Goreti Marreiros · Laura Grande ·
Juan Pedro Llerena · Luís Conceição · Hoon Ko ·
Marta Plaza · Michela Ricca
Editors

Distributed Computing and Artificial Intelligence, Special Sessions II, 21st International Conference

Springer

Editors
Goreti Marreiros
ISEP/GECAD Instituto Superior de
Engenharia do Porto
Porto, Portugal

Juan Pedro Llerena
Departamento de Ciencias de la Computación
e Inteligencia Artificial
University of Alcalá
Madrid, Spain

Hoon Ko
Sunmoon University
Asan-si, Korea (Republic of)

Michela Ricca
Department of Biology, Ecology and Earth
Sciences
University of Calabria
Rende, Italy

Laura Grande
Biotechnology, Intelligent Systems
and Educational Technology (BISITE)
Research Group
University of Salamanca
Salamanca, Spain

Luís Conceição
ISEP/GECAD Instituto Superior de
Engenharia do Porto
Porto, Portugal

Marta Plaza
Biotechnology, Intelligent Systems
and Educational Technology (BISITE)
Research Group
University of Salamanca
Salamanca, Spain

ISSN 2367-3370 ISSN 2367-3389 (electronic)
Lecture Notes in Networks and Systems
ISBN 978-3-031-80945-3 ISBN 978-3-031-80946-0 (eBook)
https://doi.org/10.1007/978-3-031-80946-0

This Springer imprint is published by the registered company Springer Nature Switzerland AG
The registered company address is: Gewerbestrasse 11, 6330 Cham, Switzerland

Lecture Notes in Networks and Systems 1151

The series "Lecture Notes in Networks and Systems" publishes the latest developments in Networks and Systems—quickly, informally and with high quality. Original research reported in proceedings and post-proceedings represents the core of LNNS.

Volumes published in LNNS embrace all aspects and subfields of, as well as new challenges in, Networks and Systems.

The series contains proceedings and edited volumes in systems and networks, spanning the areas of Cyber-Physical Systems, Autonomous Systems, Sensor Networks, Control Systems, Energy Systems, Automotive Systems, Biological Systems, Vehicular Networking and Connected Vehicles, Aerospace Systems, Automation, Manufacturing, Smart Grids, Nonlinear Systems, Power Systems, Robotics, Social Systems, Economic Systems and other. Of particular value to both the contributors and the readership are the short publication timeframe and the world-wide distribution and exposure which enable both a wide and rapid dissemination of research output.

The series covers the theory, applications, and perspectives on the state of the art and future developments relevant to systems and networks, decision making, control, complex processes and related areas, as embedded in the fields of interdisciplinary and applied sciences, engineering, computer science, physics, economics, social, and life sciences, as well as the paradigms and methodologies behind them.

Indexed by SCOPUS, EI Compendex, INSPEC, WTI Frankfurt eG, zbMATH, SCImago.

All books published in the series are submitted for consideration in Web of Science.

For proposals from Asia please contact Aninda Bose (aninda.bose@springer.com).

Preface

The 21st International Symposium on Distributed Computing and Artificial Intelligence 2024 (DCAI 2024) acts as a crucial forum for the exchange of ideas between scientists and professionals from both academia and industry. This interaction is vital for developing systems that address contemporary societal needs. Over the last ten years, research in Intelligent Distributed Systems has made significant strides, leading to the creation of many effective applications. Presently, technologies like the Internet of Things (IoT), Industrial Internet of Things (IIoT), Big Data, Blockchain, and distributed computing continue to advance due to substantial research and technical efforts from universities and businesses alike. Artificial Intelligence (AI) plays a pivotal role in this landscape, driving innovation and enhancing the capabilities of these technologies. AI techniques, such as machine learning and neural networks, are essential for analyzing vast amounts of data generated by IoT and IIoT devices, optimizing operations, and making predictive decisions. AI's integration into distributed systems has enabled more efficient resource management, improved security protocols, and the development of intelligent applications that can adapt to dynamic environments.

With computing systems ranging from personal laptops to edge, fog, and cloud computing systems, this conference offers the perfect venue to showcase innovative techniques applied to complex challenges within these domains. The combination of AI and distributed computing not only facilitates the creation of smart environments and autonomous systems but also supports advancements in healthcare, transportation, smart cities, and numerous other fields. By fostering collaboration and knowledge sharing, DCAI 2024 aims to drive the next wave of technological breakthroughs that will shape the future of intelligent distributed systems.

This year's technical program will present both high quality and diversity, with contributions in well-established and evolving areas of research. Specifically, 138 papers were submitted, by authors from 43 different countries (Algeria, Angola, Argentina, Austria, Belgium, Brazil, Bulgaria, Chile, Colombia, Croatia, Cyprus, Czechia, Denmark, Ecuador, France, Germany, Greece, Hungary, India, Iran, Ireland, Italy, Japan, Jordan, Luxembourg, Mexico, Morocco, Netherlands, Norway, Pakistan, Poland, Portugal, Qatar, Romania, Russia, South Korea, Spain, Sweden, Switzerland, Tunisia, United Arab Emirates, United Kingdom, United States), representing a truly "wide area network" of research activity.

Given the conference's success, this edition features twelve special sessions covering a wide range of topics related to AI and other areas of interest. These sessions were carefully curated to address the latest advancements and challenges in fields such as machine learning, neural networks, IoT, big data, and blockchain, among others. The accepted papers from these sessions are presented in two volumes, showcasing the diverse and innovative research being conducted in these domains. This is the second volume, which includes the sessions: Advanced AI-based Data Annotation Tools (AI-DAT), Digital Heritage Contents (DHC), female technopreneurs and early career researchers in distributed computing and AI organized by the Gendered Innovation Living Labs (GILL), New perspectives and solutions in Cultural Heritage (TECTONIC), Artificial Intelligence trends in Health & Care (AItHeC) & Doctoral Consortiums), each focusing on specific themes within the broader scope of AI and its applications.

The DCAI 2024 technical program has selected 74 papers in special sessions, and, as in past editions, it will be special issues in ranked journals such as Sensors, IJIMAI, Smart Cities, Electronics, Future Internet Systems and Advances in Distributed Computing and Artificial Intelligence Journal. These special issues will cover extended versions of the most highly regarded works. DCAI 2024 Special Sessions have been a very useful tool to complement the regular program with new or emerging topics of particular interest to the participating community.

This symposium was organized by the University of Salamanca and was held from 26 to 28 June 2024 in the city of Salamanca (Spain). We would like to thank all the contributing authors, National Associations (AEPIA, APPIA, LASI), the sponsors (AIR Institute), the funding supporting of the project "Edge-IoT platform based on energy-efficient DLT technologies for the exchange of green digital tokens through smart contract execution", Reference: PDC2022-133161-C31, fnanced by MCIN /AEI/10.13039/501100011033 and NextGeneration EU/PRTR, UE, and finally, the Organization members and the Program Committee members for their hard work, which was essential for the success of DCAI 2024.

June 2024

Goreti Marreiros
Laura Grande
Juan Pedro Llerena
Luís Conceição
Hoon Ko
Marta Plaza
Michela Ricca

DCAI 2024 Organization

Advisory Board

Kenji Matsui	Osaka Institute of Technology, Japan
Yuncheng Dong	Sichuan University, China
Francisco Herrera	University of Granada, Spain
Tan Yigitcanlar	Queensland University of Technology, Australia
Tiancheng Li	Northwestern Polytechnical University, China

Program Committee Chairs

Ravikumar C. V.	Vellore Institute of Technology University, India
Pawel Sitek	Kielce University of Technology, Poland
Nasro Min Allah	Imam Abdulrahman Bin Faisal University (KSA), Saudi Arabia

Special Sessions Chairs

Rashid Mehmood	King Abdulaziz University, Saudi Arabia
Goreti Marreiros	ISEP/GECAD, Portugal

Organizing Committee Chairs

Sascha Ossowski	King Juan Carlos University, Spain
Sara Rodríguez	University of Salamanca, Spain
Guillermo Hernández González	University of Salamanca, Spain
Laura Grande Pérez	University of Salamanca, Spain

Program Committee

Adel Boukhadra	National High School of Computer Science, Algeria
Alberto Fernandez	University Rey Juan Carlos, Spain
Alfonso González-Briones	University of Salamanca, Spain

Amel Borgi	ISI/LIPAH, Université de Tunis El Manar, Tunisia
Amin M Khan	UiT The Arctic University of Norway, Norway
Ana Belén Gil González	University of Salamanca, Spain
Ana Faria	ISEP, Portugal
Ander García	Visual Communication Technologies VICOMTech
André Zúquete	University of Aveiro, Portugal
Angel Martin Del Rey	Department of Applied Mathematics, Universidad de Salamanca, Spain
Angélica González Arrieta	Universidad de Salamanca, Spain
António J. R. Neves	University of Aveiro, Portugal
António Jorge Do Nascimento Morais	Universidade Aberta, Portugal
Araceli Queiruga-Dios	Department of Applied Mathematics, Universidad de Salamanca, Spain
Arkadiusz Gola	Lublin University of Technology, Poland
Aurélie Hurault	IRIT-ENSEEIHT, France
Benedita Malheiro	Instituto Superior de Engenharia do Porto and INESC TEC, Portugal
Bozena Wozna-Szczesniak	Institute of Mathematics and Computer Science, Jan Dlugosz University in Czestochowa, Poland
Carlos Alejandro De Luna-Ortega	Universidad Politecnica de Aguascalientes, Mexico
Carlos Carrascosa	GTI-IA DSIC Universidad Politecnica de Valencia, Spain
Carmen Benavides	University of León, Spain
Cesar Analide	University of Minho, Portugal
Dalila Durães	University of Minho, Portugal
Edgardo Bucciarelli	University of Chieti-Pescara, Italy
Eleni Mangina	UCD, Ireland
Elisa Huzita	State University of Maringa, Brazil
Emilio Serrano	Universidad Politécnica de Madrid, Spain
Ester Martinez	Universidad de Alicante, Spain
Fábio Marques	University of Aveiro, Portugal
Felix Freitag	Universitat Politècnica de Catalunya, Spain
Fernando De La Prieta	University of Salamanca, Spain
Fidel Aznar Gregori	Universidad de Alicante, Spain
Florentino Fdez-Riverola	University of Vigo, Spain
Francisco A. Pujol	Specialized Processor Architectures Lab, DTIC, EPS, University of Alicante, Spain
Francisco García-Sánchez	University of Murcia, Spain

Francisco Javier Calle	Departamento de Informática. Universidad Carlos III de Madrid, Spain
Friederike Wall	Alpen-Adria-Universitaet Klagenfurt, Austria
Giner Alor Hernandez	Instituto Tecnologico de Orizaba, Mexico
Giovanni De Gasperis	DISIM, Università degli Studi dell'Aquila, Italy
Guillermo Hernández	University of Salamanca, Spain
Gustavo Isaza	University of Caldas, Colombia
Gustavo Santos-Garcia	Universidad de Salamanca, Spain
Holger Billhardt	Universidad Rey Juan Carlos, Spain
Ichiro Satoh	National Institute of Informatics, Japan
Irina Georgescu	Academy of Economic Studies, Romania
Isabel Praça	GECAD/ISEP, Portugal
Jesús García Herrero	University Carlos III de Madrid, Spain
Jona Beysens	CSEM
Jose M. Molina	Universidad Carlos III de Madrid, Spain
José Ramón Méndez Reboredo	University of Vigo, Spain
José Ramón Villar	University of Oviedo, Spain
Jose-Luis Poza-Luján	Universitat Politècnica de València, Spain
Juan Gomez Romero	University of Granada, Spain
Juan Pavón	Universidad Complutense de Madrid, Spain
Juan-Luis Posadas-Yague	Universitat Politècnica de València, Spain
Kenji Matsui	Osaka Institute of Technology, Japan
Li Weigang	University of Brasilia, Brazil
Lourdes Borrajo	University of Vigo, Spain
Luis Fernando Castillo Ossa	University of Caldas, Colombia
Luiz Romao	Univille, Brazil
Mar Pujol	Universidad de Alicante, Spain
Mariano Raboso Mateos	Facultad de Informática - Universidad Pontificia de Salamanca, Spain
Masaru Teranishi	Hiroshima Institute of Technology, Japan
Mauricio Orozco-Alzate	Universidad Nacional de Colombia, Colombia
Michal Wozniak	Wroclaw University of Technology, Poland
Michifumi Yoshioka	Osaka Pref. Univ., Japan
Miguel A. Vega-Rodríguez	University of Extremadura, Spain
Miguel Angel Patricio	Universidad Carlos III de Madrid, Spain
Naoufel Khayati	COSMOS Laboratory - ENSI, Tunisia
Nuno Silva	DEI & GECAD - ISEP - IPP, Portugal
Pablo Negre	University of Salamanca, Spain
Paulo Mourao	University of Minho, Portugal
Paulo Novais	University of Minho, Portugal
Paulo Vieira	Insituto Politécnico da Guarda, Portugal
Pawel Pawlewski	Poznan University of Technology, Poland

Pedro Sousa	University of Minho, Portugal
Philippe Mathieu	University of Lille, France
Radu-Emil Precup	Politehnica University of Timisoara, Romania
Rafael Corchuelo	University of Seville, Spain
Raffaele Dell'Aversana	Università "D'Annunzio" di Chieti-Pescara, Italy
Reyes Pavón	University of Vigo, Spain
Rui Camacho	University of Porto, Portugal
Sara Rodríguez	University of Salamanca, Spain
Sascha Ossowski	University Rey Juan Carlos, Spain
Sergi Robles	Universitat Autònoma de Barcelona, Spain
Shimpei Matsumoto	Hiroshima Institute of Technology, Japan
Stefan-Gheorghe Pentiuc	University Stefan cel Mare Suceava, Romania
Tayuka Yoshihiro	Faculty of Systems Engineering, Wakayama University, Japan
Tiago Pinto	IPP, Portugal
Vicente Julian	Universitat Politècnica de València, Spain
Worawan Diaz Carballo	Thammasat University, Thailand
Yeray MEZQUITA	University of Salamanca, Spain
Yun Zhu	Shaanxi Normal University, China
Zbigniew Banaszak	Warsaw University of Technology, Faculty of Management, Dept. of Business Informatics, Poland
Zhen Zhang	Dalian University of Technology, China
Zhu Wang	XINGTANG Telecommunications Technology Co., Ltd., China
Zita Vale	GECAD - ISEP/IPP, Portugal

Organizing Committee

Fernando De la Prieta	University of Salamanca, Spain
Javier Prieto Tejedor	University of Salamanca and AIR Institute, Spain
Ricardo S. Alonso Rincón	AIR Institute, Spain
Alfonso González Briones	University of Salamanca, Spain
Pablo Chamoso Santos	University of Salamanca, Spain
Javier Parra	University of Salamanca, Spain
Belén Pérez Lancho	University of Salamanca, Spain
Ana Belén Gil González	University of Salamanca, Spain
Ana De Luis Reboredo	University of Salamanca, Spain
Angélica González Arrieta	University of Salamanca, Spain
Angel Luis Sánchez Lázaro	University of Salamanca, Spain
Emilio S. Corchado Rodríguez	University of Salamanca, Spain

Guillermo Hernández	University of Salamanca, Spain
Liliana Durón	University of Salamanca, Spain
Laura Grande	University of Salamanca, Spain
Carmen Losada	AIR Institute, Spain
Albano Carrera González	AIR Institute, Spain
Marta Plaza Hernández	University of Salamanca, Spain
Beatriz Bellido	University of Salamanca, Spain
María Alonso	University of Salamanca, Spain
Diego Valdeolmillos	AIR Institute, Spain
Carlos Álvarez	AIR Institute, Spain

DCAI 2024 Sponsors

Sponsors

Organizers

Support from Research Associations, Institutions & Labs

Contents

DCAI - SS-AI-DAT

Special Session on Advanced AI-based Data Annotation Tools (AI-DAT)

Scope

The past, present and future of artificial intelligence (AI) is closely linked to the efficient preparation and quality of data, which directly influences the performance of algorithms such as artificial neural networks. The allocation of resources to data annotation currently represents approximately 80% of investments in AI-based solutions, highlighting the strategic importance of research into advanced data annotation tools and their alignment with industry trends. Research on advanced data annotation tools is essential to improve the efficiency and quality of the data preparation process, as well as to address everyday technological challenges related to diverse information sources.

The session is proposed by SIGMA Cognition and the Applied Artificial Intelligence Group (GIAA) from Universidad Carlos III de Madrid. SIGMA, as a company specialized in data annotation services, recognizes the strategic importance of this task. Through its SIGMA.AI and SIGMA Cognition units and the collaboration with GIAA, SIGMA plays a key role in the development of artificial intelligence-based data annotation tools. The special session "Research on advanced AI-Based data Annotation Tools" as a result of the SIGMA Cognition project HADA (Herramientas de Anotación de Datos Avanzadas) center in the research of AI techniques in annotation tools search for new developments in this area. This session aims to share and discuss new research on AI tools and algorithms in a data annotation framework, with the goal of creating advanced tools that optimize the collection of high-quality data more efficiently and accurately than existing tools. The session will also include the participation of the Signal Processing Applications Group (GAPS) of the Polytechnic University of Madrid, which has closely collaborated with Sigma in the research and development of audio annotation processes.

Topics

Some of the topics included in this session but not limited are:

Human-in-the-loop Artificial Intelligence (HITL-AI)
Image and video segmentation
Audio Processing
Sound Event Detection
Speech Anotation
Automated audio, text, image and video annotation
Pre-anotation processes in text, image and video (Data preparation/curation)
Generative AI and Synthetic Data

Organizing Committee

Juan Pedro Llerena	Universidad de Alcalá
Miguel Ángel Patricio	Universidad Carlos III de Madrid
Ester Sancho	SIGMA Cognition
Silvia Rodríguez-Jiménez	SIGMA Cognition

Program Committee

Jorge Rico	Head of NLP and NLU Technologies at Sigma Cognition
Silvia Rodríguez-Jiménez	Computer Vision Director at Sigma Cognition
Juan Manuel Perero	Head of Speech Technologies at Sigma Technologies Global
Angel Mora	Research Scientist, Computer Vision at Sigma Cognition
Jaime Corton	Computer Vision Engineer at Sigma Cognition
Fernando Espinoza	Research Scientist, Speech processing at Sigma Cognition
Jose Manuel Molina Lopez	Ph.D. in telecommunications engineering from the UPM, professor in the area of Computer Science and Artificial Intelligence at the Carlos III University of Madrid and director of the GIAA group
Miguel Ángel Patricio Guisado	PhD in Computer Science from the Universidad Politécnica de Madrid, and is a professor in the area of Computer Science and Artificial Intelligence at the Universidad Politécnica de Madrid
Luis Hernández Gómez	University Professor ETSI of Telecommunications UPM
Ramón Fernandez Castañon	Masters student at UPM

AI-Boosted Video Annotation: Exploring Pre-Labeling with Cross-Modalities

Juan Gutiérrez-Navarro[1,2(✉)], Ángel Mora-Sánchez[2], Silvia Rodríguez-Jiménez[2], and J.L. Blanco-Murillo[1]

[1] Information Processing and Telecommunications Center Escuela Técnica Superior de Ingenieros de Telecomunicación, Universidad Politécnica de Madrid, Av. Complutense 30, 28040 Madrid, Spain
juan.gutierrez@alumnos.upm.es, jl.blanco@upm.es
[2] Sigma Cognition, Avenida de Manoteras 6, 28050 Madrid, Spain
{amora,srodriguez}@sigmacognition.ai

Abstract. Annotation in large-scale video datasets requires significant resources. To enhance the efficiency of this process, we suggest employing pre-trained cross-modal models within the Human-in-the-Loop (HITL) paradigm. We used a synthetic video dataset to generate precise semantic annotations and assess the effectiveness of different label representations in comprehending visual information across diverse vision tasks, including fine- and coarse-grained ones. We also introduce a framework for automating pre-annotation extraction from semantically similar frames. Our approach presents promising avenues for efficiently annotating video data, crucial for developing robust Machine Learning (ML) systems.

Keywords: Video Annotation · Cross-modal · Human-in-the-Loop

1 Introduction

Data annotation is crucial for Machine Learning (ML). It facilitates the creation and validation of algorithms and models while it conditions the learning outcomes and process. The annotation process involves assigning labels to data. The goal is to enhance its comprehensibility for machines, supporting the training of supervised models, and validating unsupervised and self-supervised ones.

The data annotation market has grown significantly, driven by the increasing demand for new models and data-centric algorithms [16]. Robust and reliable models largely depend on annotation quality and availability, which are conditioned by the size and complexity of the annotation task. Efforts to integrate machine learning benefits into the annotation process remain limited. Recent advancements in ML performance and processing time offer promising opportunities to improve annotation efficiency, consistency, and robustness through the application and reuse of pre-trained ML models.

The type and nature of labels largely depend on the annotation's purpose and the nature of the data. Video annotation poses additional challenges compared

G. Marreiros et al. (Eds.): DCAI 2024, LNNS 1151, pp. 5–15, 2025.
https://doi.org/10.1007/978-3-031-80946-0_1

to image annotation due to the dynamic nature of the subject and the temporal dependence between frames. Cross-modal representations have the potential to serve as the cornerstone that alleviates the burden of connecting monomodal data with their corresponding text labels, bringing together the two pieces of information: the data and the label. Supplementarily, agnostic models have demonstrated great potential to operate in open domains [15]. The combination of both shall power future annotation.

Currently, video annotation involves a significant amount of human intervention, leading to errors, inconsistencies amongst annotators and annotations, high costs, time consumption, and an inability to measure annotation quality accurately. By incorporating artificial intelligence tools into the annotation process within the Human-in-the-Loop (HITL) paradigm, we aim to reduce the number of frames to annotate and make each human input more valuable, reducing time and the possibility of annotation errors. By leveraging human intelligence to handle the nuances and complex decisions, which are the weak points of fully automated annotation, we aim to harness machines' ability to process large amounts of data. Ultimately, the goal is to manually annotate a reduced number of frames, known as keyframes, which somehow concentrate most of the information in the video. From this, we aim to develop annotation schemes based on this frame annotation that generalize and propagate well to the entire video, minimizing human intervention throughout the process.

The complete annotation comprises several phases: pre-annotation, annotation, and post-annotation. We primarily focus on pre-annotation for automatizing, which involves preprocessing the data using specific ML models to simplify the annotator's work in the subsequent stages. Pre-annotation streamlines the human annotator's work by covering direct annotations and providing assistance to make their task easier. The main goal is to provide consistent annotation results along the video and concentrate information.

In this study, we investigate agnostic cross-modal representations and perform a preliminary evaluation of the performance of the image encoder, specifically the CLIP [14] visual transformer. Based on these, we propose a framework to automate pre-annotation extraction from groups of frames sharing semantic content. Our contributions include a pipeline for video frame grouping based on CLIP, a method for identifying prototypes from clustered frames.

2 Related Work

AI-assisted Video Annotation. Various methodologies have been proposed for AI-assisted video annotation in recent years. One approach utilizes a unified framework incorporating bounding boxes, automatic temporal interpolation, frame selection for manual annotation, and a guiding mechanism based on previous annotations [9]. Another method employs spatio-temporal feature similarity and dimensionality reduction to accelerate the annotation process significantly [12]. On the other hand, as a more general approach to video classification in [2], evaluate models for video classification, leveraging spatiotemporal tokens and pre-trained image models to achieve robust performance, particularly with limited data. Nonetheless, most of these techniques focus on scene elements rather than on the overall meaning and significance of the scene. The latter is significantly more difficult to attain from frame sequences, as well as to interpret.

Highly Capable Multimodal Models. Recent advancements in large multimodal models have demonstrated impressive proficiency in understanding video content, with the latest and most notable example being Gemini [18]. Gemini models utilize a technique that samples 16 equally spaced frames from each video clip to address various video-related tasks. When integrated with state-of-the-art Large Language Models (LLMs), this approach significantly enhances comprehension in the video domain. Moreover, promising approaches have been proposed by OpenAI Sora [11], which involve reducing the dimensional embeddings to a latent space of lower dimensions. This approach forms video representations by aggregating visual patches. This technique holds remarkable potential for extracting accurate real-time descriptions from videos.

Anomaly Detection. While our research doesn't center directly on anomaly detection, it's noteworthy to discuss several studies in this domain as they closely align with ours regarding utilizing pre-trained image and text models for analyzing video contexts. This approach involves interpreting video content for frame classification, akin to annotation tasks. In [7], the potential of incorporating text descriptions within unlabeled video datasets is explored. LLM generates these text descriptions, which are then employed to identify abnormal frames by assessing their similarity with the encoded input frame, another approach in [20]. The authors refine the CLIP-derived cosine similarity with an unlabeled dataset and introduce text-conditional similarity as a metric to improve performance. Another weakly supervised video anomaly detection strategy is presented in [19], where the frozen CLIP model is utilized directly without pre-training or fine-tuning. One aspect involves coarse-grained binary classification based on visual features, while another exploits language-image alignment. Lastly, [6] utilizes the ViT-encoded visual features from CLIP to extract discriminative representations, contrasting with conventional features like C3D or I3D in the domain.

3 Preliminary Study: Visual Information Compression

To create a pipeline for the natural semantic grouping of video frames based on CLIP, we first explore the efficacy of CLIP capabilities validated against existing literature. Using a synthetic video dataset, we aim to assess the effectiveness of different label representations in understanding visuals across various tasks.

We conducted a preliminary study on the performance of the video Encoder across various vision-related tasks to gain insights into the model's behavior concerning individuals, interacting objects, and environments. We selected a state-of-the-art **cross-modal encoder**. Contrastive Language-Image Pretraining (CLIP) [13] consists of two Encoders trained in parallel. A Vision Transformer (ViT), or ResNet, is used for image embeddings, and a transformer is

Fig. 1. Snapshots from the GTA Indoor Motion (GTA-IM) Dataset.

used for language embeddings. CLIP can achieve performance comparable to standard classification models with access to extensive training examples.

Regarding the test, we selected a publicly available, **synthetic video dataset**. Due to the limited availability and quality of annotated video datasets, we created annotations using the GTA Indoor Motion dataset (GTA-IM) [4]. Synthetic data from a model can include multiple annotated video frames with textual descriptions essential for establishing a ground truth and extremely relevant due to the scarcity of datasets that include video and text annotations. It includes one million high-definition RGB-D frames.

Utilizing detailed pixel-level annotations (instance masks) from the GTA-IM dataset, we selected several classes (i.e., individuals and objects) marked in depth masks. We extracted precise bounding box coordinates. The process depicted in Fig. 1 served as our primary source of annotations, featuring textual labels.

Table 1. Results of performance evaluation tasks on the GTA-IM dataset.

Class type	Task	AP
Fine-grained	Detection of persons	< 50%
	Count number of instances	< 10%
	Proximity to the point of view	< 40%
	Within frame position	< 50%
Coarse-grained	Scene description	< 80%
	General description	< 70%

Frames from the GTA-IM dataset underwent preprocessing for a standard resolution of 240×240 ppx. We defined self-constructed classes to perform fine- and coarse-grained classification. The former task was adapted based on the proximity of elements. Results are summarized in Table 1. Tasks are as follows:

Detection of people. The model discerned between two options to identify the presence or absence of people. We measured similarities between each frame and the texts "a photo of a person" and "a photo".

Count number of instances. We generated contextual variations to quantify similarity from prompts following the form "a photo of {quantity} {instance}", where quantity consisted of selected adjectives (e.g., “a”, “much”, “many”, “one”, “two”, etc.). The count was not required to be exact; more than three objects was considered valid if the response was “many” or “much”.

Viewpoint proximity. The template “a photo of {instance} {proximity}” was used to generate all variations, where proximity could be “near” or “far”. The ground truth label was derived from the size of the bounding box and a series of manually tuned thresholds.

Within frame position. Frame deposition diagrams for a 3×3 grid (i.e., top-bottom vs. left-right) were pre-specified to handle object positional assignments. The reference label assigned to each object was compared with the template “a photo of {instance} in the {position}”.

Scene description. We used location descriptors such as “kitchen”, “living room”, “outdoor”, etc., to obtain texts and compare them for all frames.

General description: A database of actions was generated from all instances, environments, and actions to produce texts such as “a man nearby in a living room”, “a man far away in a kitchen”, “a man wearing a black shirt”, etc., using GPT-2 [14]. These texts were then manually linked with the videos.

Attending to Table 1, the zero-shot model displayed poor performance in fine-grained detection tasks, with results akin to random chance. Specifically, the model's detection rate was less than 50%, indicating significant limitations in accurately discerning the presence or absence of elements within frames. Our evaluation underscores the critical importance of addressing fine-grained and coarse-grained tasks in real-life scenarios, mainly as observed objects draw closer. Despite attempts to adapt fine-grained tasks to a coarse-grained approach based on proximity, performance remained below satisfactory across various tasks. The count accuracy for instances was less than 10%, and proximity determination achieved an accuracy of less than 40%. These findings highlight video analysis's inherent complexity and challenges in practical contexts.

4 Clustering and Prototype Selection

4.1 Problem Formulation

We aim to address the efficient generation of semantic annotations for video frame content using textual labels. Here, “semantic annotations” refer to meaningful information extracted from the visual content, including objects, scenes, or depicted concepts. Our goal in annotation is to group similar frames, leveraging the innate properties of the representations (as described in [3]), and to select the most representative one from each group, termed a “semantic keyframe”. This concept draws inspiration from video codec keyframes, which encapsulate the information of a group of frames.

4.2 Natural Clustering of Embeddings

When clustering by semantic content, frames with similar visual content are labeled or linked with the same textual label. It's essential to check if CLIP embeddings representing frames form clusters or paths for further analysis in video analysis. The t-SNE representations in Fig. 2 depict the embeddings from videos in datasets DAVIS2017 [1] (left) and HMDB51 [8] (right). For DAVIS2017, each category corresponds to a single video, showing how frames from the same video cluster together. In HMDB51, each category includes ten videos, and although the clustering is less distinct, frames from different videos within the same category still tend to cluster to varying degrees.

The observed trend is consistent with the expected natural clustering of frames from the same video displaying similar content from visual and semantic levels. In videos, consecutive frame embeddings show minimal variation, causing frames from the same video to form trajectories, indicating temporal coherence. These trajectories visually represent the underlying relationships and patterns across different videos. We leveraged image-image similarities within sequential video frames using the cosine distance metric.

In our experiments, the algorithm that worked best for clustering was the Spherical k-Means [5] due to the hypersphere geometry of CLIP's latent space [13]. However, this leads to a new challenge of determining the number of clusters. Determining the optimal number requires selecting a quality clustering metric, such as the Silhouette Score, and evaluating multiple cluster configurations.

The Silhouette coefficient condenses within-cluster and between-cluster variation. The assigned score is the mean of all silhouette coefficients for all points, aiding in determining the optimal number of clusters. A silhouette close to 1 suggests that the data point is appropriately clustered, while a silhouette score close to -1 indicates that the data point is likely assigned to the wrong cluster.

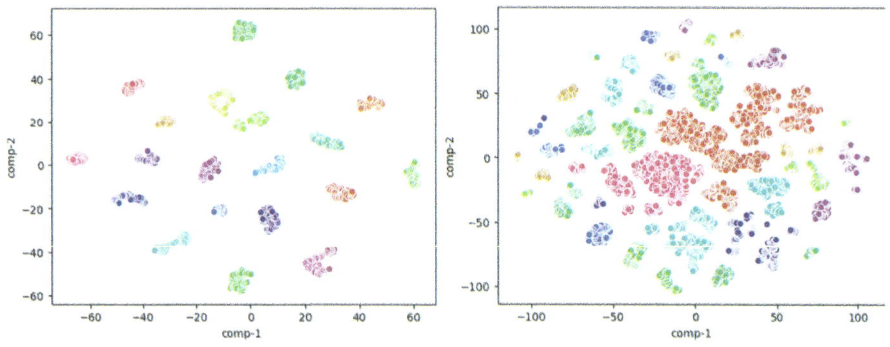

Fig. 2. t-SNE projections for color-coded categories on DAVIS2017 (20 videos, 1 per category, left) and HMDB51 (100 videos, 10 videos per category, right).

4.3 Semantic Prototype Selection

Extracting a prototype from a video chunk involves identifying and isolating a representative sample or instance (snapshot) from the video based on its content. The term "prototype" typically refers to a specific example or archetype within the video that encapsulates the essence of the analysis or extraction endeavor.

A standard method for efficiently representing a video is through keyframes. These are selected based on content variation and clustering within the video. As explained before, frames naturally group into clusters in a latent space, and a single representative is chosen. After encoding frames into embeddings and clustering them based on a given criterion, keyframes can be chosen. Prototypes for keyframes might be the cluster center or a weighted mean. A concise set of keyframes is obtained by selecting frames with embeddings closest to the cluster center, effectively capturing content variation and meaning.

Our choice for the mean value is motivated by the inner properties of the latent space and the well-known k-means clustering properties. Essentially, the mean vector is descriptive enough of the underlying category of the video, even for anomaly detection. In this context, we assume that a single average value per video does not represent the video. Thus, we include an arbitrary number of prototypes depending on the video, assuming that normal and abnormal (or more) classes could be assigned to each segment instead of one per video.

4.4 Experiences

Synthetic Video Analysis. We conducted tests using the "2020-06-12-23-54-30" video from GTA-IM, which comprises 3797 frames depicting these actions: a person walking backward through the house, a person sitting on a sofa, a person walking outside, and a person lying on a bed. These actions take place in these locations: hallways, living rooms, bedrooms, the exterior of the house, and the kitchen. The lighting conditions are bright indoors to dark outdoors, with three scene shifts occurring throughout the video.

As previously described, the results of clustering the frames and extracting keyframes are depicted in Fig. 3. We will now analyze whether the keyframes accurately represent the manually described content of the video in terms of action, location, and lighting. From the automatically selected keyframes, several conclusions can be drawn: Keyframe 0 contains a person walking backward in a house hallway with lighting, Keyframe 1 contains a person sitting on the sofa with good lighting following a scene shift, Keyframe 2 contains a person walking outside the house with nighttime lighting, Keyframe 3 contains a person in the kitchen after a scene change, with lighting, and Keyframe 4 contains a person in the bedroom, lying on the bed during the third scene change.

The content extracted from the keyframes summarizes the video content with the minimum number of frames. Furthermore, the frame classification allows us to delineate the temporal space corresponding to each semantic cluster. These techniques show more promise than random frame sampling as a video prototype.

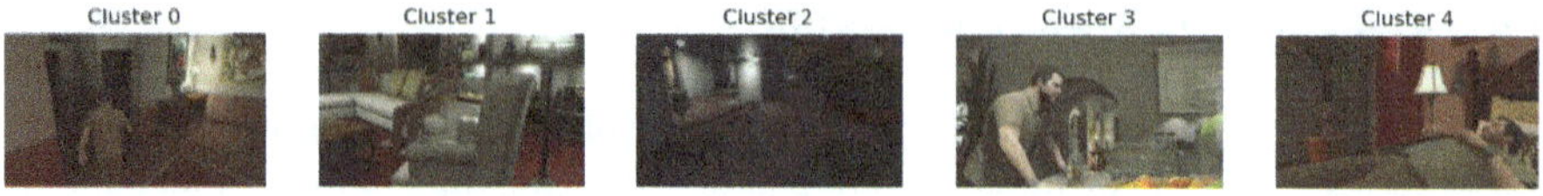

(a) Keyframes extracted from the GTA-IM video.

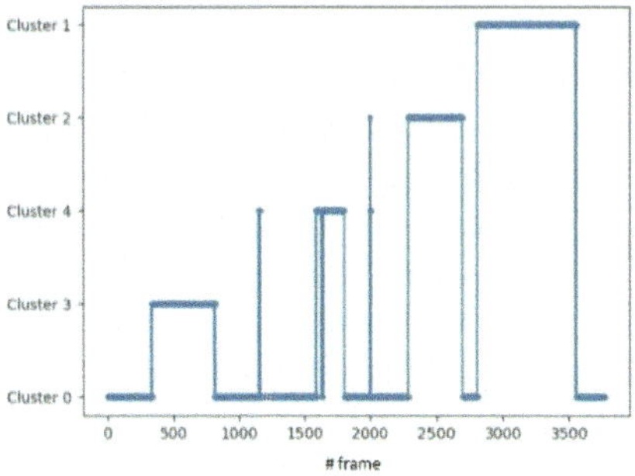

(b) Frames clustering along the video.

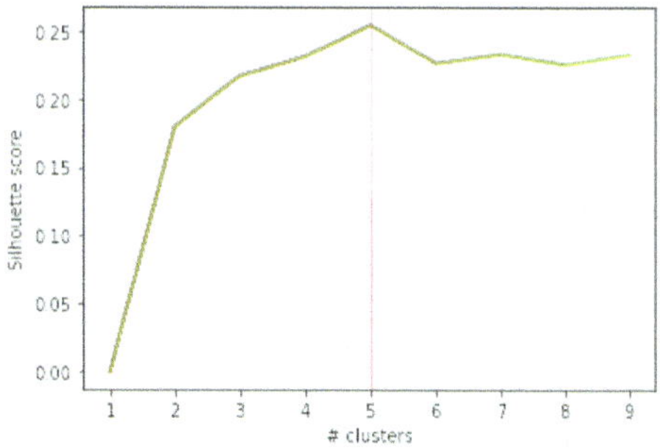

(c) Silhouette score of embedding clusters.

Fig. 3. Analysis of situations in a GTA-IM video. (a) Includes the keyframes we automatically selected. (b) Includes the clustering results across time (x-axis) relative to the selected keyframes (y-axis). (c) contains the Silhouette scoring for different clusters, the optimal being reached for 5 clusters for this video.

Real-World Video Analysis. We utilized "Normal_Videos_246_x264" from the UCF-Crime dataset [17], comprising 2 min and 46 s of security camera footage within an office environment. This real-world video captures dynamic scenes of moving individuals, showcasing various positional changes. Extracting keyframes from this video allows us to discern distinct situations.

5 A Framework for Automated Video Pre-Annotation

We have developed a systematic pipeline for automated video pre-annotation, leveraging key concepts from our research. The pipeline consists of several phases designed to process and annotate video content.

Frame Encoding. All video frames evaluated using CLIP's pre-trained visual encoder to ensure robust and comprehensive representation of visuals.

Clustering. Next, we group the encoded frames into a finite set of clusters using the spherical k-means algorithm. We utilize cosine distance as the distance metric between embedding vectors, allowing us to effectively measure the similarity between frames. This clustering process helps organize the frames, facilitating subsequent annotation tasks.

Quality Assessment We evaluate the clustering using the Silhouette Score metric to refine the clustering process to maximize performance.

Textual Labeling on Keyframes. Once the frames are clustered, we select keyframes from each cluster to represent the cluster's content. These keyframes are annotated with textual labels, leveraging CLIP's cross-modal capabilities to match visuals with textual descriptions. This step enhances the interpretability of the annotations and provides valuable context for understanding the video.

Annotation Propagation: Finally, we propagate the annotations from the selected keyframes to all frames within the same cluster. This ensures that all frames within a cluster receive consistent and accurate annotations, streamlining the annotation process and improving overall efficiency. Additionally, it is possible to employ language models to generate coherent and contextually appropriate textual interpretations of the visual content represented by the keyframes [10].

These pre-annotations have been used in another contribution currently pending evaluation wherein the produced automatic annotations have been tested in an actual annotation trial. We tested whether these pre-annotations reduce annotation times while maintaining annotation quality. We prove how our automatic keyframe selector can provide annotators with an overview of the video's content to decide whether the critical content has been correctly selected.

The advantage of this procedure is that, until now, annotators had to sequentially review individual frames for their annotation, which resulted in significant time loss when there was no relevant content to annotate. With this new approach, annotators can observe and directly access the annotated elements on a timeline, with the keyframes summarizing the content of each segment. The HITL framework ensures that annotators can efficiently merge clusters or correct misinterpretations, significantly speeding up the annotation process.

6 Conclusions

In this contribution, we used CLIP to produce automatic video annotations to streamline the process for human annotators. Our novel pipeline for automated pre-annotation combines CLIP's visual encoder with k-means natural clustering. This approach offers a streamlined solution for annotators, providing a comprehensive overview of video content and potentially reducing annotation times while contributing to consistency across annotations.

The procedure is simple, consistent, and direct, yet our investigation revealed several key insights into video using CLIP. Firstly, the lack of high-quality annotated video datasets presents a significant bottleneck in the development of video understanding algorithms. We turned to synthetically generated data to address this challenge, utilizing the GTA-IM dataset.

Secondly, our preliminary experiments shed new light on CLIP's abilities for zero-shot classification tasks. CLIP demonstrates impressive performance across domains but struggles with fine-grained distinctions within frames. This highlights the need for further refinement of CLIP, particularly for video tasks.

Thirdly, in exploring content aggregation techniques, we uncovered promising methods for extracting information from video footage. Keyframe extraction proved effective in capturing key moments and situations automatically.

Our research sets the background for forthcoming investigations. Improving CLIP's efficacy in video annotation, grinding pre-annotation methods, and confirming our conclusions on extensive datasets are desirable paths for exploration.

Acknowledgments. Research framed under HADA project "Research on advanced data annotation tools" (2021/C005/00146323), funded by EU Next Generation (through Ministry of Economy and Digital Transformation & RED.ES).

Disclosure of Interests. Juan Gutiérrez-Navarro received research funding at UPM from a project funded by Sigma Cognition.

References

1. DAVIS: Densely Annotated VIdeo Segmentation. https://davischallenge.org/. Accessed 20 May 2024
2. Arnab, A., Dehghani, M., Heigold, G., Sun, C., Lučić, M., Schmid, C.: Vivit: a video vision transformer. In: Proceedings of the IEEE/CVF International Conference on Computer Vision, pp. 6836–6846 (2021)
3. Bengio, Y., Courville, A., Vincent, P.: Representation learning: a review and new perspectives (2014)
4. Cao, Z., Gao, H., Mangalam, K., Cai, Q.-Z., Vo, M., Malik, J.: Long-term human motion prediction with scene context. In: Vedaldi, A., Bischof, H., Brox, T., Frahm, J.-M. (eds.) ECCV 2020. LNCS, vol. 12346, pp. 387–404. Springer, Cham (2020). https://doi.org/10.1007/978-3-030-58452-8_23
5. Hornik, K., Feinerer, I., Kober, M., Buchta, C.: Spherical k-means clustering. J. Stat. Softw. **50**, 1–22 (2012)
6. Joo, H.K., Vo, K., Yamazaki, K., Le, N.: Clip-TSA: clip-assisted temporal self-attention for weakly-supervised video anomaly detection. In: 2023 IEEE International Conference Image Processing (ICIP), pp. 3230–3234. IEEE (2023)
7. Kim, J., Yoon, S., Choi, T., Sull, S.: Unsupervised video anomaly detection based on similarity with predefined text descriptions. Sensors **23**(14), 6256–6273 (2023)
8. Kuehne, H., Jhuang, H., Garrote, E., Poggio, T., Serre, T.: HMDB: a large video database for human motion recognition. In: 2011 International Conference on Computer Vision, pp. 2556–2563. IEEE (2011)
9. Kuznetsova, A., Talati, A., Luo, Y., Simmons, K., Ferrari, V.: Efficient video annotation with visual interpolation and frame selection guidance. In: Proceedings of IEEE/CVF Winter Conference on Applications of Computer Vision, pp. 3070–3079 (2021)
10. Li, Y., Chen, X., Hu, B., Zhang, M.: LLMs meet long video: advancing long video comprehension with an interactive visual adapter in LLMs (2024)
11. OpenAI: Video generation models as world simulators (2024). https://openai.com/research/video-generation-models-as-world-simulators. Accessed 20 May 2024
12. Poorgholi, S., Kayhan, O.S., van Gemert, J.C.: t-EVA: time-efficient t-SNE video annotation. In: Del Bimbo, A., et al. (eds.) ICPR 2021. LNCS, vol. 12664, pp. 153–169. Springer, Cham (2021). https://doi.org/10.1007/978-3-030-68799-1_12
13. Radford, A., et al.: Learning transferable visual models from natural language supervision (2021)
14. Radford, A., Wu, J., Child, R., Luan, D., Amodei, D., Sutskever, I., et al.: Language models are unsupervised multitask learners. OpenAI blog (2019)

15. Ribeiro, M.T., Singh, S., Guestrin, C.: Model-agnostic interpretability of machine learning (2016)
16. Sigma-AI: Data annotation. https://sigma.ai/what-is-data-annotation/. Accessed 20 May 2024
17. Sultani, W., Chen, C., Shah, M.: UCF-crime. https://paperswithcode.com/dataset/ucf-crime. Accessed 20 May 2024
18. Team, G., Anil, R., et al.: Gemini: a family of highly capable multimodal models. arXiv preprint arXiv:2312.11805 (2023)
19. Wu, P., et al.: VadCLIP: adapting vision-language models for weakly supervised video anomaly detection. In: Proceedings of AAAI Conference Art International, vol. 38 (2024)
20. Zanella, L., Liberatori, B., Menapace, W., Poiesi, F., Wang, Y., Ricci, E.: Delving into clip latent space for video anomaly recognition (2023)

How Does Speech Quality Impact the Data Transcription Process?

Fernando M. Espinoza-Cuadros(✉), Rafael Ginard-Aguilera, and Juan M. Perero-Codosero

Sigma Cognition, Avenida de Manoteras 6, 28050 Madrid, Spain
{fmespinoza,rginard,jmperero}@sigmacognition.ai
https://sigmacognition.ai/

Abstract. Speech quality plays a crucial role in the audio transcription process as it influences various factors that could significantly impact the annotation results. These factors include transcription speed, annotation confidence, and the number of playbacks, among others. Most existing subjective (e.g., Mean Opinion Score (MOS)) and objective (e.g., Perceptual Evaluation Score Quality (PESQ)) speech quality measurements do not consider factors that could hinder the transcription process. In this work, we first analyze the relationship between subjective speech quality measurement MOS and factors that may impact the transcription quality. Findings show that this measure poorly correlates with the speech quality perceived by the annotator. Then, we explore the use of a novel subjective measurement, called Speech Quality Score (SQS), whose scale aims to encompass the most relevant factors involved in the audio transcription process. Additionally, we explore two different approaches to predict the SQS measure. The results indicate a linear correlation of ($r = 0.66$) and ($r = 0.34$) between the ground truth and the predicted SQS values on the VoxConverse and RTVE2020 datasets, respectively.

Keywords: Speech quality measurements · Audio transcription · Speech intelligibility

1 Introduction

Communication between people is usually affected by acoustic conditions, and sometimes the spoken content is understood with several difficulties; that is, intelligibility is affected. In light of this, we can assume that intelligibility is one of the factors that determines speech quality [1]. There are many factors that can affect speech intelligibility; from factors inherent to the nature of the audio (e.g., resolution, signal-to-noise-ratio), to factors inherent to the nature of the voice (e.g., accent, pronunciation, tone) [2]. Therefore, it is a challenge to cover the broad spectrum of such factors related to speech intelligibility.

In recent years, the aforementioned speech quality issues have also emerged in data annotation workflows, being considered a challenge that hinders precise

G. Marreiros et al. (Eds.): DCAI 2024, LNNS 1151, pp. 16–25, 2025.
https://doi.org/10.1007/978-3-031-80946-0_2

annotations [3]. Annotators often handle utterances containing different acoustic characteristics according to the environment (e.g., noise, music, etc.), the recording devices (e.g., mic type, distance, etc.), among others, making it hard to transcribe the spoken message or annotate specific sound events.

In the context of audio annotation, it is not trivial to determine the impact of the speech quality to determine how it could affect the annotation and transcription task. There are several measurements that could provide a faint idea of the quality and intelligibility of a speech recording.

Speech quality can be assessed by subjective and objective measures. On the side of subjective measures, there is the aforementioned Mean Opinion Score (MOS), which appeared as a standardized listening test ITU-T P.808 [4] based on human ratings to evaluate the speech quality from audio samples. On the other hand, many objective measures of speech quality and speech intelligibility have been proposed under the standardized ITU-R recommendations, such as the Speech Intelligibility Index (SII) [5] related to the listener's audibility, or the Perceptual Evaluation of Speech Quality (PESQ) [6] for telephony speech.

Another very well-known measure used is the Word Error Rate (WER), which allows the comparison between transcriptions in the automatic speech recognition field. This measure is often used to determine the transcription quality in annotation tasks as long as a reference transcription is available.

Nevertheless, these speech quality measures were implemented under the domain of quality of experience and telecommunications engineering, representing the opinion about the speech quality in the transmission system, however, they do not consider factors that could degrade annotation performance (e.g., transcription speed, annotation confidence, and the number of audio replays).

In this paper, our aim is to contribute to the exploration of the impact of speech quality in the human transcription process. For this purpose:

1. We first study the relationship between existing speech quality measures and factors related to transcription accuracy. For that purpose, a new set of annotations was created for the analysis.
2. We explore the use of a measure (SQS), which aims to represent the different factors involved in the speech annotation process that degrade the performance of human transcriptions.
3. We investigate the use of deep learning models to predict the SQS measure by using a new set of annotations that was created in this research.

This paper is organized as follows. Section 2 describes the importance of speech quality in the annotation context, along with a description of datasets and proposed models for SQS prediction. Experiments, results, and discussion are presented in Sect. 3. Finally, Sect. 4 concludes with a final summary and future works.

2 Speech Quality in the Annotation Context

The naturalness of the audio recording (e.g., resolution, signal-to-noise ratio) and speaker-related factors (e.g., accent, pronunciation, tone) make it harder to get high-quality transcription. In fact, this issue is a concern of the annotation companies to develop high-quality datasets. Nevertheless, a method to determine the speech quality of an audio sample before the annotation process would be useful to know the effort to transcribe it. Data preparation is a multi-step process involving annotation and revision roles. According to [7], a good practice is to assign specific samples according to the expertise of the team, increasing the productivity by 25% and the quality of the project by 32% on average. In this way, more experienced annotators could receive those hard-to-annotate audio samples, and even an extra service of revision could be proposed to guarantee the quality requirements. Thus, it would increase both the annotator performance and the annotation quality as it would prioritize those audio samples with the highest quality.

Under this context, we propose to categorize the audio recordings according to the factors that could degrade human transcription quality, i.e., aspects that could make the transcription process difficult, such as poor audio quality or unintelligible speech. To facilitate this, we propose four levels of speech quality in the annotation context: Poor, Bad, Average, and Good. These levels, referred as Speech Quality Score (SQS), are determined on the mean value of subjective judgments using a 4-point Absolute Category Rating (ACR) scale. The ACR scale assigns a score ranging from 1 (Poor) to 4 (Good), as shown in Table 1. The purpose of such levels is to represent the transcription difficulty arising from various factors such as speech naturalness, noise artifacts (i.e., background noise, channel noise), and speaker-related characteristics. It is important to note that these quality levels were established based on typical acoustic conditions encountered during annotation processes, as high-quality audio samples are not usually expected.

Table 1. Description of the speech quality levels in terms of SQS measure.

Quality	Description	Score
Poor	The speech is unintelligible. After listening to the audio more than three times, understanding it becomes difficult	1
Bad	The beginnings and endings are not clear or cut off, and parts of the speech are unintelligible	2
Average	The speech in the audio requires at least three playbacks to be understood	3
Good	The speech is mostly normal with a good rhythm and is easy to understand, although there is some background noise or loudness present	4

2.1 Speech Quality Evaluation Data

The evaluation set consists of 1,020 audio samples at 16 kHz in Spanish, extracted from the RTVE2020 Database [8]. It contains different parts of video recordings of TV programs that represent a real use case of data annotation, with a large amount of adverse acoustic conditions and challenging scenarios. Samples from evaluation set were manually selected to cover the broad spectrum of a variety of acoustic conditions. As a result, speech clips of 10 s were obtained.

The goal is to define an annotation process to obtain a new set of speech quality labels regarding the annotation aspects (i.e., SQS labels) at the time of listening and transcribing the audio samples. The annotation process was carried out using Sigma's proprietary tool for audio annotation, where each audio sample was transcribed and rated by two in-house annotators. In the case of speech quality rating, the ACR values range from 1 (Poor) to 4 (Good), as shown in Table 1. As a result, we obtained a total of 2,040 ACR values. Finally, the SQS values were obtained as the mean of those ACR annotations (i.e. 1,020 values). Information related to annotators performance (i.e. annotation time, number of playbacks) are obtained from the annotation tool.

In order to compare the SQS and MOS measures, the MOS-related ACR values were also annotated on the same evaluation set according to ITU-T P.808 [4] with five possible ratings per sample. to mitigate some bias in the annotation process, two additional in-house annotators were selected to perform this task.

2.2 Model Proposal for SQS Prediction

Inspired by studies related to MOS prediction through non-intrusive methods (i.e., do not require clean reference sample to predict the quality), in particular, deep learning models operating in an end-to-end fashion [9–11], as well as models utilizing speech representations as input for quality prediction [12–14], we propose two distinct approaches for predicting SQS measurements: 1) Finetuning the NISQA model [9], which is an end-to-end model based on a CNN self-attention model to predict speech quality, and trained using more than 13,000 audio samples (i.e., NISQA Corpus) with its respective subjective evaluation in terms of MOS [9]; and 2) using speaker embeddings as speech quality predictor, in particular ECAPA2 embeddings [15].

In the first approach, we employ transfer learning to re-train the NISQA model [9] using as a starting point the frozen weights[1] of the model (see Fig. 1a). In the second approach, ECAPA2 embeddings are first extracted from a pre-trained model[2]. Subsequently, these embeddings are utilized as input to train a deep learning model, whose architecture consists of one dense layer, three convolutional blocks, LeakyReLu activation function, and dropout (see Fig. 1b).

[1] https://github.com/ConferencingSpeech/ConferencingSpeech2022/tree/main/baseline.

[2] https://huggingface.co/Jenthe/ECAPA2.

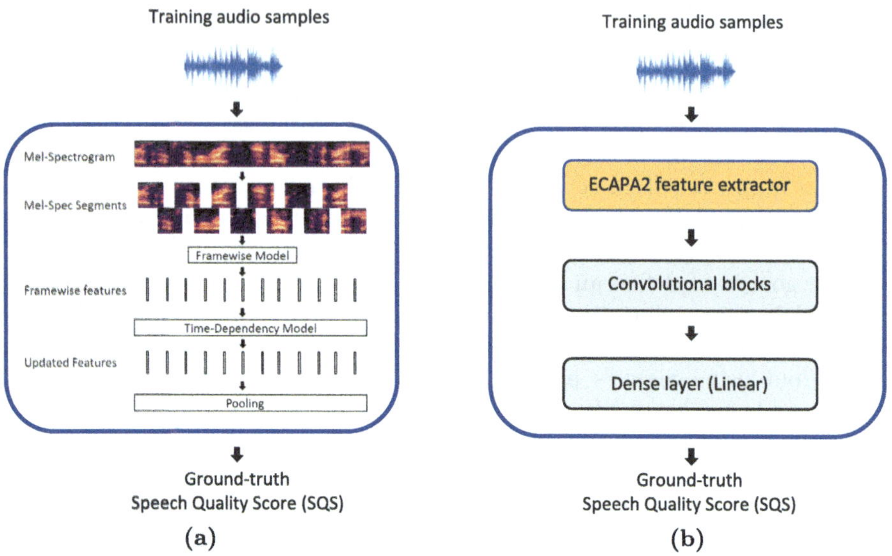

Fig. 1. Scheme of model training for SQS prediction by means of (a) Finetuning NISQA model (Adapted from [9]) and (b) ECAPA2 embeddings as input to a DNN model.

For both training and evaluation purposes, SQS values were annotated for a subset of 1,048 audio samples from the VoxConverse dataset [16], following the protocol described in the previous section. This dataset consists of video recordings of TV programs in English that contain similar acoustic conditions to the RTVE2020 Database. From the VoxConverse dataset, speech clips of 10 s duration were extracted through segmentation of the original audio samples. A subset of these segmented samples was selected based on specific speech quality criteria. Subsequently, 2,096 ACR ratings were obtained from this subset, and the SQS values were computed as the mean of these ACR annotations.

3 Experiments

In this section, we first analyze the relationship between both subjective measures MOS and SQS, as well as their relationship with the factors related to transcription quality. In the same line, the analysis is also performed for the transcription quality measure WER. This measure is estimated by the comparison between the reference transcription, i.e., made by an expert team that carried out an exhaustive quality process to ensure the reliability of the transcription, and the hypothesis transcription, i.e., resulted from the audio annotation process described in Sect. 2.1.

Secondly, we present the results for SQS prediction by proposed models. The evaluation metrics used in this study are Pearson correlation (r) with a value of significance level (p), and Mean Squared Error (MSE).

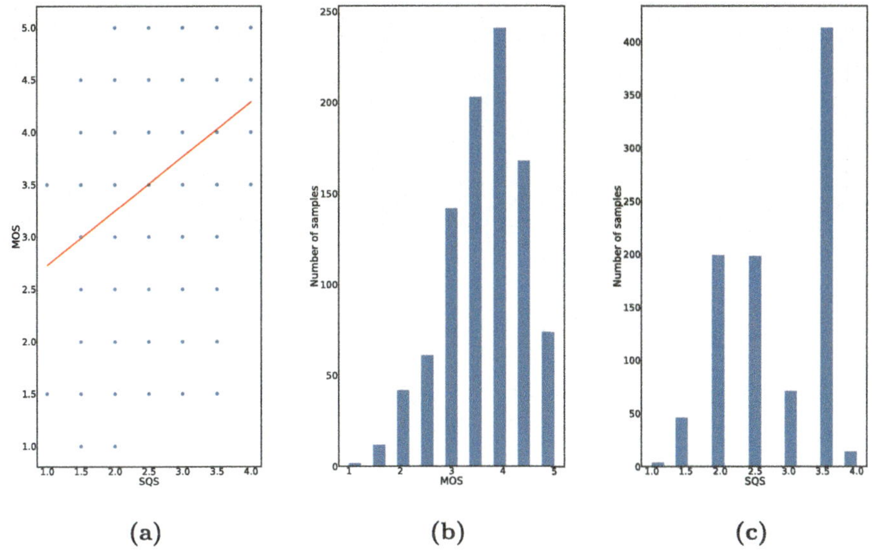

Fig. 2. (a) Scatter plot between MOS and SQS for the 1,020 audio samples with a correlation of $r = 0.4$ ($p < 0.01$). (b) A histogram of the MOS values with a mean of 3.67 and a mean standard deviation of two MOS-related ACR ratings for each sample of 0.47. (c) A histogram of the SQS values with a mean of 2.83 and a mean standard deviation of two SQS-related ACR ratings for each sample of 0.74.

For SQS prediction, the VoxConverse subset of 1,048 speech samples along with their ground truth SQS values were divided into training (80%), validation (10%), and testing (10%), respectively. To avoid false discoveries due to overfitting problems or insufficient data for model retraining, we employed the k-fold cross-validation technique (with $k = 5$) during both the training and evaluation phases. This approach enabled us to measure performance on each fold and derive the final performance metrics as the mean across all folds. Additionally, to gauge the generalization of the model, we assessed its performance on the RTVE2020 subset.

3.1 Comparison of Subjective Speech Quality Measures

Information related to speech quality is collected by both subjective measures to some extent, as shown in Fig. 2a, where both SQS and MOS measurements exhibit a moderate correlation ($r = 0.4$). However it seems that information related to the transcription process is more effectively captured by the SQS measurement, as shown in Figs. 3 and 4, where annotation factors such as transcription time ($r = -0.54$) and number of playbacks ($r = -0.56$) show the strongest correlation with the SQS measurement. Additionally, the WER shows a higher correlation ($r = 0.38$) with SQS than with MOS. This suggests that WER could partially account for speech quality in the annotation context, as

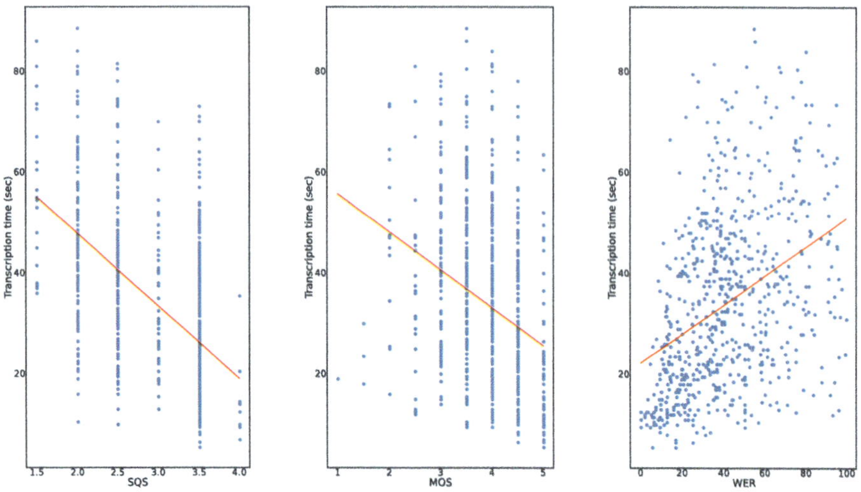

(a) (b) (c)

Fig. 3. (a) Scatter plot between the Transcription time and SQS measure with a correlation of $r = -0.54$. (b) Scatter plot between the Transcription time and MOS measure with a correlation of $r = -0.31$. (c) Scatter plot between the Transcription time and WER with a correlation of $r = 0.38$. Correlation values are significantly different from 0 ($p < 0.001$).

it reflects the effort of transcribing an audio sample under specific acoustic conditions, which may influence annotation accuracy. For example, lower levels of speech quality (e.g., Poor or Bad), might be correlated with an increased number of missing words or deletions, resulting from poor intelligibility of spoken content in the audio.

In terms of inter-annotator agreement, a higher consensus among annotators is observed for the MOS measurement, indicated by the lower level of dispersion compared to ACR values associated with MOS, as shown in Fig. 2b and Fig. 2c, in contrast to the ACR values for the SQS. We attribute this observation to the ease of achieving high agreement levels on quality in the MOS compared to the SQS, as the former relies solely on the perception of speech quality without considering factors that could impact annotation quality. Nevertheless, it is important to acknowledge that annotator expertise and others related factors should also be considered, as they may introduce bias into the assessment of the speech quality in this context.

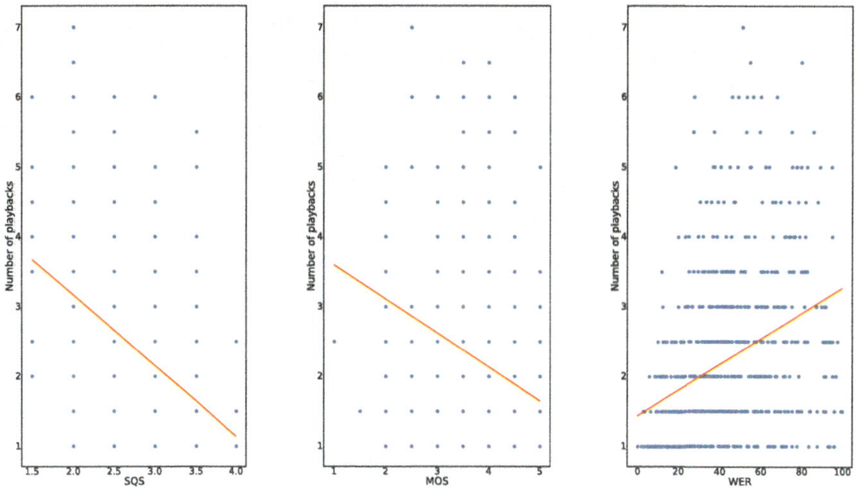

(a) (b) (c)

Fig. 4. (a) Scatter plot between the Transcription time and SQS measure with a correlation of $r = -0.55$. (b) Scatter plot between the Transcription time and MOS measure with a correlation of $r = -0.29$. (c) Scatter plot between the Transcription time and WER with a correlation of $r = 0.35$. Correlation values are significantly different from 0 ($p < 0.001$).

3.2 Speech Quality Prediction

Subjective speech quality in the annotation context (SQS) represents the effort of transcribing an audio under certain acoustic conditions that might degrade the transcription quality. However, this process is time-consuming and expensive due to the large number of participants required to conduct listening tests and provide perceptual ratings. To overcome this issue, several studies have developed DNN-based systems to predict the MOS measurement [9,10,12] with high levels of accuracy. Following a similar approach, we proposed two different approaches for predicting SQS measurements: 1) Finetuning the NISQA model, and 2) using ECAPA2 speaker embeddings [15] as speech quality predictor (Table 2).

Table 2. SQS prediction results, in terms of Pearson correlation (r) and MSE, for the proposed models, evaluated in VoxConverse and RTVE2020 datasets.

	VoxConverse		RTVE2020	
Model	r	MSE	r	MSE
Finetuned NISQA	0.66	0.52	0.34	1.06
ECAPA2	0.51	0.58	0.21	0.57

Moderate correlations ($r = 0.66$, $r = 0.51$) achieved by both approaches confirmed the effectiveness for SQS values prediction, which could be used as a substitute for human listeners to evaluate the speech quality within an annotation pipeline. Regarding the RTVE2020 dataset, weak correlations ($r = 0.34$, $r = 0.21$) might be due to differences in acoustic conditions and language between both evaluation datasets.

4 Conclusions

We present an exploratory analysis of a proposed subjective speech quality measurement, called Speech Quality Score (SQS), suitable to be used within the audio data annotation process. This measure involves the most relevant characteristics that could impact transcription quality. The comparison with the subjective quality measurement (i.e., MOS) and transcription quality metric (i.e. WER) shows that the SQS measure involves factors that could degrade the annotation quality, showing a higher correlation between those factors and the SQS. This emphasizes the necessity for an appropriate subjective measure within the annotation context, as most speech quality measures primarily address speech quality under the context of the transmission systems or the naturalness of synthetic voice.

Additionally, we proposed two different methods for predicting SQS measurements. Our experimental findings demonstrate the possibility of predicting SQS values by deep learning models, where predicted values are moderately correlated with human ratings.

In future work, we will explore different lines to address, such as looking for additional annotation-related factors, increasing the dataset with a high number of raters, as well as integrating the SQS prediction model in our annotation process, which would allow us to quantify its impact in a real large-scale project.

Acknowledgements. This research leading to these results is framed under the HADA project: "Research on advanced data annotation tools" (2021/C005/00146323), funded by the EU Next Generation through the public business entity attached to the Ministry of Economy and Digital Transformation, RED.ES. In addition, this work used data from the RTVE2020 database. This dataset was transferred by RTVE Corporation with the aim of contributing to the development of speech technologies in Spanish language.

References

1. Kondo, K.: Estimation of speech intelligibility using perceptual speech quality scores. Speech Lang. Technol. 155–174 (2011)
2. Chua, E., Fedor, J., Collins, C., Malenfant, A.: Quantifying speech intelligibility based on crowdsourcing (2017)
3. Springbord. The importance of audio annotation in 2023: What is it? (2023). https://www.springbord.com/blog/the-importance-of-audio-annotation-in-2023-what-is-it/. Accessed 05 April 2024

4. ITUT Rec. P. 808, subjective evaluation of speech quality with a crowdsourcing approach. ITU-T, Geneva (2018)
5. American National Standards Institute. American National Standard: Methods for Calculation of the Speech Intelligibility Index. Acoustical Society of America (1997)
6. ITU-T Recommendation. Perceptual evaluation of speech quality (PESQ): an objective method for end-to-end speech quality assessment of narrow-band telephone networks and speech codecs. Rec. ITU-T P. 862 (2001)
7. SigmaAI. The fundamentals of audio annotation (2023). https://sigma.ai/fundamentals-audio-annotation-services/. Accessed 05 April 2024
8. Lleida, E., et al.: Rtve2020 database description (2020). *URL:*http://catedrartve.unizar.es/reto2020/RTVE2020DB.pdf
9. Mittag, G., Naderi, B., Chehadi, A., Möller, S.: NISQA: a deep CNN-self-attention model for multidimensional speech quality prediction with crowdsourced datasets. arXiv preprint arXiv:2104.09494 (2021)
10. Reddy, C.K.A., Gopal, V., Cutler, R.: DNSMOS: a non-intrusive perceptual objective speech quality metric to evaluate noise suppressors. In: ICASSP 2021-2021 IEEE International Conference on Acoustics, Speech and Signal Processing (ICASSP), pp. 6493–6497. IEEE (2021)
11. Yang, Z., et al.: Fusion of self-supervised learned models for MOS prediction. arXiv preprint arXiv:2204.04855 (2022)
12. Oliveira, E.C., Junior, A.C., Gris, L.R.S., Soares, A.S., Galvão Filho, A.R.: Evaluation of speech representations for MOS prediction. In: Ekstein, K., Pártl, F., Konopík, M. (eds.) Text, Speech, and Dialogue. TSD 2023. LNCS, vol. 14102, pp. 270–282. Springer, Cham (2023). https://doi.org/10.1007/978-3-031-40498-6_24
13. Williams, J., Rownicka, J., Oplustil, P., King, S.: Comparison of speech representations for automatic quality estimation in multi-speaker text-to-speech synthesis. arXiv preprint arXiv:2002.12645 (2020)
14. Tseng, W.-C., Huang, C., Kao, W.-T., Lin, Y.Y., Lee, H.: Utilizing self-supervised representations for MOS prediction. arXiv preprint arXiv:2104.03017 (2021)
15. Thienpondt, J., Demuynck, K.: ECAPA2: a hybrid neural network architecture and training strategy for robust speaker embeddings. In: 2023 IEEE Automatic Speech Recognition and Understanding Workshop (ASRU), pp. 1–8. IEEE (2023)
16. Chung, J.S., Huh, J., Nagrani, A., Afouras, T., Zisserman, A.: Spot the conversation: speaker diarisation in the wild. arXiv preprint arXiv:2007.01216 (2020)

Enhancing Image Annotation Through Attention Mining: A Grounded SAM Approach

Jaime Cortón-González(✉), Ángel Mora-Sánchez, and Silvia Rodríguez-Jiménez

Sigma Cognition, Avenida de Manoteras 6, 28050 Madrid, Spain
jcorton@sigmacognition.ai

Abstract. In the domain of image annotation, the involvement of human annotators presents a series of intricate challenges tied to the complexities of visual perception. The manual labeling process demands an understanding of context and visual intricacies, all susceptible to human subjectivity. Furthermore, the presence of visual distractions compromises annotation quality, not only impeding annotation precision but also escalating time expenditures as annotators navigate through visual noise to discern pertinent details. Traditional image annotation pipelines underscore these challenges in favor of automatic or semi-automatic annotation, emphasizing the critical necessity for innovative approaches in annotation tasks where the human annotator role is fundamental. Within this context, the Grounded SAM model, emerges as a potent tool for text-prompt-based panoptic segmentation. This paper proposes a novel annotation pipeline employing Grounded SAM and LaMa cleaner models to augment the indispensable role of human annotators by enhancing annotation efficiency through natural language-based attention mining for visual distractions elimination and preannotation techniques. The effectiveness of distraction elimination is demonstrated through an annotation task involving human annotators, with half of the images processed through our pipeline and the remaining unmodified. With our current approach and with the current data analyzed, image annotation times of 70% of the annotators were reduced by 15.88%, while global annotation time was reduced by a 6.93%.

Keywords: image annotation · visual distractions · attention mining · computer vision

1 Introduction

Within the image annotation domain, the reliance on human annotators introduces a set of challenges tied to the nuances of visual perception. The process of manually labeling images demands a keen understanding of context and visual intricacies that can be susceptible to human subjectivity. Moreover, the quality of annotations is often compromised by the presence of visual distractions,

G. Marreiros et al. (Eds.): DCAI 2024, LNNS 1151, pp. 26–35, 2025.
https://doi.org/10.1007/978-3-031-80946-0_3

leading to inaccuracies and inconsistencies in the labeling process. Distractions impede annotation precision and increase time spent as annotators struggle to discern relevant details amidst visual noise. As we dig into the traditional image annotation pipelines, these challenges become palpable, underscoring the critical need for innovative approaches. In this context, Grounded SAM model [4], which is a combination of the Grounding DINO model [5] developed by IDEA-Research and Segment Anything (SAM) model [6] developed by Meta AI, emerges as a powerful model for text-prompt-based panoptic segmentation. In this paper, we propose a novel annotation pipeline that uses Grounded SAM and LaMa cleaner [7] models as building blocks with the purpose of keeping the valuable human annotator role by increasing its annotation efficiency through distraction elimination and preannotation techniques using natural language attention mining. As our experiment shows, the process of removing distractions is effective especially for certain types of images, tasks and degree of experience of the annotators.

2 Related Work

2.1 Image Annotation

According to the Cognilytica's 2020 white paper on AI Data Engineering Lifecycle Checklist [8], 50% of the total time spent in any AI Project goes to Data Cleaning (25%) and Data Labeling (25%) making those steps one of the most important ones along the project. This importance stems from the necessity of precise annotations, serving as the foundational truth for training models to accurately recognize and comprehend visual information. The meticulous annotation of images necessitates a careful deployment of resources, such as human annotators, computational infrastructure, and extensive manual efforts. The scarcity of good quality annotated image datasets, coupled with the escalating complexity and diversity of real-world visual data, poses substantial challenges. The labor-intensive nature of annotation tasks, relying on human expertise for precision, further exacerbates the strain on resources and time. In the context of advancing AI model's complexity and the escalating demand for labeled data, the development and implementation of efficient annotation methodologies, automation tools, and strategic resource management are imperative. Addressing these challenges is fundamental for optimizing the overall efficiency of AI projects, highlighting the intricate interplay between the importance of data labeling and the resource and time constraints inherent in image annotation.

There are few strategies aimed at optimizing annotation pipelines involving human annotators. The majority of these approaches focus on fully or partially replacing human annotators with automatic or semi-automatic annotation models [9,10]. This prevalent trend tends to overlook the significant role that human annotators play in certain annotation tasks that require high expertise. Nevertheless, there are some research developments centered in approaches to support the human annotator. Those methods consist on the use of just point annotations (click supervision) to train object detection models [11,12], or to create semantic segmentation models [13]. There are also some pipelines that eliminate the role of the human annotator by changing it to a human validator role, so that once the bounding box annotation is done by an automatic model, the human validator points the correct and incorrect annotations [14]. Moreover, active learning (AL) strategies try to reduce the amount of annotation payload delivered to single (single-head models) or multiple (multiple-head models) human annotators by selecting an optimal-most-informative subset of unlabeled images based on certain methods such us, approximate bayesian inference [15], informativeness of instances [16] or the classical uncertainty sampling [17]. More recent works of Xinpeng Wang et al. [18] and Connor Baumler et al. [19] explore the variety of annotation criteria for different annotators and use it as a building block for their pipeline. In the first case they use this variety to generate a disagreement aware active learning approach while in the second case it consists on an annotator aware approach, where the model chooses who to ask for new annotations based on the unique labeling properties of that annotator. Recently, the power of using a foundation models, like Grounded SAM [4], has been demonstrated as a key component for zero-shot labeling pipelines.

2.2 Saliency and Visual Distractions

Image annotation requires the usage of human annotators for a long period of time that depends on the annotation task difficulty and extension as well as the annotator expertise. Visual distractions introduce significant time delays over the annotation and impact on annotation quality and efficiency. Saliency is closely related to this behavior of human perception because salient regions can direct attention within an image. In the context of annotation, saliency is the inherent prominence or significance of visual elements and, a visual distraction is a salient visual element diverting annotators' attention from the annotation task and introducing ambiguity. Therefore, innovative solutions have been designed to mitigate the influence of distractions and enhance the accuracy of annotated datasets. The importance of visual attention mining has been widely studied, with several models trying to predict it on an image. This prediction can be based on traditional techniques, such as, background feature analysis [20] or fourier analysis [21] and it can use deep learning modern techniques like multi-scale multi-path fusion network with cross-modal interactions [22] or joint learning and densely-cooperative fusion (JL-DCF) architectures [23].

In particular, the DeepGaze model proposed by M. Kümmerer et al. [24] is the one that this study will focus on, being one of the golden standards for human saliency prediction as stated by the MIT300 Saliency Benchmark [25] and the MIT/Tuebingen Saliency Benchmark [26]. DeepGaze model, uses a complex architecture that allows the analysis of the interplay of scene content and scanpath history. This model will be used for the validation of human visual perception in terms of salient and distracting elements over an image. To remove distracting elements from images, the LaMa model by R. Suvorov et al. [7] excels in inpainting complex structures of images.

3 Methodology

3.1 Pipeline Description

The goal of this new pipeline is to optimize the annotation process by eliminating visual distractions, thus improving the efficiency of human annotation. This is achieved by enabling the use of natural language to identify and eliminate potential distractions. The pipeline generates pre-annotations from raw image input and cleans visual distractions, allowing human annotators to focus on validation and correction.

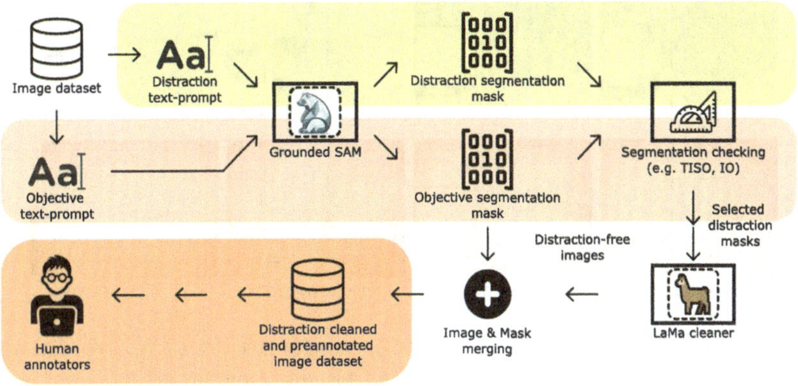

Fig. 1. Enhanced annotation pipeline: The image dataset is processed by Grounded SAM model to produce two segmentations (DES and OES). After segmentation checking, distracting elements are removed using the LaMa cleaner model if criteria are met. The result is a distraction-cleaned, pre-annotated dataset for faster final annotation refinement and correction (see Sect. 3.1), models' logos adapted from [1,2].

The initial setup involves a large dataset requiring annotation. Each annotation task has a specific goal, such as detecting or segmenting elements in the image. Based on this task objective, human experts will identify distracting and relevant elements using natural language, eliminating the need for programming skills. With that description, the distraction (DTP) and the objective text-prompts (OTP) will be generated. Subsequently, the DTP and OTP are inputted into the Grounded SAM model along with the image dataset, producing two segmentations per image: distracting elements segmentation (DES) and objective elements segmentation (OES). After obtaining two segmentations per image, a checking process analyzes the DES independently and compares it with the OES. Based on this analysis, the algorithm determines which DES are inputted into the LaMa cleaner model, considering the total image surface occupation (TISO) and inter-segmentation overlapping (IO). TISO is the user-defined percentage of DES occupation over the entire image, while IO sets a user-defined threshold for acceptable overlapping percentage between DES and OES. The final dataset will include then pre-annotations from the OES, with some images having distracting elements removed based on TISO and IO conditions. Human annotators then correct or add annotations where the Grounded SAM model was inaccurate (See Fig. 1 for segmentation task).

3.2 Experiment Explanation

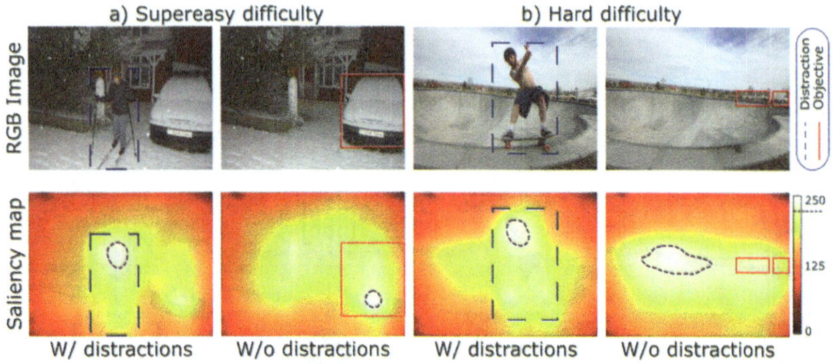

Fig. 2. Image difficulty visual explanation. The area of interest for the annotation task is marked in red and the crop is zoomed in for visualization purposes.

In order to evaluate the impact of the visual distraction elimination in the context of image annotation the following experiment was designed: **Annotation task definition**. Initially the annotation task was defined as a simple vehicle detection, where each annotator will have to draw a rectangular bounding box over all the vehicles of 4 or more wheels; **Selection of annotators**. Human annotators participating in this study were not professional annotators but they had different experiences within the computer vision/image annotation domain. This way we can classify them into 2 groups based on the expertise: With and without computer vision skills; **Selection of images**. All the images will be extracted from Microsoft Common Objects in Context (COCO) Dataset [3]. With the intention of generating a fair comparison, the images to annotate were selected so that the salient objects are not the objects of interest for the annotation task, making it some short of camouflaged object detection task. Images selected contained mainly people doing different activities as salient elements near the camera plane while the vehicles to annotate were in a secondary plane.

Data Preparation. The images were classified in 5 categories of difficulty (ultrahard, hard, medium, easy and supereasy. See Fig. 4 for the details) based on the scale of the annotations and the number of annotations to perform per

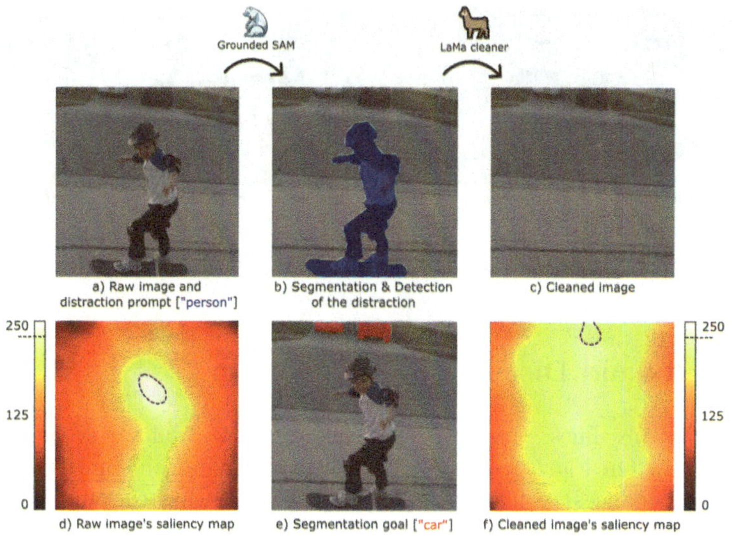

Fig. 3. A text-prompt describing the distraction elements (a) for this segmentation task (e) and a raw image (a) are used as input for Grounded SAM, after that, the segmentation masks produced are inputted to LaMa cleaner for its elimination within the background context. Over the second line, the saliency map of the raw image (d) and the cleaned image (f) are displayed, showing the attention shift towards the objects of interest for this segmentation task. The saliency map goes from 0 to 255, with contour lines shown at a value level of 240. Models' logos extracted from [1,2] and raw images extracted from [3].

image as well as the background complexity and proximity. Before the annotation experiment started, an attention mining strategy was followed by analyzing several saliency maps between image pairs (with/without distractions), this analysis confirmed our division in difficulty categories. On instances labeled as hard and ultrahard, the attention peak switched from the distraction object to the complex background instead of switching to the object of interest (See Fig. 2), while for the rest of categories the attention peak shifted towards the objects of interest (See Fig. 3). With that 5 categories in mind, the separation of the images in 2 sets of 20 images was done, ensuring that way that both annotation groups were having similar difficulty and that one annotator will not be annotating the distraction-free modified image and the original raw image at the same task. **Annotation process and evaluation**. Each annotator will have 20 images, where 10 will be distraction free images and 10 will be unmodified images. Finally, the annotation time was measured for each annotator and for each image annotated separately. The annotation tool used for this project was Label Studio labeling software [27].

Fig. 4. Examples of the 5 difficulty categories in ascendant difficulty order from left to right.

4 Results and Discussion

On the following lines, the annotation experiment results will be presented and discussed. On Fig. 5 (Supereasy, Easy, and Medium instances) and Fig. 6 (Supereasy instances) you can see barplots representing the annotation time difference by user comparing the annotation over images with and without distractions. The bars are displayed as the percentage that the median time difference between the annotation on images with and without distractions (median time with distractions minus median time without distractions) represents over the total mean annotation time of that user. Time is measured as seconds per bounding box annotated. Finally, the mean time percentage is displayed for the positive % bars (green dashdotted line) and negative % bars (red dotted line) as well as for both of them together (black dashed line). First 5 users correspond to annotators with computer vision skills and last 5 users correspond to annotators without computer vision skills (C.V.S). The analysis conducted in this study has revealed that for a 70% of the annotators, the removal of visual distractions present on images implied a reduction by a 15.88% in the annotation time per bounding box. For the whole annotation team, thanks to the visual distractions removal, the mean annotation time was reduced by a 6.93% (See Fig. 5). Moreover, considering just the supereasy difficulty images (difficulty category where the attention shift towards the objects of interest was more prominent), this time reduction was bigger, specifically an 8.92% for the whole group of annotators and having an 80% of annotators with positive impact instead of the previous 70%. Moreover, the annotation times without distractions present more data points in the lower tail and less on the upper tail of the distribution compared to the annotations with distractions.

In the displayed plots (Figs. 5 and 6), annotations on difficult images (ultrahard and hard) were excluded to maintain focus on distraction removal's efficacy. Such distractions had minimal impact on annotation times (median reduction: just 2.83% per user considering all difficulty categories). This variation in annotation times, influenced by annotators' diverse skills, warranted exclusion to avoid statistical noise coming from secondary distractions and undesirable attention shifts (See Fig. 2).

Observations revealed that distraction removal was less effective in images with complex backgrounds and prominent distracting objects (See Fig. 4). Anno-

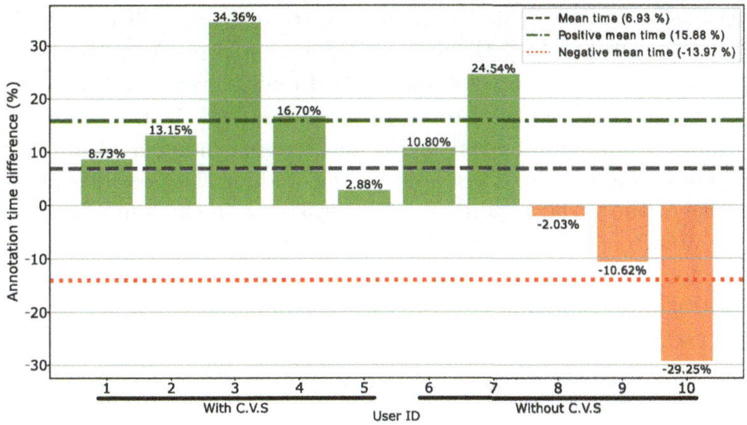

Fig. 5. Barplot representing the annotation time difference percentage by user for the supereasy, easy and medium instances.

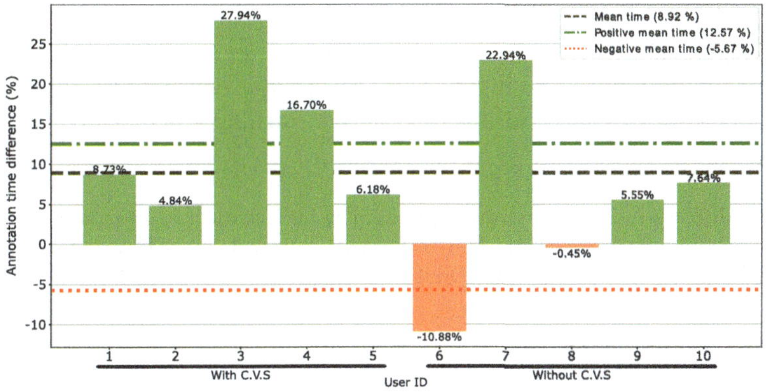

Fig. 6. Barplot representing the annotation time difference percentage by user for just the supereasy instances.

tators spent significant time searching for additional objects amidst such distractions. Conversely, distraction elimination worked better in images with foreground distracting objects and moderately complex backgrounds. Removing such distractions promptly redirected annotators' attention to the primary object of interest.

5 Conclusions

The proposed pipeline make use of the previously defined model to generate a promising framework that addresses these human-centric challenges by integrating natural language-based attention mining mechanisms to enhance the annotation process mitigating the impact of visual distractions. Grounded SAM's contribution is integral to this comprehensive framework, collectively elevating the quality and efficiency of the image annotation process, with the pipeline serving as the focal point of our approach. Finally, even with this promising initial results, there is plenty of room for improvement. As next steps it would be important to perform an study with an increased number of annotators as well as images to annotate. This way, we will reduce the possible noise generated because of the variable annotator's expertise with the annotation tool and we will be able to evaluate how the annotators behave in the long run after annotating for certain period of time. It would be interesting to include certain pupil tracking algorithm to asses how the annotators eye fixation behaves on the distraction-free images in order to evaluate the exact eye scanning path along the image in comparison with the images with distractions. The proposed pipeline implies a broader process that will include pre-annotations of the objects of interest, once the distraction elimination is validated, it will be necessary to perform a final study in which annotation times are measured between images processed with the full pipeline and raw images. With our current approach and with the current data analyzed, image annotation times of 70% of the annotators will be reduced by 15.88%, while the total annotation time will be reduced by a 6.93% and that's why in the future, it would be very interesting to have this promising pipeline linked to highly human-expertise-dependant image annotation domains such as medical imaging, active learning schemes or fine-grained visual recognition tasks.

Acknowledgements. Research framed under HADA project "Research on advanced data annotation tools" (2021/C005/00146323), funded by EU Next Generation (through Ministry of Economy and Digital Transformation & RED.ES).

Disclosure of Interests. Authors have no conflict of interest to declare.

References

1. Tainhe, R.: IDEA-Research. https://github.com/IDEA-Research. Accessed 05 Mar 2024
2. Suvorov, R., et al.: Image Inpainting, Resolution-robust Large Mask Inpainting with Fourier Convolutions, WACV 2022. https://github.com/advimman/lama. Accessed 06 Mar 2024
3. Lin, T., et al.: Common Objects in Context, Microsoft COCO (2015)
4. Ren, T., et al.: Assembling Open-World Models for Diverse Visual Tasks, Grounded SAM (2024)
5. Liu, S., et al.: Marrying DINO with Grounded Pre-Training for Open-Set Object Detection, Grounding DINO (2023)
6. Kirillov, A., et al.: Segment Anything. (2023)
7. Suvorov, R., et al.: Resolution-robust Large Mask Inpainting with Fourier Convolutions (2021)
8. Cognilytica AI Data Engineering Lifecycle Checklist. https://www.cloudera.com/content/dam/www/marketing/resources/whitepapers/ai-data-lifecycle-checklist-cloudera-whitepaper.pdf?daqp=true

9. Adnan, M., Rahim, M., Rehman, A., Mehmood, Z., Saba, T., Naqvi, R.: Automatic image annotation based on deep learning models: a systematic review and future challenges. IEEE Access. **9**, 50253–50264 (2021)
10. Cheng, Q., Zhang, Q., Fu, P., Tu, C., Li, S.: A survey and analysis on automatic image annotation. Pattern Recogn. **79**, 242–259 (2018)
11. Papadopoulos, D., Uijlings, J., Keller, F., Ferrari, V.: Extreme clicking for efficient object annotation (2017)
12. Papadopoulos, D., Uijlings, J., Keller, F., Ferrari, V.: Training object class detectors with click supervision (2017)
13. Bearman, A., Russakovsky, O., Ferrari, V., Fei-Fei, L.: What's the Point: Semantic Segmentation with Point Supervision (2016)
14. Papadopoulos, D., Uijlings, J., Keller, F., Ferrari, V.: We don't need no bounding-boxes: training object class detectors using only human verification (2017)
15. Gal, Y., Ghahramani, Z.: Dropout as a bayesian approximation: representing model uncertainty in deep learning. In: Proceedings of the 33rd International Conference on Machine Learning, vol. 48, pp. 1050–1059 (2016). https://proceedings.mlr.press/v48/gal16.html
16. Zhang, M., Plank, B.: Cartography Active Learning (2022)
17. Lewis, D., Gale, W.: A Sequential Algorithm for Training Text Classifiers. In: SIGIR 1994, pp. 3–12 (1994)
18. Wang, X., Plank, B.: ACTOR: Active Learning with Annotator-specific Classification Heads to Embrace Human Label Variation (2023)
19. Baumler, C., Sotnikova, A., Daumé III, H.: Which examples should be multiply annotated? Active learning when annotators may disagree. In: Findings of the Association For Computational Linguistics: ACL 2023, pp. 10352–10371 (2023). https://aclanthology.org/2023.findings-acl.658
20. Feng, D., Barnes, N., You, S., McCarthy, C.: Local Background Enclosure for RGB-D Salient Object Detection. In: Proceedings of the IEEE Conference on Computer Vision and Pattern Recognition (CVPR) (2016)
21. Li, J., Levine, M., An, X., Xu, X., He, H.: Visual saliency based on scale-space analysis in the frequency domain. IEEE Trans. Pattern Anal. Mach. Intell. **35**, 996–1010 (2013). https://doi.org/10.1109/TPAMI.2012.147
22. Chen, H., Li, Y., Su, D.: Multi-modal fusion network with multi-scale multi-path and cross-modal interactions for RGB-D salient object detection. Pattern Recogn. **86**, 376–385 (2019). https://www.sciencedirect.com/science/article/pii/S0031320318303054
23. Fu, K., Fan, D., Ji, G., Zhao, Q.: JL-DCF: joint learning and densely-cooperative fusion framework for RGB-D salient object detection. In: 2020 IEEE/CVF Conference on Computer Vision and Pattern Recognition (CVPR), pp. 3049–3059 (2020)
24. Kümmerer, M., Bethge, M., Wallis, T.: DeepGaze III: modeling free-viewing human scanpaths with deep learning. J. Vision. **22**, 7 (2022). https://doi.org/10.1167/jov.22.5.7
25. Bylinskii, Z., et al.: MIT Saliency Benchmark
26. Kümmerer, M., et al.: MIT/Tübingen Saliency Benchmark. https://saliency.tuebingen.ai/
27. Tkachenko, M., Malyuk, M., Holmanyuk, A., Liubimov, N.: Label Studio: Data labeling software. https://github.com/heartexlabs/label-studio Open source software. https://github.com/heartexlabs/label-studio

Can Large Sound Event Detection Models Be Accurately Adapted to Specific Acoustic Scenarios?

Ramón Fernández-Castañón[1(✉)], Fernando M. Espinoza-Cuadros[2], Juan M. Perero-Codosero[2], Ester Sancho-Lozano[2], and Luis A. Hernández-Gómez[1]

[1] ETSIT - Universidad Politécnica de Madrid, Avenue Complutense 30, 28040 Madrid, Spain
r.fcastanon@alumnos.upm.es

[2] Sigma Cognition, Avenida de Manoteras 6, 28050 Madrid, Spain

Abstract. Sound Event Detection (SED) is an audio signal processing field aiming at the automatic identification of sound events within an audio stream. SED systems are gaining increasing relevance on a variety of applications including context-aware chatbots, environmental surveillance, urban noise management, healthcare monitoring, etc. The fast development of Deep Learning (DL) together with the availability of large datasets, such Audioset, has led to the emergence of a new generation of SED models that can identify a broad range of sound classes (generally 527 Audioset classes); we will refer them as Large SED (LSED). Besides this technological advance, adapting LSED to specific applications is still challenging mainly because it requires an important effort for accurate labelling of specific sound events in a representative set of audios. To address this challenge, this paper explores three learning strategies to LSED adaptation: 1) zero-shot learning, just using supervised semantic mapping between the LSED classes to application-specific classes, 2) transfer learning, training a DL model to map from all LSED classes to application-specific classes, and 3) active learning, DL mapping only over problematic subset of sounds. DL mapping was explored using seq2seq models (BLSTM and Self-Attention Transformer Encoder). Several experimental setups are presented to show how different LSED adaptation strategies can provide different balances between sound detection accuracy and audio labelling cost.

Keywords: Sound Event Detection · Deep Learning · Zero-shot Learning · Transfer Learning · Active Learning · Strong Labels · Semantic Mapping · Seq2Seq Models

1 Introduction

1.1 Sound Event Detection Systems

Sound Event Detection (SED) is an emerging field in audio signal processing that focuses on the automatic identification and classification of various sound events within an audio stream. A sound event is any audible occurrence that can be distinguished from the

G. Marreiros et al. (Eds.): DCAI 2024, LNNS 1151, pp. 36–45, 2025.
https://doi.org/10.1007/978-3-031-80946-0_4

background noise, such as a dog barking, a car horn, or a door closing. The importance of SED for many applications is to provide capabilities to interpret and interact with our environment, providing a more intuitive and immersive user experience. The complexity of automatic detection of acoustic events is very high due to the complex way in which sound sources interact and overlap in different acoustic scenes. Automatic acoustic event detection and classification can be addressed from two different perspectives: *audio tagging* and *sound event detection* [1]. In Audio Tagging (AT) the goal is to detect the presence of sound events within an audio clip without its temporal localization, while Sound Event Detection (SED) recognizes and temporally localizes sound events. In this paper we will consider SED systems.

SEDs have experienced a relevant progress over the last years thanks to organizations such as DCASE[1] (Detection and Classification of Acoustic Scenes and Events) promoting research and innovation in the field. Over recent years, Deep Learning (DL) has greatly impacted SED systems in a similar way as in other audio processing areas, like speech recognition or music information retrieval. Training DL models requires large number of annotated audios, and collecting and labeling sound events is a very costly and time-consuming task. The recent availability of large audio datasets, mainly Audioset,[2] has somewhat alleviated this problem. Audioset is a collection of over 2 million human-labeled 10-s sound clips from YouTube videos, covering 632 sound classes. Adapting pre-trained models to a particular task mitigates the needs for such large number of audios. Several pre-trained models have been developed along the years, starting from Convolutional Neural Network (CNN) architectures such as YAMNet, VGGish [2] or PANN [3] and progressing to Convolutional Recurrent Neural Networks (CRNNs) that uses CNNs to extract local features followed by LSTMs to model the temporal context and dynamics of sounds [4]. But it is with the irruption of Self-Attention and Transformers when more powerful pre-trained SED systems have appeared [5], we will refer them as Large SED or LSED. More recently, self-supervised learning (SSL) as BEATS (Bidirectional Encoder representation from Audio Transformers) offers excellent possibilities for fine tuning SED systems [6].

Despite these significative advances, we believe that there is still a gap to apply them into real-world applications. Practical applications usually require detecting a small set of sound classes (a few tens) and only a small budget is available to record and label audios for adapting a pre-trained LSED system. Although some research challenges have been focusing on these situations (e.g. DESED (Domestic Environment Sound Event Detection), ESC-50 (Environmental Sound Classification), TUT SED, etc.) their results cannot be extrapolated to real deployments scenarios.

To reduce this research-industrial gap, in this paper we explore a process based on two main steps. First, we propose and evaluate a simple baseline SED system based on a supervised semantic mapping between Audioset ontology and application-specific sound classes. Second, we will train specific DL models to map the output of the LSED system to the application-specific sound classes. It is important to highlight that audio labelling for SED training is particularly costly as it requires strongly annotated labels, or *strong labels*, where sounds are labeled including precise start and end times for each

[1] https://dcase.community/.
[2] https://research.google.com/audioset/.

sound event. Therefore, in our experiments we will consider the compromise between sound event detection accuracy and cost for labelling training audios.

2 Adapting LSEDs to Specific Applications

The proposed LSED adaptation process assumes that, once the set of sound classes for a specific SED application have been defined, there is a budget to record a set of representative audios and manually annotate them with strong labels. Then we leverage on an open source state-of-the-art LSED system, trained on Audioset, to process these audios and provide activation heatmaps for 527 sound classes. That is, for each audio **x**, an activation heatmap can be represented as a matrix H^{TxC}, where every element $h_{t,c}$ corresponds to the *posterior* probability of the presence of a sound class c at a given time frame t. T is the number of time frames corresponding to the whole audio duration and C is the number of Audioset classes C $\in \{1, \ldots 527\}$. Once we have this information available, our next step will be to manually create a mapping between application sound classes and Audioset sounds.

2.1 Supervised Semantic Mapping

Manual sound mapping is not easy, but even describing sound classes is not trivial due to ambiguities in audio characteristics, their sources and acoustic scene. Supposing we are given a clear description of application specific sound classes, manual semantic mapping presents many difficulties, being *subjectivity* the main one. Different individuals may have different interpretations of subtle distinctions in sibling and parent–child relations in Audioset ontology [7] and several biases, inconsistencies and incorrect assignments may appear. To alleviate this, once a first manual semantic mapping is available, we perform a statistical analysis on a relatively small dataset of audio recordings including a balanced representation of each application sound class. All audios are first processed through the LSED system to obtain their corresponding activation heatmaps H^{TxC}. The scores in theses heatmaps will indicate the probability of Audioset sound classes at each specific time frame. Audios must also be manually annotated to obtain ground truth strong labels for each one of 527 Audioset classes. Using these H^{TxC} heatmaps and strong labels, a statistical analysis will study time overlaps between Audioset classes with high activation levels and event intervals in ground truth strong labels. Given an Audioset sound A and an application sound B, not mapped to A, we search for situations where Audioset activations of A (above a given threshold) occurs during ground truth time segments of B. We measure overlaps in terms of frame percentages. For the detected overlaps a manual supervision needs to be done to decide whether: a) a missed sound class has occurred, that is, the Audioset class A has not been included in the semantic mapping for B, or b) the frequent activation of A and B classes during overlapped time segments may point at high confusable sounds when A and B are clearly different sound classes.

2.2 Baseline SED

Once the supervised semantic mapping has been obtained, the pre-trained LSED can be easily used in a zero-shot way (i.e. no training data) to create what we will call application-specific baseline SED, or simply baseline SED. Defining S^{TxD}, as the application heatmap, where D is the number of application-specific sound classes. The $s_{t,d}$ scores or *posterior* probabilities of a sound event class d at time frame t can be easily obtained by taking the maximum score of the subset of mapped Audioset sound classes in LSED heatmap H^{TxC}. That is:

$$s_{t,d} = \max_{k \in g(d)} h_{t,k} \quad \text{for } t = 1, \ldots, T \text{ and } d = 1, \ldots, D \tag{1}$$

where $g(d)$ represents a subset, among all Audioset classes C, that are semantically mapped to the application-specific sound class d. We will refer this approach as semantic max pooling over Audioset heatmaps.

$$S^{TxD} = f_{semax}\left(H^{TxC}\right) \tag{2}$$

SED Performance Evaluation. The resulting SED baseline system can be tested using the same dataset collected for our statistical analysis. However, the evaluation of SED systems is very challenging as the different nature of sounds presents a high variability both in activation scores and accuracy. Furthermore, due to subjectivity in the interpretation of the temporal structure of sound events, inaccuracies in labeling strong labels are very common [8]. Consequently, when evaluating a SED system, both the tolerance in the detected start and end times, as well as the tolerance for human annotators must be considered. Hopefully the research community has recently proposed a SED evaluation framework [9] that considers both annotator and system tolerances to define event detection ROC curves, called Polyphonic Sound Detection ROC (PSD-ROC) curves, plotting TPs (True Positives) and FP (False Positives) rates per unit of time for different operating points. A single performance metric PSDS score is also defined providing more comprehensive insights in system performance, considering several working points, than classical F1 scores. PSD-ROC curves and PSDS score will be used during our experimental evaluations in Sect. 3.

2.3 Seq2Seq Models for LSED Adaptation

To improve the zero-shot baseline SED system, in this section we propose applying deep neural networks (DL) to map between LSED H^{TxC} and application S^{TxD} heatmaps:

$$S^{TxD} = f_{\theta}\left(H^{TxC}\right) \tag{3}$$

where $f_{\theta}(.)$ will be a DL model with weights $\{\theta\}$. Considering that H^{TxC} heatmaps are internal representations of a pre-trained LSED system, training a DL model on top of it can be considered a transfer leaning approach. Additionally, we consider that seq2seq (sequence-to-sequence) DL models are well-suited for mapping an input sequence H^{TxC} to an output sequence S^{TxD}, where T represents the time dimension, and C and D represent the feature dimensions of the input and output sequences, respectively. To explore labelling cost vs. event detection accuracy, experiments in Sect. 3 will consider two main dimensions: different DL models and different mapping tasks with different needs of labeling data. A simple BLSTM (Bidirectional Long Short-Term Memory) and a more complex Self-Attention Transformer Encoder were considered. Different mapping strategies requiring different labelling costs can be formulated considering that application heatmaps S^{TxD} can be obtained concatenating two heatmaps: 1) S^{TxD^1}, where D^1 are classes obtained using supervised semantic mapping (semantic max pooling) with 2) S^{TxD^2} obtained after training a seq2seq model to map the rest of D^2 classes, with $D = D^1 + D^2$. That is:

$$S^{TxD} = \text{concat}\begin{cases} S^{TxD^1} = f_{semax}\left(H^{TxC^1}\right) & D^1 \rightarrow C^1 \text{semantically mapped} \\ S^{TxD^2} = f_\theta\left(H^{TxC^2}\right) & D^2 \rightarrow C^2 \text{ seq2seq mapping} \end{cases} \tag{4}$$

where C^1 and C^2 corresponds to Audioset classes used for mapping application-specific classes D^1 and D^2 respectively. Using this formulation, higher accuracy, but also high labelling cost, can be expected when $D^1 = 0$ and $D^2 = D$; that is when all application classes D are mapped using a seq2seq model. While lower accuracy and labeling costs may be expected when using a) semantic mapping (requiring no training data) for D^1 classes performing well (above a given threshold), and b) only training the seq2seq model to model a subset D^2 of lower performance classes. Note that different possibilities in Eq. (4) can come not only when using different classes in C^1 (depending on the performance threshold over the baseline system), but also because C^2 can be defined in different ways. For example, it can be the whole set of 527 Audioset classes, it can only include those classes in Audioset that are mapped to D^2, or it may include all classes in C that are mapped to D.

3 Experimental Case Study

3.1 Case Study: Database and LSED System

We have defined a scenario with the following 27 application sound classes: *Car*, *Train*, *Boat_water* (sounds of water vehicles), *Motorcycle*, *Helicopter*, *Gunshot*, *Fireworks*, *Burst_pop_boom*, *Alarms*, *Door*, *Dishes*, *Frying*, *Blender*, *Typing*, *Bells*, *Dog*, *Cat*, *Bird*, *Farm_animals*, *Wind*, *Water*, *Fire*, *Respiratory_sounds*, *Human_locomotion*, *Clapping_applause*, *Loud_unintelligible_speech*, and *Reverberation_echo*. For our experiments we rely on audios extracted from Audioset strong labels train and evaluation partitions [8]. Using already annotated audios we can easily design and evaluate the proposed LSED adaptation approaches with different cost/performance balances.

We selected the Hierarchical Token-Semantic Audio Transformer, HTS-AT [10] as a state-of-the-art LSED. HTS-AT introduces a hierarchical transformer structure with a token semantic CNN layer after the final transformer block that will provide the Audioset heatmap H^{TxC}.

3.2 Statistical Analysis and Supervised Semantic Mapping

An initial manual mapping between 527 Audioset classes and 27 application sounds from a human expert was supervised using the statistical analysis in Sect. 2.1. A subset of 1890 audios (70 audios for each one of the 27 classes) from Audioset strong-labels evaluation partition was used. These audios were processed through HST-AT to get sound classes activation scores. We then looked for cases where an Audioset sound A overlaps during a significant percentage of time with application sound B in ground truth intervals, and A and B have not assigned in the semantic mapping. Some mistakes during the manual mapping were detected: Audioset classes *Sirens* and *Buzzer* where not mapped to *Alarms*, Audioset *Fowl* was not assigned into *Farm Animals*, and Audioset classes *Firecracker* and *Fireworks* were not assigned into *Burst pop boom*.

In other cases, confusable or interacting sound classes were identified. This is the case for Audioset class *Water tap, faucet* that was found to frequent activates when application class *Dishes* is annotated. This is not an error or mistake in the mapping, it is just that usually the sound of one event typically occurs in the presence or company of another one. In this case they both could be related in a higher category of Kitchen sounds.

3.3 Baseline SED Evaluation

The *supervised semantic mapping* (Sect. 3.2) was used to define the baseline SED that was evaluated using the psds_eval[3] tool on the same 1890 audios used for our statistical analysis; this set of audios will be the test dataset for all our experiments in this paper. PSDS scores and PSD-ROC curves for individual classes (left plot) and whole system (right) are presented in Fig. 1. As it can be seen in the left plot of Fig. 1 there is a high variability in the performance for different sound classes. This is an expected behavior in SED due to the different complexities and nature of sounds. In our next experiments (Sect. 3.4) though we will try to improve all the sound classes, we will also explore the possibilities of putting annotation efforts only on those more problematic ones. To that aim, from the detailed analysis of ROC curves, the following 5 sound classes with the lower performance were selected: *Boat_water* (sounds from water vehicles), *Motorcycle*, *Fire*, *Respiratory_sounds* and *Loud_unintelligible_speech*.

3.4 Improving the Baseline SED Using Seq2Seq Models

Two adaptation scenarios were explored. In a *full mapping* scenario, we assume that the cost of having a relatively high number of annotated audios is acceptable and thus we can train seq2seq models that map all classes from C to D. That is, in Eq. (4) $D^1 = 0$, i.e. no semantic mapping is used, thus $D^2 = D$, and $C^2 = C$. While in what we call *Lower performance mapping*, seq2seq is only used to map the 5 sound classes that presented the lowest performance in the baseline SED; consequently, the required number of annotated audios may be smaller than in *full mapping*.

3.5 Comparing Mapping Strategies

For *full mapping*, seq2seq models must map all classes from C to D. That is, in Eq. (4) $D^1 = 0$, i.e. no semantic mapping is used, thus $D^2 = D$, and $C^2 = C$. In our experiments

[3] https://pypi.org/project/psds-eval/.

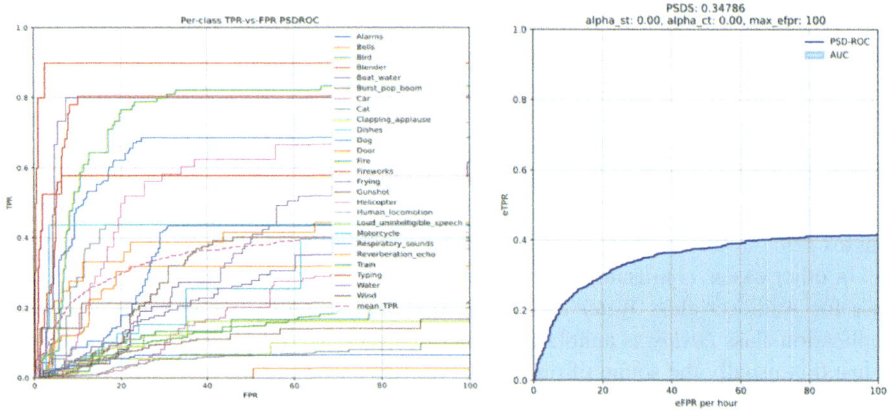

Fig. 1. PSD-ROC curves and PSDS score for Baseline SED.

we first trained the seq2sqe model using all available 49,574 Audioset audios, to have a reference for top accuracy at highest annotation cost. Then we considered more restrictive labelling situations with 5,400, 2,700 and 1,350 audios, corresponding to annotation efforts of 50, 100 and 200 audios for each one of 27 sound classes, respectively. As described in Sect. 2.4, two seq2seq models were evaluated: a BLSTM model and a Self-Attention Transformer encoder. The BLSTM model was quite simple: including a single Bidirectional LSTM layer with dimensionality 256, followed by a time-distributed dense layer with 27 sigmoid outputs (resulting in 1,619,483 trainable parameter). The Self-Attention model was an 8 heads Transformer encoder with feed forward dimension 2048, followed by a top time-distributed dense layer with 27 sigmoid outputs; no positional encoding was used (11,078,034 parameters). Evaluation results in terms of PSDS scores were 0.41, 0.37, 0.34 and 0.30, for BLSTM, and 0.44, 0.41, 0.39 and 0.38 for Self-Attention when trained with 49,574, 5,400, 2,700 and 1,350 audios respectively. Figure 2 presents the corresponding PSD-ROC curves, also including the Baseline SED curve that obtained a PSDS score of 0.35.

According to these results, full mapping improves de Baseline SED for both BLTSM and Self-Attention except when BLTSM is trained with 1,350 or 2,700 audios (blue dotted and dot-dash lines in left plot Fig. 2). Also, as expected, Self-Attention always provides significative better results than BLSTM, and when Self-Attention is trained with only 5,400 audios it reaches a performance slightly better than BLTSM trained with 49,574 audios. Self-Attention results trained on 2,700 or 1,350 audios achieves a very similar performance.

In the *lower performance mapping* scenario, seq2seq models only map 5 application classes $D^2 = 5$ and semantic mapping is used for $D^1 = 22$ (see (4)). The labelling cost will be significantly lower as now we only need to annotate 5 instead of 27 sound classes. As in the full mapping scenario, three experimental setups were defined to evaluate the balance between annotation cost and performance. In this case seq2seq models were trained with: 1) 2400 audios (400 per 5 low performance classes + 400 for other classes), 2) 1200 audios (200*5 + 200) and 3) 600 audios (100*5 + 100). Only Self-Attention Transformer encoder was considered due to its clear better performance than

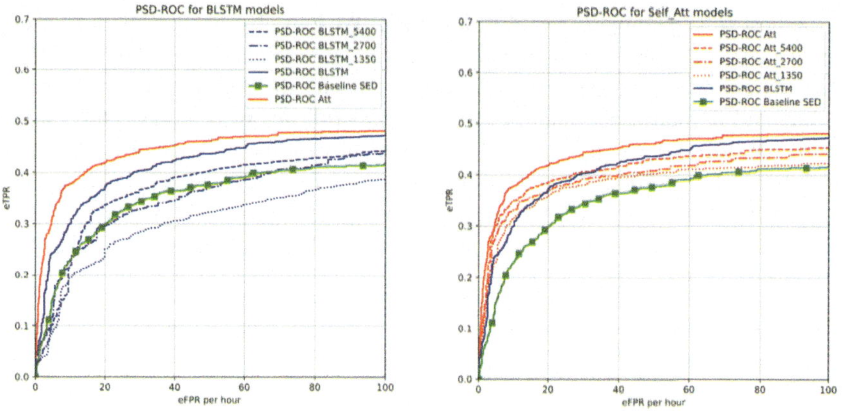

Fig. 2. PSD-ROC curves for Baseline SED, BLSTM and Self-Attention Transformer models trained with different number of annotated audios (Color figure online).

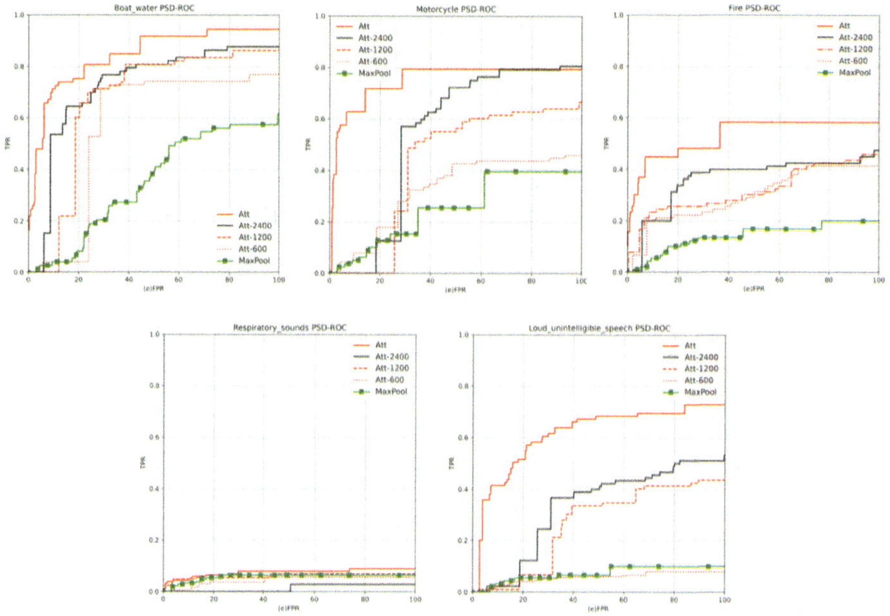

Fig. 3. PSD-ROC curves for *lower performance* adaptation.

BLSM in full mapping. The complexity of the Transformer encoder was reduced to 4 heads and feed forward dimension 64. PSDS scores for 2400, 1200 and 600 audios were very similar: 0.39, 0.39 and 0.38 respectively. Although all of them are slightly above 0.35 Baseline SED, it is not a surprising that mapping only 5 sound classes may only have a small increase on the global PSDS score. PSD-ROC curves for each one of the 5 classes are presented in Fig. 3, including ROC curves for the Baseline SED and the best full mapping using Self-Attention. We can observe that the use of class-specific mappings improves 4 of the 5 sound classes (*Boat_water*, *Motorcycle*, *Fire* and *Loud_unintelligible_speech*). PSD-ROC curves for these classes are well above the Baseline and in many cases approaching the best full mapping model. Only *Respiratory_sounds* showed no improvement and the results for the baseline and full mapping models are also very poor. After carefully inspection for this class, we concluded that it is a complex sound that in many cases appears heavily overlapped with speech events making it barely audible. This makes that the strong labels are also in many situations unreliable.

4 Conclusion and Future Lines

In this work, several approaches to adapt state-of-the-art pre-trained SED systems (LSED) to specific applications have been proposed. Results have shown that a simple baseline system, relying on manual semantic mapping, can provide a reasonable performance only requiring the labelling cost for a small dataset of audios. When higher labeling budgets are possible, we have shown how seq2seq models can increase the system performance mapping all sound or only those with low performance. Future research will consider the extension of this work fusing an ensemble of LSED systems as well extending our experiments different industrial applications. Examples of planned applications include environmental monitoring, where SED systems could detect sounds related to illegal activities, such as hunting in protected areas, helping the preservation of wildlife, and context-aware conversational agents, where the interaction could be adapted to different user's acoustic scenes.

Acknowledgments. This research leading to these results is framed under the HADA project: "Research on advanced data annotation tools" (2021/C005/00146323), funded by the EU Next Generation through the public business entity attached to the Ministry of Economy and Digital Transformation, RED.ES. UPM research in this work was also funded by the European Commission (grant ASTOUND (101071191 HORIZON-EIC-2021-PATHFINDERCHALLENGES-01).

Disclosure of Interests. The authors have no competing interests to declare that are relevant to the content of this article.

References

1. Ebbers, J., Haeb-Umbach, R.: Self-trained audio tagging and sound event detection. In: Domestic Environments. Proceedings of the 6th Detection and Classification of Acoustic Scenes and Events 2021 Workshop (DCASE2021), pp. 226–230 (2021)
2. Hershey, S., et al.: Cnn architectures for large-scale audio classification. In: Proceedings International Conference on Acoustics, Speech and Signal Processing (ICASSP), New Orleans, LA, USA (2017)
3. Kong, Q., Cao, Y., Iqbal, T., Wang, Y., Wang, W., Plumbley, M.D.: PANNs: large-scale pretrained audio neural networks for audio pattern recognition. In: IEEE/ACM Transactions on Audio, Speech, and Language Processing, vol. 28, pp. 2880–2894 (2020)

4. Lu, Z.: Sound event detection and localization based on CNN and LSTM. In: Detection Classification Acoustic Scenes Events Challenge, Tech. Rep. (2019)
5. Gong, Y., Chung, Y.A. Glass, J.: AST: audio spectrogram transformer. In: Proceedings Conference of the International Speech Communication Association (INTERSPEECH), Brno, Czech Republic, pp. 571–575 (2021)
6. Shao, N., Li, X., Li, X.: Fine-tune the pretrained ATST model for sound event detection. In: Proceedings IEEE International Conference on Acoustics, Speech and Signal Processing (ICASSP), Rhodes Island, Greece, pp. 911–915 (2023)
7. Raj, B.: Improving sound event detection with ontologies. J. Acoustical Soc. Am. **153** (3_supplement), A364-A364 (2023)
8. Hershey, S., et al.: The benefit of temporally-strong labels in audio event classification. In: IEEE Proceedings International Conference on Acoustics. Speech and Signal Processing (ICASSP), ON, Canada, Toronto, pp. 366–370 (2021)
9. Bilen, C., Ferroni, G., Tuveri, F., Azcarreta, J., Krstulović, S.: A framework for the robust evaluation of sound event detection. In: Proceedings IEEE International Conference on Acoustics, Speech and Signal Processing (ICASSP), Barcelona, Spain, pp. 61–65 (2020)
10. Chen, K., Du, X., Zhu, B., Ma, Z., Berg-Kirkpatrick, T., Dubnov, S.: HTS-AT: a hierarchical token-semantic audio transformer for sound classification and detection. In: Proceedings IEEE International Conference on Acoustics, Speech and Signal Processing (ICASSP), Singapore, pp. 646–650 (2022)

Innovative Quality Metrics for Enhanced Interpretation of Instance Segmentation in Complex Image Scenarios

Juan Pedro Llerena[1(✉)], Miguel A. Patricio[2], José Manuel Molina[2], Ángel Mora-Sánchez[3], and Silvia Rodríguez-Jiménez[3]

[1] Computer Science Department, Universidad de Alcalá, Alcalá de Henares, Spain
jp.llerena@uah.es

[2] Applied Artificial Intelligent Group, Universidad Carlos III de Madrid, Colmenarejo, Spain

[3] Sigma Cognition, Avenida de Manoteras 6, 28050 Madrid, Spain

Abstract. Manual data annotation involves human annotators labeling and reviewing data according to predefined criteria. It can be time-consuming and expensive, so automated data annotation based on artificial intelligence algorithms has gained popularity. However, the classical integrated artificial intelligence models have some limitations, especially in complex scenarios characterized by occlusions, low resolution, and illumination variability. This paper introduces new evaluation methods for image instance segmentation, focusing on the need for easy-to-understand quality metrics in the face of the complexity of image annotation. Considering the difficulties of manual annotation processes and the diversity of computer vision models focused on instance segmentation, four quality metrics are proposed: Number of Detections, Normalized Bounding Box Area, Normalized Mask Area, and Bounding Box Occlusion Rate. These metrics attempt to overcome the difficulties of select the best model for automated labeling based on conventional metrics such as Intersection Over Union and Mean Average Precision, especially in scenarios where classical models have limitations. The application and analysis of these metrics in the OVIS database demonstrates their potential to improve the interpretation of instance segmentation models, thus facilitating a more accurate and accessible automated annotation process.

Keywords: Image Annotation Tools · Image Quality Metrics · Image Instance Segmentation and Image Detection

1 Introduction

Image annotation tasks are of vital importance for industry and research. Generally, the annotation process is manual, and the tools used often assist annotators to easily generate lightweight files that specify regions of interest or related categories among others.

Some image annotation tools as [1, 2] integrate artificial intelligence models, which aim to speed up the hard and repetitive process of annotation, helping to reduce the cost, time-consuming and minimizing human subjectivity.

G. Marreiros et al. (Eds.): DCAI 2024, LNNS 1151, pp. 46–55, 2025.
https://doi.org/10.1007/978-3-031-80946-0_5

Nowadays there are a great variety of computer vision models in charge of image segmentation at different levels. However, the selection of a model is not necessarily trivial, as some models are specialized in accuracy, others in speed, etc. [3]. To evaluate image segmentation models, there are quality metrics such as Intersection over Union (IoU) and Mean Average Precision (mAP) among others, which can be difficult to understand when the scene casuistry to be segmented is varied for an annotator.

Some key challenges for segmentation systems include, viewpoint variation, occlusions, lighting conditions, textured background, or shape variety among others [3].

Therefore, detectors must be able to recognize objects considering these difficulties, i.e. be invariant to these challenges.

To address these challenges in a generalist way in terms of objects, there are many classical databases such as COCO [4] or DAVIS [5] among others [6–8] that include people as part of the objects. However, there are highly challenging databases such as Moments in Time (MiT) [9] or Occluded Video Instance Segmentation (OVIS) [10] that pose a challenge in terms of detection due to their contextual complexity.

This paper proposes four innovative quality metrics to address the difficulties of interpreting object segmentation models in images, especially in complex situations involving occlusions, low resolution or distant objects, and variations in illumination. These factors significantly complicate the task of accurately segmenting objects, especially people, both indoors and outdoors, and present limitations for existing computer vision systems, especially with respect to instance segmentation.

This paper is organized as follow: Sect. 2 introduces the proposal including quality metrics description. Section 3 shows the possibilities of cases of interest and dataset. Section 4 shows the results and discussion of the approach on specific cases. Finally, the conclusions are discussed in Sect. 5.

2 Proposal

Accurate evaluation of instance segmentation under specific real-world conditions poses significant challenges, especially due to limitations in the interpretation of classical metrics such as Intersection Over Union (IoU) and Mean Average Precision (mAP) in the automated data annotation process.

In view of this problem, we propose the development of a set of innovative quality metrics designed to provide a more comprehensive assessment of the performance of instance segmentation models for automated labeling. These new metrics seek to incorporate crucial aspects in the annotations that are difficult to interpret in traditional metrics, such as the number of detections, the detection area, or the occlusion rate.

2.1 Quality Metrics

Due to the difficulty to interpret the general metrics (IoU and mAP) by non-experts, it is proposed to develop a set of interpretable and easily explainable qualitative metrics to help determine which model performs better in different situations.

The four-quality metrics proposed are Number of Detections (ND), Normalized Bounding Box Area (NBA), Normalized Mask Area (NMA), and Bounding Box Occlusion Rate (BOR). An example of the quality metrics for an image from the OVIS database is shown in Fig. 1. For example, Fig. 1b) shows that there are 6 detections (ND) that are easily identifiable in Fig. 1 a). The following three metrics are bounded between 0–1 and their properties are detailed in the following subsections.

Number of Detections (ND)
Since detection is implicit in the segmentation of instances, it was decided to use the error in the number of detections as a quality metric (1).

This metric is of high interest in cases where the objects are in complex visibility conditions (very small objects, appear blurred in the background, are occluded, etc.).

$$error_x = \frac{x - \hat{x}_i}{x} \tag{1}$$

where x is the number of labeled objects in the image, $\hat{x}_i$ is the number of objects detected by the i-detection system.

Normalized Bounding Box Area (NBA)
With this metric, we measure the accuracy of the detections on the database. In this case, the accumulated area of a list of bounding boxes (bboxes) is calculated and normalized by dividing it by the total area of the image. This involves going through all bboxes, computing their area, dividing it by the total area of the image (height multiplied by width) and adding it to a global normalized area counter (2).

$$NBA = \sum_{\mathrm{k}=0}^{n_j} \frac{B_k}{Width_{\mathrm{j}} \cdot Height_{\mathrm{j}}} \tag{2}$$

were B_k is the area of bounding box k, and n is the number of detections in j-image. The error in this metric is the difference between the Ground Truth (GT) and the value estimated by the i-model.

Normalized Mask Area (NMA)
The purpose of this metric is to know the accuracy of the segmentation masks. In this case the area of the masks is normalized by the total area of the image. We receive a list with n_j- matrices of dimension equal to the size of the j-image $(W_j \cdot H_j)$ (each k-element of this list is associated with each k-object present in the image), and we perform the sum of 1's that are in the whole k-mask, whose sum yields the area of the mask of that object $\delta_{\alpha\beta}$. Then, the mask area is divided by the total area of the j-image and added to a counter representing the sum of the different normalized areas.

$$NMA = \sum_{k=0}^{n_j} \frac{\sum_{\alpha=0}^{W_k} \sum_{\beta=0}^{H_k} \delta_{\alpha\beta}}{W_j \cdot H_j} \tag{3}$$

The error in this metric is the difference between the GT and the value estimated by the i-model.

Bounding Box Occlusion Rate (BOR)
This metric has been developed by the OVIS research team [10], and is designed to be able to measure and discriminate different levels of occlusions. The calculation is performed using the bbox areas of the objects as input parameter, through their coordinates (4). BOR perform the union of the areas of the intersections between bboxes, and then divide it by the union of the areas of all bboxes, all in absolute value.

$$BOR = \frac{\left|\bigcup_{i,j}(B_i \bigcap B_j)\right|}{\left|\bigcup_k B_k\right|} \tag{4}$$

(a)

	GT
ND	6
NBA	0.148257
NMA	0.052740
BOR	0.496867

(b)

Fig. 1. OVIS video 10, frame 19. a) Ground Truth (GT) annotations, b) Values of the quality metrics on the GT.

3 Cases of Interest and Dataset

The proposed cases of interest focus on a combination of 10 classes (scenes), different subclasses and the combinations between them, reflecting a wide range of contexts and situations in the annotation of images and videos of people. These scenes are described in detail below:

1. **Type of space:** Defines whether the action takes place indoors or outdoors, with specific classes for indoors, outdoors, and situations where the context is unknown.
2. **Time of day:** Distinguishes between daytime scenes, nighttime scenes, and scenes where the time of day cannot be determined.
3. **Lighting conditions:** Evaluates the quality of light present and classifies it as good or bad, which directly affects the visibility and perception of elements in the image.
4. **Weather conditions:** Provides a variety of environmental contexts such as sunny, cloudy, rainy, foggy, or unknown, adding a layer of complexity to the environmental analysis of scenes.
5. **Occlusion type:** This class is particularly rich in subclasses because of the different ways in which objects or people can obstruct the view of other elements. It includes Person-to-Person (PoP), Object-to-Person (OoP), Non-Occlusion, and combinations of these such as PoP-OoP, PoP-No, and PoP-OoP-No, reflecting the complexity and variability of visual interactions in images.

6. **Camera motion:** This is classified as fixed or variable, indicating whether the camera remains static or follows the action, which has significant implications for the perception of motion and image stability.
7. **Distance from camera to people:** This class is broken down into short, medium, long, and various combinations of these (short-medium, short-long, medium-long, and short-medium-long), providing a detailed understanding of the scale and perspective of people in relation to the camera.
8. **Image capture angle:** Identifies the camera's point of view as either contrapposto, normal, or swooping, each providing a different visual interpretation of the scene.
9. **View of people:** This category allows for multiple subclasses (Reflections, Slice, No Skin, Crowded), capturing the variety of ways people can appear in the image. Multiple categories or none can be assigned per image.
10. **Image quality:** Difference between blurred and normal images, a critical distinction for analyzing the clarity and accuracy of detail in the image.

In order to have a reference dataset to evaluate the behavior of a set of images to be annotated, based on the characteristics that define it or the challenges for the segmentation of people found in it, in this work we proceeded to the annotation of these categories of scenes in a series of videos. For this new annotation, we use the OVIS database on the subset of 82 videos showing more than 2 people. The new annotation is done manually by video experts following the next process: 1) Annotation task definition; 2) Data selection and preparation; 3) Annotation process; 4) Review process to assure accuracy and consistency. Once the data is annotated, the evaluation of the quality metrics is carried out for the evaluation of performance of several models on instance segmentation of persons.

An example of categories corresponding to the scene shown in Fig. 2 b) are listed in Fig. 2 a).

1. **Type of space:** indoor
2. **Time of day:** unknown
3. **Lighting conditions:** good
4. **Weather conditions:** unknown
5. **Occlusion type:** PoP-OoP
6. **Camera motion:** variable
7. **Distance from camera to people:** long
8. **Image capture angle:** swooping
9. **View of people:** slice, no skin

(a)

(b)

Fig. 2. Example of case of interest. a) Description of the scene by categories, b) Example scene video Id = 67 frame 1.

4 Results and Discussion

The set of possible combinations of cases of interest with the ten types of scenarios and their individual peculiarities is very high, so to illustrate to the reader the behavior of the proposed quality metric, we show an illustrative case.

Consider a type of indoor scene where there are PoP occlusions and non-occlusions throughout the scene, and there are reflections. This condition is identified in the video with Id = 10 corresponding to a scene of a dance choreography with six dancers in a rehearsal room as in Fig. 3b). The complete scene has 300 frames. Figure 3c) shows the case with high occlusion level in which the annotations only identify 5 dancers.

Figure 3a) shows that throughout the video, except for frame 175, the number of people is constant and equal to 6. Figure 3b) and c) show a case with all dancers and a reflection in a mirror and a case with high occlusion.

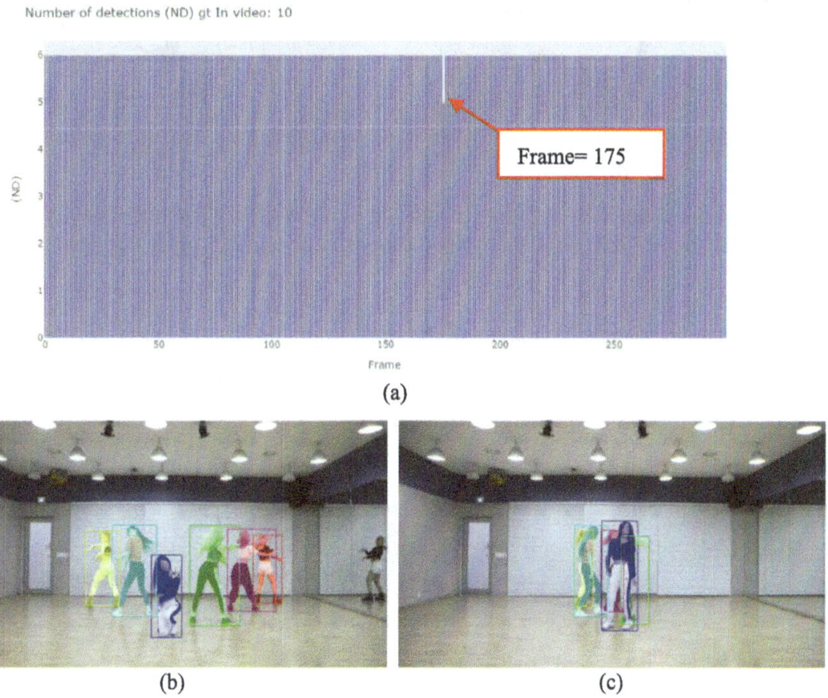

Fig. 3. Number of Detections (ND) in video Id = 10 OVIS. a) ND along the video, b) General situation, frame 204, c) High occlusion in frame 175.

The normalized areas of the bounding boxes throughout the video are shown in Fig. 4a). Figure 4b) and c) show the images of the minimum and maximum values throughout the video. It is found that in the first case, Fig. 4 b), the objects are in the background so the bboxes are "small" while in the second case, Fig. 4c), the objects are in the foreground of the scene, so their size is larger.

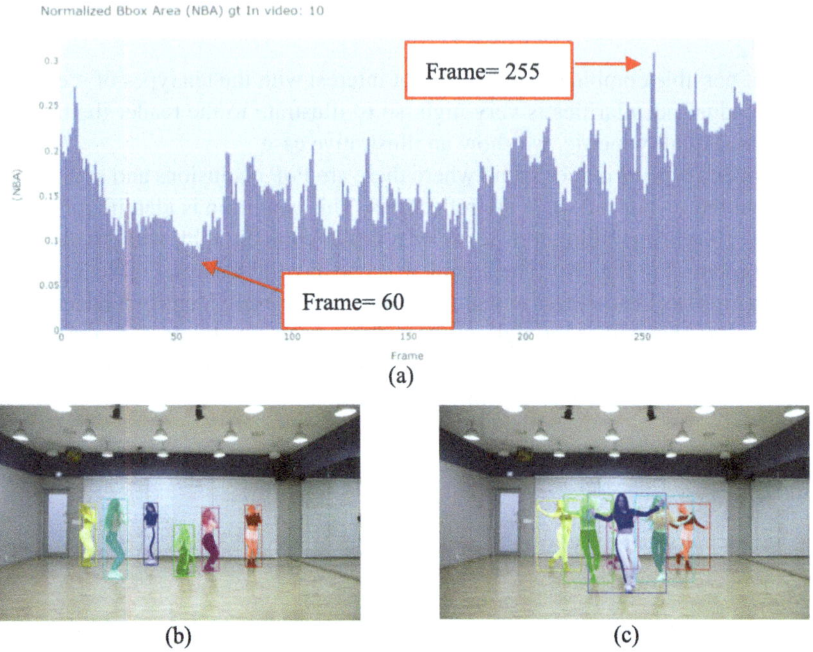

Fig. 4. Normalized Bbox Areas (NBA) in video Id = 10 OVIS. a) NBA along the video, b) NBA Minimum, Frame 60, NBA = 0.086, c) NBA Maximum, Frame 255, NBA = 0.309.

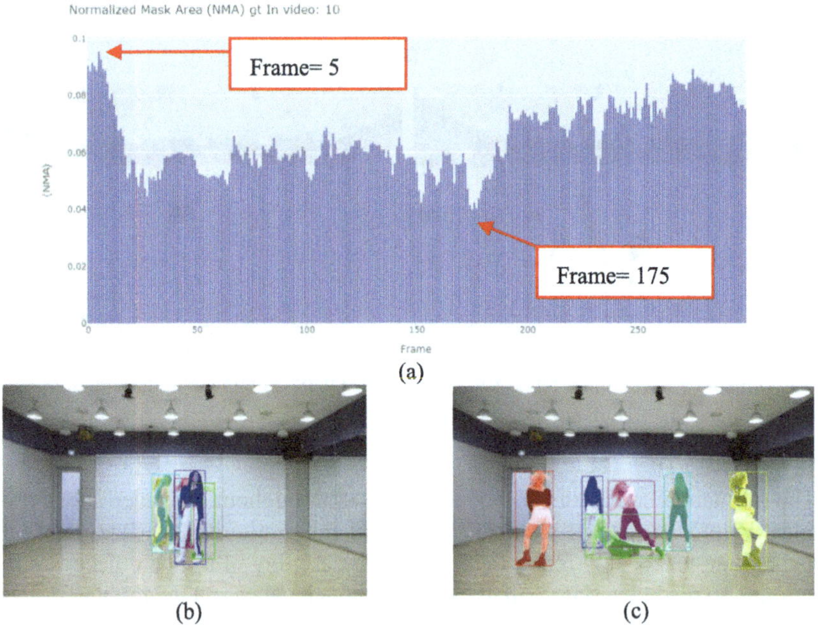

Fig. 5. Normalized Mask Areas (NMA) in video Id = 10 OVIS. a) NMA along the video, b) NMA Minimum, Frame 175, NMA = 0.096, c) NMA Maximum, Frame 5, NMA = 0.039.

Regarding the normalized mask areas, Fig. 5 a) shows their result throughout the video. The minimum value is identified in a high occlusion area where there are small parts of the objects, Fig. 5 b), while the value is maximum in a scene with "large" foreground objects, Fig. 5 c).

Finally, the BOR metric along the video Id = 10 is shown in Fig. 6 a). The higher the occlusions, the higher the BOR, this can be seen between Fig. 6 b) and c). In Fig. 6 b) the scene is free of occlusions and BOR = 0, while in the second one a high level of occlusions is shown, Fig. 6 c).

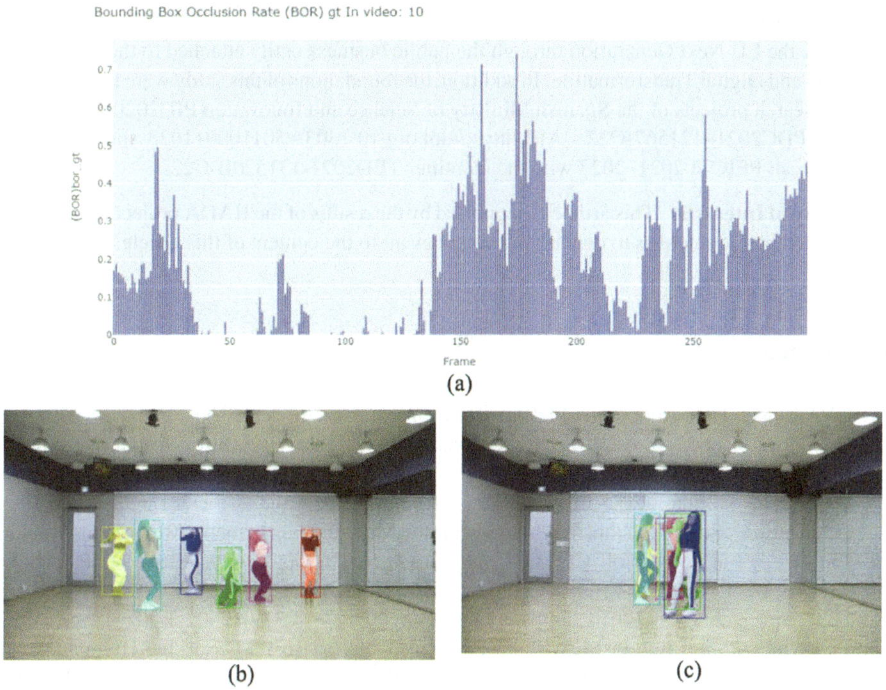

Fig. 6. Bounding box Occlusion Rate (BOR) in video Id = 10 OVIS. a) BOR along the video, b) BOR Minimum, Frame 55, BOR = 0, c) BOR Maximum, Frame 174, BOR = 0.741.

5 Conclusions

Innovative quality metrics show consistent results in scene segmentation. The number of detections is an easily understandable and computable metric, so its evaluation along the scenes allows the application of new segmentation & tracking algorithms. The Normalized Bounding Box Area (NBA) and Normalized Mask Area (NMA) identify different levels of object segmentation, but the evolution of the scenes throughout the example video shows that the results are related and consistent between them.

Finally, Bounding Box Occlusion Rate (BOR) allows to quantify the degree of occlusion between objects, which is illustrated in the examples shown to demonstrate the consistency between the scenes and the results obtained.

The promising results of the innovative quality metrics have the potential to be used with advanced instance segmentation algorithms in powerful advanced data annotation tools.

Acknowledgments. We would like to thank the SIGMA COGNITION S.L. team, in particular Esther Sancho and Pierre Plaza, for their support throughout the project and the trust placed in the GIAA team for the development "HADA project: Research on advanced data annotation tools". This research leading to these results is framed under the HADA project (2021/C005/00146323), funded by the EU Next Generation through the public business entity attached to the Ministry of Economy and Digital Transformation. In addition, the foundations of this study were funded by the public research projects of the Spanish Ministry of Science and Innovation PID2020-118249RB-C22 and PDC2021–121567-C22 - AEI/https://doi.org/10.13039/501100011033 and the project under the call PEICTI 2021–2023 with the identifier TED2021-131520B-C22.

Disclosure of Interests: This article is supported by the results of the HADA project. The authors have no competing interests to declare that are relevant to the content of this article.

References

1. CVAT, Open Data Annotation Platform. https://www.cvat.ai/ Accessed 11 May 11 2024
2. Roboflow, "Roboflow Annotate: Label Images Faster Than Ever. https://roboflow.com/annotate Accessed 11 May 11 2024
3. García, J., Molina, J.M., Llerena, J.P., Amigo, D., Sánchez, D.: Engineering UAS Applications: Sensor Fusion, Machine Vision and Mission Management. Artech House (2023)
4. Lin, T.Y., et al.: Microsoft COCO: Common Objects in Context. In: Fleet, D., Pajdla, T., Schiele, B., Tuytelaars, T. (eds) Computer Vision – ECCV 2014. ECCV 2014. LNCS, vol. 8693. Springer, Cham (2014). https://doi.org/10.1007/978-3-319-10602-1_48
5. Perazzi, F., Pont-Tuset, J., McWilliams, B., Van Gool, L., Gross, M., Sorkine-Hornung, A.: A benchmark dataset and evaluation methodology for video object segmentation. In: Proceedings of IEEE Computer Society Conference on Computer Vision and Pattern Recognition, vol. 2016-Decem, pp. 724–732, (2016). https://doi.org/10.1109/CVPR.2016.85
6. Everingham, M., Eslami, S.M.A., Van Gool, L., Williams, C.K.I., Winn, J., Zisserman, A.: The pascal visual object classes challenge: a retrospective. Int. J. Comput. Vis. **111**(1), 98–136 (2015). https://doi.org/10.1007/s11263-014-0733-5
7. Galasso, F., Nagaraja, N. S., Cardenas, T. J., Brox, T., Schiele, B.: A unified video segmentation benchmark: Annotation, metrics and analysis. In: Proceedings of IEEE Computer Society Conference on Computer Vision, pp. 3527–3534 (April 2013). https://doi.org/10.1109/ICCV.2013.438
8. Gupta, A., Dollar, P., Girshick, R.: Lvis: a dataset for large vocabulary instance segmentation. In: Proceedings of IEEE Computer Society Conference on Computer Vision and Pattern Recognition, vol. 2019, pp. 5351–5359 (June 2019). https://doi.org/10.1109/CVPR.2019.00550

9. Zhou, B., Zhao, H., Puig, X., Fidler, S., Barriuso, A., Torralba, A.: Scene parsing through ADE20K dataset. In: Proceedings of 30th IEEE Computer Society Conference on Computer Vision and Pattern Recognition, CVPR 2017, vol. 2017, pp. 5122–5130 (Janua 2017). https://doi.org/10.1109/CVPR.2017.544
10. Qi, J.,et al.: Occluded Video Instance Segmentation: Dataset and ICCV 2021 Challenge. In: NeurIPS, pp. 1–11 (2021). http://arxiv.org/abs/2111.07950

DCAI - SS-DHC

Special Session on Digital Heritage Contents (DHC)

Scope

The "Special Session on Digital Heritage Contents" features a prominent session that serves as a platform for diverse discussions on emerging trends and perspectives at the intersection of digital technology and cultural heritage preservation. The session aims to facilitate the exchange of ground-breaking ideas and research findings, expanding insights into how we can more effectively understand and preserve our history and cultural heritage in the current digital age.

Objective

This special session aims to provide a comprehensive view of the current state and future directions of the field of digital heritage contents. By promoting collaboration and knowledge exchange among international researchers, experts, and scholars, the session seeks to deepen the understanding of the impact of digital technology on cultural heritage.

Topics

Potential topics include, but are not limited to:

Digital Archiving and Preservation: Discussion on the latest trends and technological challenges in archiving and preserving cultural heritage using various methods offered by digital technology.
Utilization of Virtual Reality (VR) and Augmented Reality (AR): Research and case studies on using virtual and augmented reality technologies to provide new experiences and convey cultural heritage in innovative ways.
3D Scanning and Modeling: Exploration of techniques and technological challenges in utilizing digital 3D scanning and modeling to accurately reproduce and visualize archaeological sites and cultural heritage.
Integration of Digital Art and Cultural Heritage: Examination of the impact and interaction of digital art with cultural heritage, including discussions on collaborative possibilities for creative expression.
Digital Platforms and Education: Research and experiences related to leveraging digital platforms to innovate cultural heritage education, including discussions on various methods and the distribution of educational materials.

Organizing Committee

A Case Study on Digital Contents Using UNESCO Cultural Heritage in ASEAN

Sang-hee Wi(✉) and Hee-soo Choi

Sangmyung University, 20, Hongjimun 2-Gil, Seoul, Jongno-Gu, South Korea
weesh76@hnamil.net

Abstract. Digital heritage can increase accessibility to existing cultural heritage and change the way it is experienced through informatization and digitization of cultural heritage. And digital cultural heritage can be used for international cooperation and cultural exchange. In Korea, digital cultural heritage is being produced and utilized targeting UNESCO cultural heritage sites in the ASEAN region produced in Korea from 2000 to the present. The main purpose of these contents is digital restoration of individual cultural heritage, and the developed contents were mainly used for museum exhibitions. Applied digital technologies include 3D video, 3D scanning, VR, AR, and multimedia.

Keywords: digital heritage · UNESCO world heritage · ASEAN cultural heritage

1 Introduction

Cultural heritage is what we left behind in the past, what we live in the present, and what we pass on to future generations. Recently, digital technology is being actively integrated into the field of cultural heritage as the way the public enjoys cultural heritage changes due to the development of ICT technology.

Through the informatization and digitization of cultural heritage, it can be used for international cooperation and cultural exchange through cultural heritage.

Digital content is information produced digitally using the latest ICT technology, and is information that allows you to see, hear, and experience overseas cultural heritage without being restricted by time and space and without visiting the historic site. Accordingly, since 2000, content introducing world cultural heritage sites, including UNESCO cultural heritage sites, has been consistently produced in Korea.

Therefore, in this study, we examine examples of digital content targeting UNESCO cultural heritage in the ASEAN region among the digital cultural heritage developed in Korea over the past 20 years, and through this, seek ways for future international exchange and cooperation utilizing cultural heritage.

The research method first examines the concepts of UNESCO cultural heritage and digital cultural heritage through literature research. Second, we investigate and analyze digital content cases targeting UNESCO cultural heritage sites in the ASEAN region. Lastly, we will explore ways to develop and utilize digital cultural heritage content targeting overseas cultural heritage in the future.

G. Marreiros et al. (Eds.): DCAI 2024, LNNS 1151, pp. 61–70, 2025.
https://doi.org/10.1007/978-3-031-80946-0_6

2 Theoretical Background

2.1 UNESCO World Heritage

In 1945, UNESCO was created for mutual communication and cooperation in the fields of education, science, and culture. A representative example of UNESCO's cultural activities is the UNESCO World Heritage system. Based on the World Heritage Convention established in 1972, it is a system to share the value of heritage and preserve it for future generations by registering heritage that must be shared and preserved by humanity on the World Heritage List. Here, legacy refers to what we left behind in the past, how we live in the present, and what we pass on to future generations.

The agreement, which emphasizes the role of local communities, can address climate change, rapid urbanization, mass tourism, sustainable social and economic development, and other contemporary challenges. Today, 195 countries adhere to the Convention to identify and protect the world's most important cultural heritage.

UNESCO World Heritage is largely divided into cultural heritage, natural heritage, and mixed heritage. Among these, cultural heritage registration targets include monuments, ruins, and buildings, and Outstanding Universal Value (OUV), which is a core value, must be recognized by the advisory body and the World Heritage Committee.

There are various types of heritage and their values, such as the pyramids in Egypt, Baroque cathedrals in South America, and temples in Korea. The value of even a single piece of heritage can be perceived differently depending on time and space, and depending on groups that have different relationships with the heritage.

Therefore, UNESCO World Heritage is preserved and managed with the goal of preserving universal values. Additionally, cultural heritage must be fundamentally authentic in terms of materials and technology. Lastly, both cultural and natural heritage include various elements that can demonstrate the value of the heritage. The current World Heritage list is 1,199 (993 cultural heritage, 227 natural, and 39 mixed heritage), and 42 are scheduled to be added in 2023 [1] (Table 1).

Table 1. Classification of UNESCO World Heritage (@ UNESCO)

Name	Contents
Cultural heritage	**Monuments:** Heritage of outstanding universal value in history, art, and scholarship, among monuments, buildings, commemorative sculptures and paintings, archaeological relics and sculptures, inscriptions, troglodyte sites, and mixed sites
	Buildings: Heritage of outstanding universal value in history and art as independent or connected structures
	Historic Site: Heritage of outstanding universal value in terms of history, physiognomy, ethnology or anthropology in areas containing man-made or combined man-made and natural remains and archaeological remains

(*continued*)

Table 1. *(continued)*

Name	Contents
Natural heritage	Natural monuments containing inorganic or biological elements, geology and geological products, habitats for animals and plants, etc., which have outstanding universal value in terms of science, conservation and aesthetics
Mixed heritage	A heritage that have characteristics of both cultural and natural heritage

2.2 Digital Heritage

Since the 1990s, with the advancement of technology, digital contents has begun to be produced in the field of cultural heritage. In particular, UNESCO defined digital heritage as 'resources created through digital technology' or 'heritage converted into digital form' in the 2003 'Charter for the Preservation of Digital Heritage' [2].

Accordingly, digital heritage began to be introduced in Korea after 2000. At first, it was implemented through a web page through digital restoration of cultural heritage, but since 2010, various technologies such as 3D stereoscopic images, holograms, VR, AR, XR, AI, and multimedia have been applied. These contents are mainly used for exhibition, education, preservation, digital restoration and recording Fig. 1.

So far, projects related to Korea's digital cultural heritage have largely progressed in two directions. The first is 'digital records' such as the Cultural Heritage Administration's National Cultural Heritage Portal, the Academy of Korean Studies' Local Culture Electronic Exhibition Hall, and the National Museum of Korea's e-museum. The second is the 'digitization of individual heritage,' such as immersive content created by combining specific cultural assets with computer technologies such as 3D scanning, VR (virtual reality), AR (augmented reality), and MR (mixed reality), etc. [3].

Digital heritage is the computer-based production of cultural heritage that is valuable to preserve for future generations. Combined with digital technologies, content delivered through visualizations, interactive websites, crowdsourcing and virtual environments is transforming the way cultural heritage is accessed and experienced [4].

In particular, in the case of overseas cultural properties, direct visits may be difficult due to time and space constraints. However, in the case of digital cultural heritage, you can virtually experience historic sites through VR, AR, etc.

As such, the purpose of digital heritage is not only to preserve the heritage of the past, but also to enable future generations to use it creatively. Additionally, cultural heritage plays an important role in promoting cultural exchange and understanding around the world. Accordingly, Korea has been producing digital content targeting overseas cultural heritage, including Asian cultural heritage, for the past 20 years.

Initially, digitization of Angkor Wat (2006) were methods of digitally restoring cultural heritage and making it public on web pages. However, since 2010, it has evolved into a digital exhibition by applying various digital technologies such as 3D scanning, VR, AR such as the 'ASEAN Digital Cultural Heritage Project' (2017–2021) [5].

Fig. 1. Seokguram VR contents, 2016, (@ photo by TRIC, https://www.tric.or.kr, last accessed 2024/04/12)

3 Digital Contents of UNESCO Cultural Heritage in ASEAN

3.1 Association of Southeast Asian Nations (ASEAN)

ASEAN is an association of 10 Southeast Asian countries that promotes common development through political, economic, and security cooperation among member countries. ASEAN was founded in 1967 amid fears of communism and a thirst for economic development. And in 2007, they agreed to create the Community of Southeast Asian Nations, consisting of political, security, economic, social and cultural communities, by 2015. The member countries are Brunei, Cambodia, Indonesia, Laos, Malaysia, Myanmar, Philippines, Singapore, Thailand and Vietnam Fig. 2.

In the 21st century, ASEAN is attracting attention from around the world, to the extent that the term 'ASEAN connectivity and centrality' is mentioned. Southeast Asia, where ASEAN countries are located, has a population of over 600 million (8% of the world population) and is an economically and culturally important region [6].

As ASEAN relations improve, Korea is also promoting active exchanges across politics, economy, society, and culture. Currently, various races, religions, and cultures coexist in ASEAN. Additionally, this region has a long history, tradition and many excellent heritage sites. UNESCO World Heritage include Angkor Wat in Vietnam (1992), Luang Prabang in Laos (1995), Thang Long Imperial city in Vietnam (2010), and Borobudur Buddhist Temple in Indonesia (1991), etc. [7].

3.2 Digital Contents Using UNESCO Cultural Heritage in ASEAN

ASEAN is Korea's main export destination and one of Koreans' favorite tourist destinations. Therefore, economic, tourism, and cultural exchanges are currently taking place actively between Korea and ASEAN. In addition, Korea has consistently produced digital heritage contents using ASEAN's UNESCO World Cultural Heritages from the late 2000s to the present. Specific examples are as shown in Table 2.

Digitalization of Angkor Wat in Cambodia (2006). The Korean government considers history and traditional culture as a core element of cultural content creation and a new knowledge industry resource, and promoted the 'Cultural Archetype Digital Content Project' between 2002 and 2005 using traditional culture that contains cultural identity.

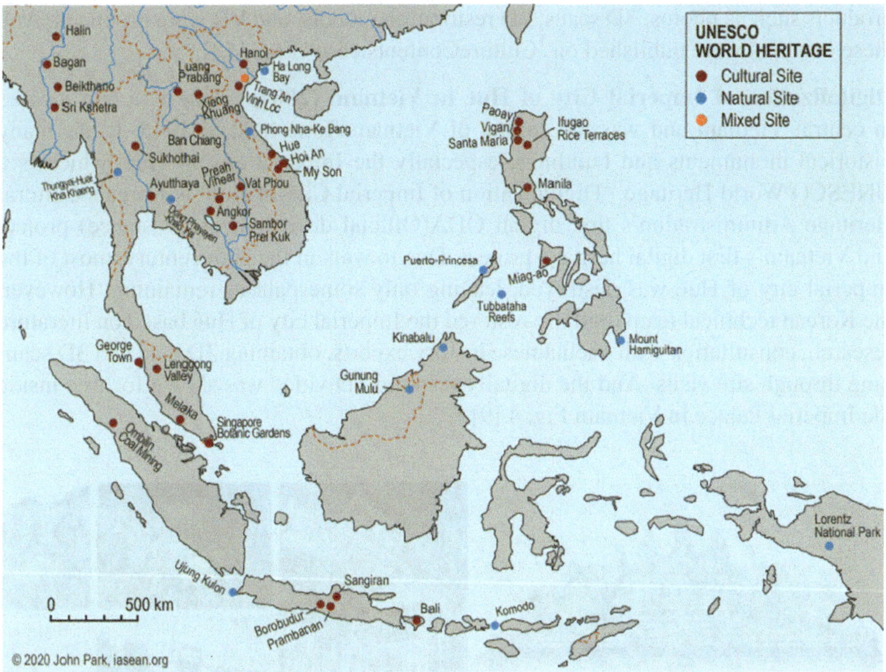

Fig. 2. Map of UNESCO World Heritage Sites in ASEAN (@ Institute of ASEAN studies, https://iasean.org/iasean0402/10613, last accessed 2024/04/12)

Table 2. Digital contents using ASEAN's UNESCO cultural heritage

	project	nation	year
1	Digitalization of Angkor Wat in Cambodia	Cambodia	2006
2	Digitalization of Imperial City of Hué in Vietnam	Vietnam	2007
3	3D stereoscopic video using Borobudur in Indonesia	Indonesia	2012
4	Virtual reality experience center in Imperial City of Hué	Vietnam	2018
5	ASEAN Digital Cultural Heritage Project	ASEAN	2017–2019
6	VR exhibition using Takht-i-Bahi Temple in Pakistan	Pakistan	2019
7	Immersive contents using the Ayutthaya Buddha statue at the Bangkok national Museum in Thailand	Thailand	2023

Angkor Wat is a royal palace built by Surya Varman II, who led the Khmer Empire to its heyday in the 12th century. The 'Angkor Wat Digitization Project' was to digitize Angkor Wat and make the Angkor ruins more accessible to the general public through digital media. In this project, ruins and artifacts from Angkor Wat were 3D scanned and digitally restored. This case is Korea's first overseas project, and various multimedia products such as photos, 3D scans, 3D restoration models, and VR were produced. And these contents were published on 'CultureContents.com" Fig. 3 [8].

Digitalization of Imperial City of Hué in Vietnam (2007). Hué is a city located in central Vietnam and was the capital of Vietnam from 1802 to 1945. It has many historical monuments and buildings, especially the Imperial city of Hué, which is a UNESCO World Heritage. 'Digitalization of Imperial City of Hué' is Korea's Cultural Heritage Administration's first digital ODA(Official development assistance) project and Vietnam's first digital heritage project. Due to wars in the 20th century, most of the Imperial city of Hué was destroyed, leaving only some palaces remaining. However, the Korean technical team digitally restored the Imperial city of Hué based on literature research, consultation with Vietnamese history experts, obtaining 3D data and 3D scanning through site visits. And the digitally restored 3D video was shown for free inside the Imperial Palace in Vietnam Fig. 4 [9].

Fig. 3. Digitalization of Angkor Wat (2006) (@ photo by CG-wave)

Fig. 4. Digitalization of Imperial City of Hué (2007) (@ photo by KAIST CTRI)

3D Stereoscopic Video Using Borobudur Temple In Indonesia (2013). Borobudur Temple ("Monk on the Hill" in Sanskrit), built on the island of Java in 824, is a representative Buddhist temple in Indonesia. This is a square 9-story temple with a height of 34.5m. A total of 1 million stones were stacked like a pagoda, and each floor is engraved with reliefs expressing the world of Buddhism. However, because it had been buried under volcanic ash for over 1,000 years, all of the pagoda's decorations and pigments were lost, and its overall shape was severely deformed. Accordingly, a Korean technical team performed the world's first digital restoration of Borobudur Temple in 2013. The work process first used a laser scanner to obtain accurate actual measurement data about the employee. And based on the opinions and materials of local scholars, we digitally restored its original appearance rather than its current dark gray color Fig. 5 [10].

Virtual Reality Experience Center Using Imperial City of Hué in Vietnam (2018). This project is VR contents that digitally restored the Imperial City of Hué, which was severely damaged during the Vietnam War and was rebuilt in 1970. The work identified the remaining architectural style and structure using a 3D scanner, and then created a realistic VR palace based on photo data of the palace in the past. This includes a VR beacon that allows you to see the burned down buildings in the palace right in front of you, a VR telescope that allows you to look down on the entire palace, a VR simulator that allows you to see the palace while flying in the sky, and VR walk-throughs that allows you to experience the inside of King Khai Dinh's tomb. And in 2019, Vietnam's first digital cultural tourism facility, the VR Center, was built Fig. 6 [11].

Fig. 5. 3D Borobudur temple (2013) (@ photo by WIPCO & TRIC)

Fig. 6. VR Imperial City of Hué (2018) (@ photo by Jin-ho Park)

ASEAN Digital Cultural Heritage Project (2017–2019). In 2017, the Ministry of Foreign Affairs of the Republic of Korea carried out the ASEAN Cultural Heritage Digital Project using the South Korea-ASEAN Cooperation Fund. This project aims to build a network for digital records and technology exchange and utilization in the field of cultural heritage through the production of digital heritage content based on Korean cultural technology. Development targets include Omar Ali Siaputin Mosque in Brunei, Bagan Historic City in the Philippines, Borobudur Temple in Indonesia, Sukhothai in Thailand, Wat Phu Temple in Laos, Angkor Wat in Cambodia, Singapore Botanical Garden, Historic City of Bagan in Myanmar, and Imperial City of Hue in Vietnam. Applied technologies include 3D scanning, 4k UHD video, aerial filming using drones, 360° VR, and interactive VR. And VR contents developed is currently being screened in the VR experience room of the ASEAN Culture House in Korea Fig. 7 [12].

Fig. 7. ASEAN Digital Cultural Heritage Project (2017–2021) (@ photo by ASEAN-KOREA Center, WIPCO)

VR Exhibition Using Takht-i-Bahi Temple in Pakistan (2019). The Buddhist ruins and group of Buddhist monasteries of Takht-i-Bahi (meaning 'Throne of Origins') were built in the early 1st century. This temple is a cultural heritage site considered the pinnacle of Gandhara art and it is still a well-preserved site of early Buddhist art and architecture. In the 2019 'Asian Stupa Road: Takht-i-Bahi' Project, the entire temple was scanned in 3D and filmed using video and drones. The missing parts of the temple have been restored in VR, while the remaining parts of the temple have been stored digitally. VR experiencers can understand the origin and production process of Asian pagodas and virtually experience the pagoda restoration process Fig. 8 [13].

Immersive Contents of the Ayutthaya Buddha Statue at the Bangkok National Museum in Thailand (2023). This case is the first time that Korea's accumulated immersive media technologies, such as VR, AR, and XR, have been applied to a museum in another country. Specifically, two types of immersive content were installed in the Ayutthaya Room of the Bangkok National Museum. One is an artifact information interactive system, and the other is media wall content construction. The artifact information interactive system targeted two cabinets from the Ayutthaya period. The 'digital magnifying glass' allows you to rotate or enlarge the person drawn on the cabinet 360 degrees. In addition, the 'Ayutthaya City Wall', drawn in the 17th century, was expressed in a wide panoramic form and constructed as a media wall Fig. 9 [14].

Fig. 8. Takht-i-Bahi temple VR(2019) (@ photo by Asia Culture Center)

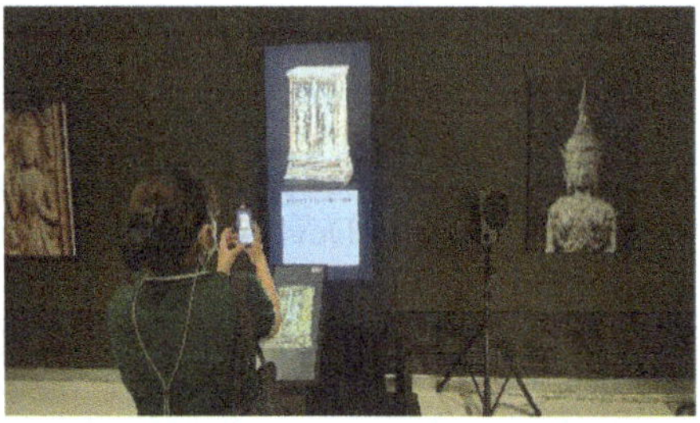

Fig. 9. Bangkok national Museum (2023) (@ photo by ZDNET korea)

3.3 Digital Contents Using UNESCO Cultural Heritage in ASEAN

Digital contents using ASEAN cultural heritage has been continuously developed and utilized in Korea from 2006 to 2023. The cases are summarized as follows.

First, the targets of content are mainly representative UNESCO cultural heritage in the ASEAN region. However, in the case of Angkor Wat in Cambodia, Hue Castle in Vietnam, and Borobudur in Indonesia, the same cultural heritage was repeatedly turned into content in different years. Currently, there are various UNESCO cultural heritage sites in the ASEAN region. Therefore, in order to widely publicize, preserve, and utilize ASEAN's cultural heritage to the public, it is necessary to diversify the content target.

Second, the development of digital application technology. Applied digital technologies are evolving from early 3D stereoscopic images to VR, AR, XR, AI, etc. For example, in 2006, the 'Angkor Wat Digitalization Project' was produced as a 3D stereoscopic video, but since 2010, experiential VR contents has been developed. And in 2023, the Bangkok National Museum is attracting attention by introducing an interactive media wall or immersive content based on Korean cultural technology.

Third, the purpose of content production is focused on the digital restoration of individual cultural heritage. The work process involves securing photos, videos, and 3D scan data through on-site visits, and then post-processing the data through computer work. Based on this, photo or point cloud-based modeling is performed, and finally photo-based 3D detailed modeling is performed.

Fourth, the digital content secured in this way was mainly used for museum exhibitions, etc., and systematic data collection or data recycling through digital archiving or platforms is insufficient. And, except for the virtual reality experience center using Imperial City of Hué (2018) and the realistic content at the National Museum in Bangkok, Thailand (2023), most digital cultural heritage was exhibited in domestic museums or cultural heritage-related exhibition halls.

Lastly, there is a lack of funding or business models. Most of these are public projects produced with support from the Korean government, and the content cannot be recycled after the free temporary exhibition. However, the VR Imperial City of Hué (2018) is the only case where a Korean company established a virtual reality center in Vietnam and secured paid revenue. Therefore, for a sustainable digital cultural heritage business, it is necessary to secure the funds necessary for content development and a business model that can generate stable profits.

4 Conclusion

ASEAN is an association of 10 Southeast Asian countries that promotes common development through political, economic, and security cooperation among member countries. Currently, ASEAN has a long history, tradition, and cultural heritage sites.

After 2000, due to the development of ICT and the increase in the cultural industry, digital content began to be introduced in earnest into the cultural heritage field. Accordingly, Korea has been producing digital heritage using UNESCO cultural heritages in ASEAN to introduce the ASEAN's culture and promote international cooperation.

UNESCO World Heritage sites, including those in the ASEAN region, are of outstanding universal value to humanity and are the basis for promoting peace, cultural diversity, creativity and intercultural dialogue.

Therefore, in order to promote digital content using ASEAN cultural heritage, it is necessary to expand to a wider range of regional historic sites, find concrete alternatives for mutual exchange, and explore profit models for the sustainability of these projects. To this end, the methods for producing and utilizing digital content are as follows.

1) 3D scanning is used to create accurate digital models of cultural heritage, and photogrammetry techniques are used to create detailed 3D models of historic sites, buildings, etc. This is then used for online exhibitions, virtual tours, and education.
2) Provides an experience that allows users to virtually explore UNESCO cultural heritage sites through VR content. AR apps provide additional information using your smartphone at historic sites or museums.For example, Angkor Wat or Borobudur can be recreated in VR so that people around the world can experience them.
3) The history, importance, and preservation status of cultural heritage are introduced through an interactive website. And through mobile apps, users can visit historical sites and get related information, and elements such as quizzes and games are added to increase the effectiveness of education.

4) We build an archive by digitizing cultural heritage-related documents, photos, and videos. We operate an online database accessible to researchers, students, and the general public, and improve global accessibility through multilingual support.
5) We use social media to promote the importance of cultural heritage and the need for preservation. We also produce storytelling content related to UNESCO cultural heritage and encourage public participation through online challenges and campaigns.
6) We introduce how to create digital heritage content through training programs and workshops, and work with schools, universities, and cultural organizations to develop digital cultural heritage education programs. Additionally, accessibility to content is improved through MOOC (Massive Open Online Course).
7) We develop content in collaboration with local governments, non-governmental organizations and UNESCO, and share resources and technologies through international joint projects. And the quality of content can be improved through collaboration with technology companies, media artists, historians, etc.

Through these methods, UNESCO cultural heritage in ASEAN on can be digitally preserved, promoted to the world, and used as an educational and tourism resource.

References

1. UNESCO. https://www.unesco.org, Accessed 12 April 2024
2. Wi, S.-H., Ahn, H.-K.: Study on the direction of digital heritage in the post-corona era, Global Culture 56, Global Cultural Contents Society, pp. 56–58 (2023)
3. Jeong, S.-M.: a theoretical study on the application of ODA to digital cultural heritage - focusing on Myanmar cultural heritage management system. Korean J. Cultural Heritage Stud. 53(4), National Research Institute of Cultural Heritage, 198~215 (2020)
4. Yoo, D.-h.: Current Status and Prospects of Digital Cultural Heritage Research, Cultural Contents Research 3, Glocal Culture Strategy Research Institute in Konkuk University, pp.95–121 (2013)
5. Park, J.-H.: Cultural Heritage in the Age of Metaverse and Artificial Intelligence, Baekje Historical Sites Academic Research Series 6, Baekje World Heritage Center, p. 41 (2023)
6. ASEAN-KOREA center. https://www.aseankorea.org, Accessed 12 April 2024
7. UNESCO. https://www.unesco.org/en/world-heritage, Accessed 12 April 2024
8. Ministry of Culture, Sports and Tourism of South Korea. https://mcst.go.kr/kor/s_notice/press/pressView.jsp?pSeq=8407, Accessed 12 April 2024
9. An, J.-H.: Digitalization project of Hue Imperial City in Vietnam, Korea Information Processing Society Review 15(3), Korea Information Processing Society, p. 80 (2008)
10. Park, J.-H.: Digital Restoration of Cultural Heritage; Borobudur Temple in Indonesia (2014). https://www.ggbn.co.kr/news/articleView.html?idxno=27412, Accessed 12 April 2024
11. Bae, T-.W.: Vietnam's last ruined royal palace revived with Korean VR technology, Korea Economic Daily (2018). https://www.hankyung.com/article/2018043058891, Accessed 12 April 2024
12. ASEAN culture house. https://www.ach.or.kr, Accessed 12 April 2024
13. Asia Culture Center. https://www.acc.go.kr/main/exhibitiion.do?PID=0202&action=Read&bnkey=EM_0000003297, Accessed 12 April 2024
14. Lee, C.-G.: The technological power of Korean new technology convergence content established at the Bangkok National Museum in Thailand, K-Museum, ZDNET korea (2023). https://zdnet.co.kr/view/?no=20230620141, Accessed 12 April 2024

The Use of Geographic Information in the Digital Heritage: The Case of Heritage GIS in Korea

Cha Yong Ku and Hee-Soo Choi(✉)

Sangmyung University, 20 Hongjimun2-Gil, Seoul, Jongno-Gu, Korea
{koostar,choice}@smu.ac.kr

Abstract. Most tangible cultural heritages have a location so they can be expressed and managed as geographic information. The Geographic Information System (GIS) is used for the preservation and management of cultural heritage. In Korea, it has been established to provide users with location information and specific data on cultural heritage. In this study, the characteristics of the Heritage Geographic Information Service (HGIS) in Korea were identified, and problems and development directions were presented. In Korea, HGIS services have been provided to preserve and manage cultural heritage since 2008, and many users are using HGIS services for the preservation and management of cultural properties. To improve and develop HGIS, the measures proposed in this study are as follows. First, the data provided by HGIS must be refined and processed to provide user-centered information. To this end, it is necessary to improve the data schema and develop integrated metadata. Second, it is necessary to provide various map expressions of the spatial distribution of heritage. The theme map expression should be improved to effectively express the spatial distribution of heritage even on a small scale. In addition, it is necessary to combine high-resolution satellite images and 3D building modeling to effectively express heritage even in virtual reality. In this study, an improvement case of the method of map expression of heritage was presented using Python's Folium and Google Earth.

Keywords: Heritage GIS in Korea · Cartographic Presentation · Geographic Information · Virtual Reality

1 Introduction

Cultural heritage refers to social and cultural dissipation that is worth inheriting from the next generation for future cultural development. Cultural heritage is classified into tangible and intangible cultural heritage depending on the presence or absence of form, and in the case of tangible cultural heritage, it occupies a space and can be expressed and managed as geographic information. This is because it is possible to preserve cultural heritage and develop tourism resources using geographic information, and more systematic management of cultural heritage can be achieved by using the Geographic Information System (GIS). In the case of cultural heritage, it is expressed in geographic

G. Marreiros et al. (Eds.): DCAI 2024, LNNS 1151, pp. 71–80, 2025.
https://doi.org/10.1007/978-3-031-80946-0_7

coordinates and has location information as a created place and exists as a cube of time and space. Location information of cultural heritage is expressed in the form of points, lines, and polygons according to the spatial dimension, and attribute information includes a detailed explanation of cultural heritage, historical records, and utilization plans. Therefore, cultural heritage can be said to be a representative case by using a geographic information.

Many countries around the world manage cultural heritage using GIS technology. Petrescu (2007) summarized the current status of using GIS technology to manage cultural heritage in 22 countries around the world. He emphasized that GIS is an essential technology in cultural heritage management. GIS has the advantage of being able to make spatial predictions and decisions through spatial relations or spatial distribution pattern analysis using location information and attribute information of geographic information, rather than simply building and managing geographic information as a database. In addition, since geographic information is expressed in the form of a map, it has the advantage of changing the spatial distribution characteristics according to the scale of the map.

In Korea, a number of studies have been conducted to integrate and analyze cultural heritage into the GIS. Moon and Park (2003) presented a model that developed a map representing the location of simple relics into a database for studying the method of cultural expression. Jang (2008) conducted a study to integrate the cultural heritage management system conducted at the local level of Korea into the national level using GIS. As a result of these efforts, the Heritage Geographic Information Service (HGIS) was established with the Cultural Heritage Administration at the national government of Korea. This system provides various types of geographic information such as state-designated heritage, local designated heritage, national registered heritage, local registered heritage, and standards for preservation of the phenomenon. In addition, detailed descriptions of each cultural heritage are provided as attribute information.

In this study, the HGIS is introduced as an example of building and managing cultural heritage using GIS technology in Korea, and problems and improvement directions in the service operation process are addressed. The specific service contents of the HGIS operated in Korea were identified, and problems of the current service were identified. In addition, a future improvement direction applying map expression method and virtual reality technology was suggested. Through this, it is intended to present a plan to utilize geographic information in digital cultural heritage in Korea and to suggest future directions.

2 Importance of GIS in Cultural Heritage Management

In countries with long historical backgrounds, numerous cultural assets are distributed throughout the country, and GIS is used to investigate and manage them. This is because through GIS, location information, zone information, and detailed explanatory information related to cultural heritage can be quantified and built into a database, and spatial analysis can be performed using various search conditions and spatial relationships. Accordingly, in Japan and the United Kingdom, cultural heritage GIS is established to preserve and manage cultural properties and to establish a development plan to minimize damage to relics. In the case of the United States, the cultural property prediction model is used to improve efficiency in the cultural property investigation process and support the establishment of a rational development plan (Kang, *et al.* 2009). Therefore, GIS plays a big role in the preservation and management of cultural properties.

GIS provides effective support tools not only for the preservation and management of cultural assets, but also for the user's rational decision-making process. This is because GIS is a system that supports users' decision-making through spatial analysis and modeling. This decision support system enhances the role of helping decision-makers make rational decisions from various alternatives. In the management of cultural properties where conflicts between preservation and development arise, GIS can be used to compare and analyze the spatial results caused by various alternatives, and to make optimal decisions. In addition, rational and effective decisions can be made by providing prediction points for the distribution of cultural properties when developing and investigating cultural properties.

3 The Heritage Geographic Information Service in Korea

The Cultural Heritage Administration of Korea has established a geographic information utilization system that combines national heritage information with map information. Location information and attribute information on national heritage are established as a database and are provided through the Internet, thereby supporting the use of cultural heritage contents based on location information. The HGIS allows anyone to easily access the service and use geographic information through the Internet homepage (https://gis-heritage.go.kr/). In addition, photos, drawings, videos, and explanations of national heritage are also provided along with location information, so various types of contents on national heritage can be used.

The HGIS is a service that displays the location on the map and informs the allowable range for each area in order to literally preserve and manage national heritage. This service provides location information and theme maps of national heritage based on the V-World and edited cadastral map of the Ministry of Land, Infrastructure and Transport. National heritage is divided into national-designated heritage and local-designated heritage, as well as national-registered heritage and local-registered heritage. National-designated heritage is a national heritage selected after deliberation by a national-designated committee for cultural heritage with great historical, artistic, academic, and landscape value. Local-designated heritage is a cultural heritage designated by the local government in recognition of its value among cultural heritage that is not registered as a national-designated heritage. A national registered heritage is a cultural heritage that has been produced more than 50 years ago among cultural heritage that has not been designated by the national or local government and is a modern and contemporary cultural heritage registered by the national government because it requires special preservation or utilization. Local registered heritage is a modern and contemporary cultural heritage registered in recognition of its preservation value by local governments, even though it was not registered in the national government. The HGIS manages and provides the location information of these national heritage as polygonal geographic information. In addition, the scope of the acceptance criteria for preserving the status quo is provided around these national heritage sites. The acceptance criteria for preserving the status quo are the legally designated scope to preserve the historical and cultural environment and prevent damage to cultural heritage caused by construction work. Therefore, the HGIS provides not only geographical location and attribute information of cultural heritage, but also spatial protection of cultural properties in the urban development process.

The HGIS can perform various search functions for national heritage. The location can be expressed on a map by using the name of a national heritage, or you can also search for attribute information of the national heritage on the map. Figure 1 shows the case of HGIS screenshot. Figure 1(a) is a screen showing the thematic map of national heritage near Gyeongbokgung Palace, and Fig. 1(b) is a result screen when searching for "Gyeongbokgung". The search results appear as a list as shown in the figure, and the list number and location are displayed on the map. When you press a number on the map, detailed information about the heritage is displayed on the left. The contents of detailed information are the title and location, type classification, designation date, photo information, and notification history. The notification history provides the contents of the notification of the Cultural Heritage Administration on the acceptance criteria, etc., along with notices and drawings. These are very important information in judging policy due to the extent to which the development permitting standards for the preservation of heritage can be reached. In addition, photographic information is important, because the preservation and management status of the heritage can be checked through photographic information. Therefore, it is important to have a lot of photographic information.

(a) Thematic map (b) Search result

Fig. 1. Screenshots of HGIS

The HGIS provides a separate themed content service to service various contents related to cultural heritage. Themed content service provides various contents as map information such as the National Heritage 360 Panorama Service, Geological Heritage Fossil Stone Sample Service, Natural Monument Ecology Map, World Heritage Map of Korea, Korean Temple Site, and Dolmen on the Map.

In the case of the geological heritage sample information service, it is largely divided into two types: the sample status by institution and the sample status by type. First, when sample information for each institution is selected and a specific institution is selected by a preservation management institution, the total number of samples and image thumbnails are displayed along with the brief information of the institution. Detailed information will appear when you click on the image thumbnail. If you select all sample information below on the thumbnail screen, the electronic map disappears, and the image thumbnails of all samples are displayed in the form of galleries. Similarly, if you select a thumbnail, detailed information is displayed. If sample information for each type is selected and a specific type is selected from the sample types below, the sample status by the holding institution appears in the form of a bar graph. At the same time, the distribution status of each sample type is shown on the map as a circular graph. These services induce people's interest in cultural heritage and allow them to utilize various cultural contents.

4 Future Improvement Directions for HGIS

4.1 Problems and Improvement of Data and Service in HGIS

The spatial information service for cultural properties in Korea started with the establishment of spatial information for buried cultural properties in 2002, and in 2004, the spatial information for buried cultural properties and designated cultural properties was integrated to provide the current HGIS service. Currently, there are about 40,000 users of the spatial information service every month, and 140,000 cases of information are provided to state agencies and the public. Since the HGIS service was launched in 2008, the service has been improved by reflecting user feedback while providing spatial information services for about 15 years. Nevertheless, HGIS has several problems.

First, it is urgent to improve the appearance of different services depending on the search result item in the main search window. The detailed information and maps in the designated heritage item and the detailed information and map contents in the cultural heritage distribution map item do not match. For example, if you search for "Toppyeong-ri" in the search window of the main page, 12 data are searched. The search results were searched by items such as designated heritage, criteria for allowing changes in the status quo, burial heritage site, cultural heritage distribution map, and address. If you look at the search results of designated heritage, you can see the Seven-story Stone Pagoda in Toppyeong-ri, and on the left side you can see detailed information such as location and type classification, designated date, photo, and notification history. However, if the Seven-story Stone Pagoda in Toppyeong-ri is selected from the cultural heritage distribution map item, contents different from those found in the designated heritage will be developed. On the left side, detailed information is displayed, but the location, type classification, and photos are shown, but the number of photos is too small and the old photo comes out. Also, even the location mark is not revealed on the map. It seems that the metadata of the cultural heritage distribution map produced by the Cultural Heritage Administration in the past was imported and used as it was. Even after Jungwon-gun was integrated into Chungju-city, the name of the tower comes out as Seven-story Stone Pagoda in Jungwon-Toppyeong-ri. The Chinese character notation comes out as the Seven-story Stone Pagoda in Chungju-Toppyeongri. They do not match each other.

There is certainly a problem with this result within the same service. This is because if the service content varies depending on which information the user views, the reliability of the information will be problematic. The currentization of cultural heritage distribution maps is urgently needed. Since the purpose of this service is to focus on designated and buried heritage, a map of cultural heritage distribution including non-designated heritage may be urgently needed, but the use of data established in the past as it is must be improved urgently.

Second, since the purpose of this service is the preservation management of the state, the service should be focused on preservation management. In the detailed information service on heritage, general information can be easily obtained anywhere, but this information is important because information on heritage and heritage reserves subject to preservation management can only be obtained here. However, currently, only location information is displayed on the service, and the reason for the designation of the area is not explained. It is a situation where it is necessary to check the notice one by one, but if there is a lot of notice history, it is difficult to check each one. Therefore, it is necessary to supplement metadata.

Third, there is a problem with the geological sample service. The critical problem with the sample service is in the metadata. The reason why a geological sample is important is not because the holding organization is important, but where the sample was collected, and the information cannot be found anywhere in the sample service. The importance of an electronic map can be said to show the geographic location of geological samples, but the information is missing. In addition, there is no reason for different services by holding organization and sample type to be conducted. This is because the sample type is important, so there is no problem in finding the holding organization information even if the service is based on it.

Fourth, above all, the current service has a problem in that it overly subdivides the database and applies it to the service as it is. If designated heritage, burial heritage, or geological sample, all fall within the category of national heritage, it is desirable to develop integrated metadata information, serve it layer by layer on an electronic map, and unify the service method. The same is true of the electronic map for each subject included in the theme map. Perhaps, in the case of a preservation management map, the service is based on the aspect, but in the case of a theme map, since the service is based on the point, it is necessary to unify these matters.

4.2 Improvement of Cartographic Representation for HGIS

The HGIS provides the location of all national heritage sites as polygon type. Since polygon type occupies an area, it is effective when expressing detailed locations and areas on a large-scale map. Therefore, it is suitable when the zoom level expresses the detailed space occupation status of national heritage at the street level. However, geographic information has a characteristic that the effect of being expressed varies depending on the scale. In other words, it is important to express detailed locations on a large-scale map, but it is effective to generalize geographic information to express the overall spatial distribution or status on a small-scale map.

Although the current HGIS provides detailed geographic information at the street level of a large scale, the polygon expressing the location of the national heritage is reduced as it is at the city level of a small scale, so the spatial distribution characteristics of the national heritage do not appear. The location of the national heritage cannot be expressed when reduced at the small-scale. Therefore, it is necessary to improve the map expression method to express the spatial distribution characteristics of the location information of national heritage expressed in the form of polygons at the zoom level of a small scale.

Figure 2 is an example of expressing location information at a small-scale national level using Python's Folium library for national-designated cultural heritage. Figure 2(a) is an example of expressing the location of national-designated cultural heritage in polygonal form. As shown in the figure, when geographic information in the form of polygon is expressed on a small scale, the spatial distribution characteristics of national heritage are not effectively expressed. Figure 2(b) is an example of expressing the location of national-designated cultural heritage as points. When the location of a national heritage is expressed as points, the spatial distribution characteristics of national heritage can be expressed to some extent, and in particular, areas where national heritage is spatially concentrated can be expressed. Figure 2(c) is an example of expressing the spatial distribution characteristics of national heritage as a cluster map in order to effectively express the spatial distribution characteristics of national heritage. A cluster map is a map expressed by the number of clustered points after grouping geographic information in the form of points according to distance. If the zoom level of the cluster map is enlarged, clusters are formed according to the distance in an enlarged state and expressed as a map. Therefore, the spatial distribution characteristics of national heritage according to distance can be effectively expressed. Figure 2(d) is to express the location of national heritage in the form of a heatmap. Heatmap is an expression of the spatial density of geographic information, and when the density is high, it is expressed as a red cluster. Therefore, the degree of spatial density of national heritage can be effectively expressed. In this way, the spatial distribution characteristics of national heritage can be effectively expressed in a small-scale map. Therefore, the current HGIS provides geographic information services to large-scale maps, but if the spatial distribution characteristics are expressed for small-scale maps, it can be more effectively expressed as a map.

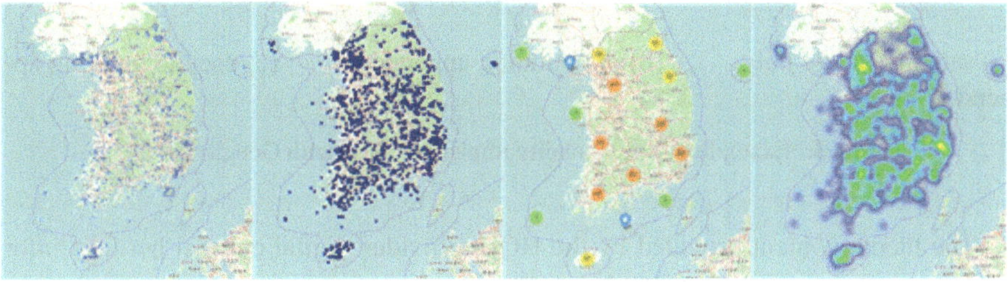

(a) polygon (b) point (c) cluster map (d) heat map

Fig. 2. Cartographic representation of national designated heritage at a small scale

4.3 Improvement Directions for Virtual Reality

The HGIS provides geographic information of national heritage in the form of a map. If virtual reality is implemented using such geographic information, national heritage information can be provided more interestingly. In Korea, Song and Cho(2008) implemented virtual reality by expressing cultural heritage information using 3D scanning on a high-resolution satellite image. Dore and Murphy(2012) effectively visualized digital cultural relics by using 3D building information modeling and GIS.

Google Earth is a representative platform that uses geographic information to implement virtual reality. Google Earth provides location information based on high-resolution satellite images, making it a suitable platform for virtual reality and digital twin. In addition, Google Earth provides geographic information content in the form of a 3D building model with satellite images in the background. In the case of representative cultural heritages, such a 3D model is also provided. Therefore, using Google Earth can effectively implement the virtual reality of cultural heritage. Figure 3 is a Google Earth screen targeting Gyeongbokgung Palace in Korea. Figure 3(a) is a Google Earth screen with only high-resolution satellite images in the background, and Fig. 3(b) is a screen representing cultural heritage located in Gyeongbokgung Palace in a 3D building model. By applying a 3D building model as shown in the figure, it is possible to effectively express the virtual reality of cultural heritage. In addition, a 3D building model can be created and added to Google Earth using Sketchup software. Figure 3(c) is an example of creating a pseudo building in Sketchup and then expressing it at an arbitrary location in Google Earth. As shown in the figure, a pseudo building that does not currently exist can be produced as a 3D model using Sketchup software and stored as a KML file to be implemented in virtual reality on Google Earth.

(a) satellite image only (b) 3D building appended (c) pseudo building appended

Fig. 3. Example of virtual reality implementation with Google Earth

The theme content provided by the HGIS provides digital cultural heritage information on various topics. Among them, the 360-degree panoramic image can be used as a stereoscopic image provided by Virtual Reality. By using a simple VR App, a panoramic image can be converted into a VR stereoscopic image, and the converted image can implement virtual reality using a simple VR headset. Figure 4 is an example of VR implementation of a panoramic image. Figure 4(a) is a panoramic image for Korean seedlings provided by the HGIS, and Fig. 4(b) is an example of converting it into a VR stereoscopic image. VR can be implemented through a simple conversion process using various cultural heritage contents as shown in the figure. If Virtual Reality is applied to the HGIS in this way, various cultural heritage contents can be provided more interestingly.

(a) panoramic image (b) stereoscopic image

Fig. 4. Example of implementation of stereoscopic images from panoramic images

5 Conclusion

Since tangible cultural heritage occupies a spatial location, it is possible to express and manage cultural heritage in digital form using geographic information. This is because geographic information does not only provide simple location information of cultural heritage, but also provides specific attribute information on cultural heritage. In addition, the proximity and spatial distribution pattern of cultural heritage can be expressed by using the spatial characteristics of geographic information.

In Korea, national heritage has been built and managed as a GIS database, and various information is provided to users through the HGIS. In addition to the geographical location and detailed data of national heritage, protected areas and management conditions are provided to prevent and preserve damage to historical sites during urban development. However, the HGIS has several problems. Different results are appearing in the search process for cultural heritage, and the search results are also expressed as they are without processing the original data. The location of the national heritage is expressed in detail on a large-scale map, but the spatial distribution characteristics are not expressed on a small-scale. In this study, the problems of the HGIS were identified and measures to improve them were suggested. First, the data provided by the HGIS should be refined and processed to serve more user-centered metadata. In addition, a plan to express the spatial distribution pattern of national heritage is needed even at a small scale. In addition, in order to use HGIS in a virtual space in the future, it is necessary to apply 3D building modeling technology based on high-resolution images such as Google Earth. If these GIS technologies are combined, the usability of the HGIS can be further improved.

References

Dore, C., Murphy, M.: Integration of HBIM ad 3D GIS for digital heritage modeling. Digital Documentation, 22–23 October 2012, Edinburgh, Scotland (2012)

Petrescu, F.: The use of GIS technology in cultural heritage. In: XXI International CIPA Symposium, 01–06 October 2007, Greece, Athens (2007)

Jang, M.-H.: A study on the distribution map designs for the construction of cultural properties in GIS. J. GIS Associat. Korea **16**(3), 303–315 (2008)

Kang, D., Ko, I., Kim, K., Kim, B.: Utilization and orientation of cultural GIS in foreign cases. In: Preservation and Management of Cultural Heritage Using GIS, pp. 200–234. Sahoi-Pyungron Press, Seoul, Korea (2009)

Chae, M.B., Wook, P.H.: A study on model of constructing the cultural database system. J. GIS Associat. Korea **11**(2), 185–201 (2003)

Song, S., Cho, M.: GIS and satellite image development using 3D cultural properties information system. In: 2008 Proceedings of HCI Korea, pp. 595–600 (2008)

HGIS Homepage. http://gis-heritage.go.kr,Accessed 10 April 2024

Technology for 3D Recording of Cultural Heritage

Sungheuk Jung(✉), Seungki Lee, Doyeon Choi, and Seungwoo Lee

TerraPix Co., Ltd., Cheongju 28644, Republic of Korea
terrapix@naver.com

Abstract. Currently, LiDAR, photogrammetry, photometric stereo, and 3D Gaussian Splatting technologies are used as 3D recording methods for cultural heritage. Measurement methods and equipment can be determined according to the purpose of 3D recording, which can range from simple to precise 3d scanning of cultural heritage. For precise measurement, precise 3D scanners and photogrammetry are used, either alone or in combination with photometric stereo technology, if detailed measurements are required. Precise data for large areas can be acquired by using drone photogrammetry and drone LiDAR. Low-poly models can be easily produced using a LiDAR or photogrammetry application running on a smartphone. The 3D Gaussian Splatting technique can be used for background or rough 3D recording for realistic content, such as games and movies. In this study, we investigated recent cultural heritage 3D recording technologies and compared these methods according to the type, purpose of use, and precision.

Keywords: Culture heritage · LiDAR · Photogrammetry · Photometric stereo · 3D Gaussian Splatting

1 Overview of 3D Recording Methods

1.1 LiDAR

LiDAR, a 3D scanning technology, is used to acquire 3D surface information for an object using light or lasers. Long-range 3D scanners are used to measure large-scale areas, buildings, and natural landscapes, and precision 3D scanners are used for small objects, such as relics, tombstones, and pottery [22, 24, 25, 31, 37, 39]. LiDAR sensors are included in recent iPhones and iPads for simple 3D scanning [1, 2, 16].

This approach has several advantages over other techniques, including the ability to quickly acquire dense 3D point cloud data and its simplicity. However, long-range and precision 3D scanners are relatively expensive. Additionally, although efficient quantitative measurement is possible, obtaining high-resolution textural information for an object is more feasible using photogrammetry.

Technologies have been developed for simple recording, such as Matterport and Cupix. A simple 3D point cloud for an area can be generated using 360° panoramic images and LiDAR data acquired from 360° camera, such as Insta360 or 3D scanner,

G. Marreiros et al. (Eds.): DCAI 2024, LNNS 1151, pp. 81–90, 2025.
https://doi.org/10.1007/978-3-031-80946-0_8

and a site can be recorded using a VR tour method based on 360° panoramic images. However, depending on the equipment used, the precision is lower than that of LiDAR surveying or photogrammetry, and there are limitations in post-processing [8, 10, 17, 21, 38].

Drone LiDAR technology is used when it is difficult to work with 3D scanners on the ground (e.g., in large areas, coastal cliffs, forests, and high buildings). Data can be acquired from a LiDAR sensor mounted on a drone and combined with ground LiDAR and photogrammetry to improve precision and texture resolution [36] (Figs. 1, 2, 3, and 4)

Fig. 1. Long-range 3D scanner (Leica RTC360)

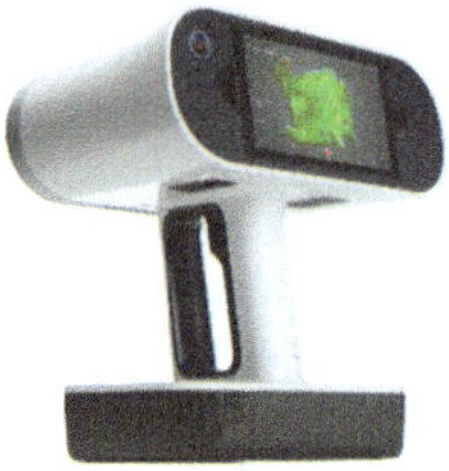

Fig. 2. Precision 3D scanner (Artec Leo)

Fig. 3. Matterport (www.matterport.com)

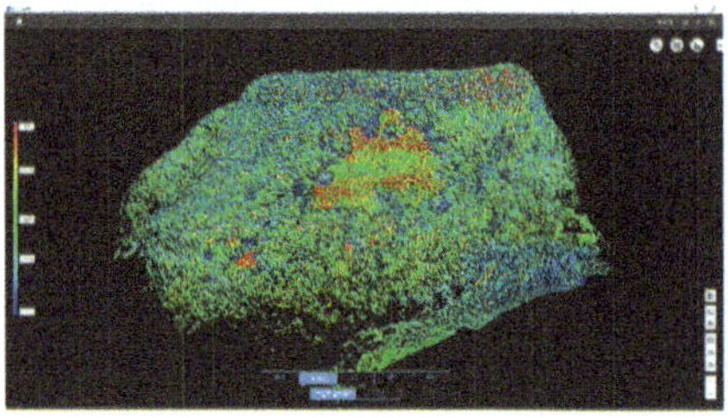

Fig. 4. Drone LiDAR data

1.2 Photogrammetry

The resolution of digital cameras and the performance of computer hardware for data processing have improved significantly, and photogrammetry tools have been developed. Recently, photogrammetry is not only used in cultural heritage preservation but also in other fields, such as GIS, civil engineering, urban development, and environmental sciences [4, 6, 12, 13, 15, 19, 32, 33, 35].

Photogrammetry can easily and quickly capture an object using cameras, such as DSLR, Mirrorless, or smartphone cameras. Data are processed using commercial software, such as Metashape, Capture Reality, or 3DF Zephyr, and open-source tools, such as COLMAP or Meshroom. Images are processed in the align, 3D reconstruction, and texturing procedures and used to reconstruct a 3D model.

Since a general camera is used, the setup cost is less than that for LiDAR technology, and qualitative measurement is possible because detailed textures of an object are acquired. In addition, this approach enables the acquisition of information that is not visible to the human eye using various sensors, such as multi-spectral, thermal, hyperspectral, RGB sensors [3, 28, 34].

Photogrammetry can be divided into terrestrial and aerial photogrammetry depending on the exposure location. It is effective in acquiring data for large buildings, natural heritage, and large terrain by using drones. Additionally, data quality can be improved through fusion processing with LiDAR point cloud data [30, 41] (Figs. 5, 6 and 7).

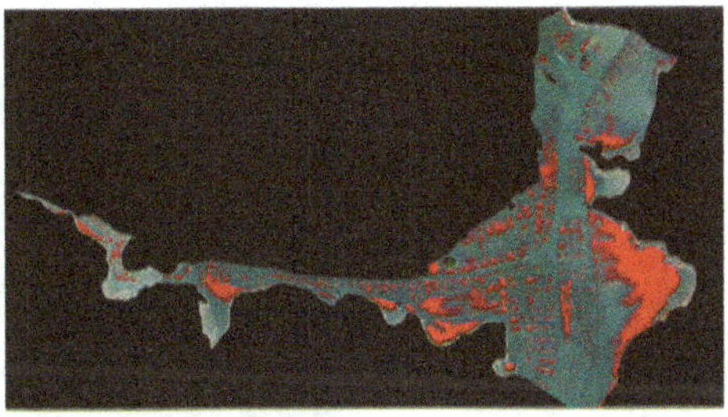

Fig. 5. Classification map by multi-spectral sensor

Fig. 6. Photo-realistic model by photogrammetry

Fig. 7. Accuracy Comparison (Artec Leo, Artec Spider and photogrammetry method)

1.3 Photometric Stereo

Photometric stereo is a method for recording the 3D surface shape of an object as a 2D image through inclined light around the object and implementing it in 2.5D or 3D with a 2D normal map. This technology has been used for a long time; however, its utilization is increasing recently through convergence with photogrammetry technology. It is suitable for recording objects, such as letters or drawings carved on a tombstone, detailed traces, and the surface of a paper or cloth [11, 29, 40] (Figs. 8 and 9).

Fig. 8. Photometric stereo rig

Fig. 9. Normal map from Photometric stereo method

1.4 3D Gaussian Splatting

Since the introduction of 3D Gaussian spreading using machine learning technology, the technique has been used in many studies and services, such as Luma AI, Polycam, and Scaniverse. It is possible to render 3D shapes easily from video and image sources, extract them as point cloud or mesh data, and use them as 3D assets in game engines, such as Unity3D and Unreal Engine.

Ease of use and fast processing times are benefits of the approach; however, data quality is still insufficient. Therefore, it can be used to record the surrounding environment but not to produce precise models. It is suitable for objects that are difficult to capture using LiDAR or photogrammetry, such as small trees, grass, and flowers. In addition, it is expected to be used in disaster management, which requires rapid processing.

Since active research is ongoing, the accuracy and precision are expected to increase significantly in the future, and the usability of cultural heritage records is expected to increase [7, 9, 20] (Figs. 10 and 11).

Fig. 10. Rendered image on the platform (Luma AI)

Fig. 11. Point cloud data from 3D Gaussian Splatting

2 Application of Technologies

2.1 Architectural Heritage

For architectural heritage, terrestrial photogrammetry, drone photogrammetry, and LiDAR surveying are the main approaches, depending on the size and level of detail. 3D records are feasible; however, external and internal records of complex buildings are time-consuming, not only with respect to field work but also data post-processing and asset modeling [5, 18].

2D drawings, such as floor plans, elevation, and cross-sectional views, can be easily produced from orthogonal projection images using AutoCAD, Adobe Illustrator, or ArcGIS and BIM modeling can be performed from point cloud data using Autodesk Revit, Graphisoft ArchiCad, and other software [23, 26] (Figs. 12 and 13).

Fig. 12. Ortho image

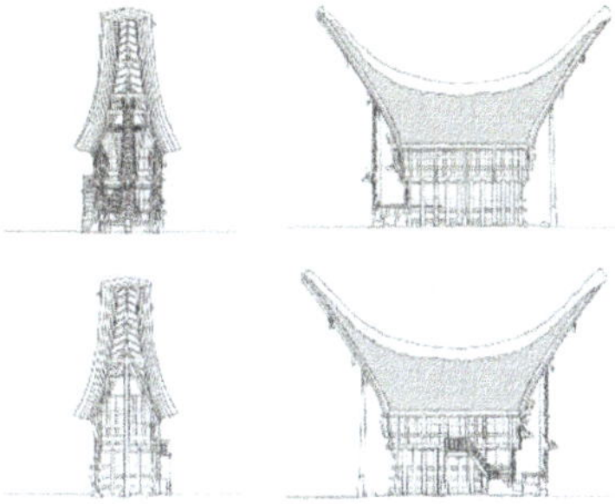

Fig. 13. 2D Drawings

2.2 Natural Heritage

It is difficult to record trees, plants, and flowers at natural cultural heritage sites. Large branches of a tree can be sufficiently 3D scanned by photogrammetry and LiDAR surveying, and the size and shape of the entire tree are recorded and utilized as point cloud data.

However, since small branches and leaves move, even for slow wind speeds, it is very difficult to obtain measurements. Furthermore, when converting 3D scan data into solid mesh data, thin small branches and leaves result in incomplete models.

Therefore, cultural heritage trees, plants, and flowers are 3D scanned to measure the overall shape, height, width, and perimeter depending on the purpose of recording. In addition, the overall shape can be recorded using the 3D Gaussian Splatting technique, and the photometric stereo technique can be used to accurately record leaves or tree trunk surfaces (Figs. 14 and 15).

Fig. 14. 3D mesh model (LiDAR)

Fig. 15. 3D textured model(LiDAR + Photogrammetry)

2.3 Archaeological Relics

Precision 3D scanners and photogrammetry are mainly used for precise recording of archaeological relics. In cultural heritage research, Artec Spider and Leo scanners are widely used to acquire high-density 3D scan data in a relatively short period of time. However, in the case of 3D scanners, this method is insufficient for acquiring high-resolution textures, and images for textural analyses are sometimes obtained separately.

Photogrammetry is used to produce an asset or photorealistic model that requires high-resolution textures, and the photometric stereo technique can be used for fine-scale archaeological relics. Photogrammetry is suitable for recording thin carved letters, drawings, and surface roughness, and the photometric stereo technique can generate a 2D normal map and use it to reproduce a 2.5D shape. Although precise 3D recording along with photogrammetry is used, a substantial number of photographs are required, and the processing time is longer than that for the general method [14, 27] (Figs. 16 and 17).

2.4 Historical Sites

To record historical sites and old cities, aerial photogrammetry is suitable. Aerial photogrammetry by traditional manned aircrafts and drone photogrammetry has been widely used in recent years. The method should be determined according to the survey area, economic feasibility, and efficiency.

Fig. 16. 3D relief map and visualization

Fig. 17. 3D relief map(Photogrammetry)

Measurement methods include LiDAR and photogrammetry. Aerial LiDAR surveying is an appropriate method to obtain a DTM (Digital Terrain Model) of a surface where trees have been removed.

Terrestrial LiDAR and photogrammetry are used for precise recording of small sites, such as tombs and building sites. 3D Gaussian Splatting technology may be efficient for simple 3D records and content production of old cities and alleys that are not for survey purposes (Figs. 18 and 19).

Fig. 18. 3D map of a large area

Fig. 19. Digital surface model of a city

3 Results

Measurement methods and equipment can be determined according to the purpose, which can range from simple to precise 3d scanning of cultural heritage. For precise measurement, precise 3D scanners and photogrammetry are used, either alone or in combination with photometric stereo technology, if detailed measurements are required. Precise data for large areas, such as historical sites, natural heritage sites, and old cities, can be acquired by using drone photogrammetry and drone LiDAR.

Low-poly models that do not require high resolution can be easily produced using a LiDAR or photogrammetry application running on a smartphone. The 3D Gaussian Splatting technique can be used for background or rough 3D recording for realistic content, such as games and movies.

References

1. Abdelhamid, T.G.: Digital techniques for cultural heritage and artifacts recording. Resourceedings **2**(2), 72–112 (2019)
2. Askar, C., Sternberg, H.: Use of smartphone lidar technology for low-cost 3D building documentation with IPhone 13 Pro: a comparative analysis of mobile scanning applications. Geomatics **3**(4), 563–579 (2023)
3. Baltsavias, E.P.: A comparison between photogrammetry and laser scanning. ISPRS J. Photogramm. Remote Sens. **54**(2–3), 83–94 (1999)
4. Campana, S., Remondino, F (eds.): 3D Recording and Modelling in Archaeology and Cultural Heritage: Theory and Best Practices. BAR publishing (2016)
5. Choi, J.-W., Kim, H.-Y., Kim, D,-W.: A study on cost-effective 3D city simulation techniques and applications based on the PC environment. J. Korea Planning Assoc. **38**(7), 245–256 (2003)
6. Choi, S.Y.: A study on the improvement of uav aerial photogrammetry policies. J. Korean Assoc. Cadastre Inform. **19**(3), 125–139 (2017)
7. Chu, C.-U., et al.: Recent trends of 3D reconstruction technology. Electr. Telecommun. Trends 22(4), 1–11 (2007)
8. Cortinhal, T., Kurnaz, F., Aksoy, E.E.: Semantics-aware multi-modal domain translation: From lidar point clouds to panoramic color images. In: Proceedings of the IEEE/CVF International Conference on Computer Vision (2021)
9. Costa, J.M.G.L.B. Optimization of filling systems for low pressure by Flow 3D. Diss (2015)
10. Cui, T., et al.: Line-based registration of panoramic images and LiDAR point clouds for mobile mapping. Sensors **17**(1), 70 (2016)
11. Dessì, R., et al.: Recent improvements in photometric stereo for rock art 3D imaging. Digital Appli. Archaeol. Cultural Heritage **2**(2–3)., 132–139 (2015)
12. Eulitz, M., Reiss, G.: 3D reconstruction of SEM images by use of optical photogrammetry software. J. Struct. Biol. **191**(2), 190–196 (2015)
13. Fryer, J., Mcintosh, K.: Enhancement of image resolution in digital photogrammetry. Photogramm. Eng. Remote Sens. **67**(6), 741–749 (2001)
14. Heritage, English. 3D laser scanning for heritage: Advice and guidance to users on laser scanning in archaeology and architecture. English Heritage (2007)
15. https://www.heritage.go.kr, https://www.heritage.go.kr/heri/html/HtmlPage.do?pg=/threeD/3dInfomation.jsp, Accessed 01 Mar 2024

16. https://www.makeuseof.com/, https://www.makeuseof.com/uses-lidar-sensors-other-than-iphones/, Accessed 1 Mar 2024
17. Hu, Q., Zhang, Z., Hu, W.: RangeLDM: Fast Realistic LiDAR Point Cloud Generation. arXiv preprint arXiv:2403.10094 (2024)
18. Hwang, B.Y., Park, J.K., Lee., T.H.: A study on the comparison of building data using 3D scanning. J. Korea Academia-Industrial Cooperation Soc. **22**(2), 50–56 (2021)
19. Janiszewski, M., et al.: Rapid photogrammetry with a 360-degree camera for tunnel mapping. Remote Sensing **14**(21), 5494 (2022)
20. Jeona, M., Kim, D., Chai, Y.: A comparison of 3D reconstruction through the passive and pseudo-active acquisition of images. J. Broadcast Eng. **21**(1), 3–10 (2016)
21. Johnson, R.A., Solis, A.: Using photogrammetry to interpret human action on Neolithic monument boulders in Ireland's Cavan Burren. J. Archaeol. Sci. Rep. **8**, 90–101 (2016)
22. Kim, H.J., et al.: Trends in high-resolution 3D data generation technologies. Electr.Telecommun. Trends **37**(3), 64–73 (2022)
23. Kim, I., et al.: Development of an application to generate 2D drawings in automation using open BIM technologies. Korean J. Comput. Design Eng. **21**(4), 417–425 (2016)
24. Kim, J., Kwon, K.K., Lee, S.I.: Trends and applications on LiDAR sensor technology. Electr. Telecommun. Trends **27**(6), 134–143 (2012)
25. Kim, M. H., Rushmeier, H., Ffrench, J., Passeri, I., Tidmarsh, D.: Hyper3D: 3D graphics software for examining cultural artifacts (2014)
26. Kim, S.-A., et al.: A Study on Setting Up Work Conditions for Improving Productiviyt of BIM-based Cost Estimation. Korean J. Construction Eng. Manag. **17**(1), 56–65 (2016)
27. KIM, S., et al.: A Relative Study of 3D Digital Record Results on Buried Cultural Properties. MUNHWAJAE Korean J. Cultural Heritage Stud. **55**(1), 175–198 (2022)
28. Kovanič, Ľ., et al.: Review of photogrammetric and Lidar applications of UAV. Appli. Sci. **13**(11), 6732 (2023)
29. Kuparinen, T., Kyrki, V., Toivanen, P.J.: Measurement noise in photometric stereo based surface reconstruction. VISAPP (2) (2008)
30. Leipner, A., et al.: Multi-camera system for 3D forensic documentation Forensic Sci. inter. **261**, 123–128 (2016)
31. Randles, B., et al.: The accuracy of photogrammetry vs. hands-on measurement techniques used in accident reconstruction. No. 2010-01-0065. SAE Technical Paper (2010)
32. Remondino, F., Guarnieri, A., Vettore, A.: 3D modeling of close-range objects: photogrammetry or laser scanning. In: Videometrics VIII. Vol. 5665. SPIE (2005)
33. Remondino, F., Stylianidis, E.: 3D recording, documentation and management of cultural heritage, vol. 2. Whittles Publishing, Dunbeath, UK (2016)
34. Takhtkeshha, N., et al.: Multispectral light detection and ranging technology and applications: a review. Sensors **24**(5), 1669 (2024)
35. Torkan, M., et al.: High-resolution photogrammetry to measure physical aperture of two separated rock fracture surfaces. J. Rock Mech. Geotech. Eng. (2023)
36. Tsai, M.-D., et al.: Exploring airborne LiDAR and aerial photographs using machine learning for land cover classification. Remote Sensing 15(9), 2280 (2023)
37. Wang, X., et al.: The evolution of LiDAR and its application in high precision measurement. IOP Conf. Ser. Earth Environ. Sci. **502**(1) (2020)
38. Williamson, B.M., et al.: Creating a 360-Degree RGB-D Sensor System for Augmented Reality Research

39. Zhang, X., et al.: A large-scale microelectromechanical-systems-based silicon photonics LiDAR. Nature 603(7900), 253–258 (2022)
40. Zheng, Q., Shi, B., Pan, G.: Summary study of data-driven photometric stereo methods. Virtual Reality Intell. Hardware **2**(3), 213–221 (2020)
41. Zong, W., et al.: The method of multi-camera layout in motion capture system for diverse small animals. Appli. Sci. **8**(9), 1562 (2018)

Agile Heritage: Revolutionizing Digital Content Development in Cultural Preservation and Beyond

Sangheon Kim(✉)

SangMyung University, Seoul 03016, Republic of Korea
enigma92@smu.ac.kr

Abstract. This paper investigates the integration of agile methodologies, artificial intelligence (AI), and the metaverse in digital heritage content development, highlighting a transformative approach to preserving and disseminating cultural heritage. Through agile processes and AI, projects gain adaptability and efficiency, while the metaverse offers immersive experiences, making heritage accessible globally. Ethical considerations and interdisciplinary collaboration are essential for ensuring respectful and accurate cultural representations. Future research should focus on ethical frameworks, sustainability, and inclusivity within the metaverse. This integration promises a dynamic engagement with heritage, underscoring the role of technology in bridging historical understanding across generations.

Keywords: Digital Heritage · Metaverse · agile process · content development

1 Introduction

Digital heritage content development is pivotal in preserving and sharing human civilization's vast legacies. This field utilizes digital tools to document, safeguard, and disseminate cultural heritage. With the rise of the metaverse, these assets can be interacted with in immersive virtual environments. This is crucial due to threats like environmental changes, urban growth, and conflict. Digital technologies enable realistic, interactive experiences of cultural sites through virtual and augmented reality, enhancing education and emotional connection. Digital heritage thus becomes a key preservation strategy, creating lasting archives and making heritage universally accessible, fostering global solidarity. This paper will discuss the essential role of digital heritage content in maintaining cultural memory and promoting worldwide understanding, ensuring the endurance and vitality of human history and culture for future generations.

2 Process of Digital Heritage Content Development

The foundation of cultural conservation in the digital age is rooted in the development of digital heritage content. This crucial preservation process merges technology with heritage management to safeguard historical artifacts against degradation. Selection of

G. Marreiros et al. (Eds.): DCAI 2024, LNNS 1151, pp. 91–99, 2025.
https://doi.org/10.1007/978-3-031-80946-0_9

artifacts for digitization requires judgment, prioritizing those of historical significance or at risk of loss.

The digitization process transforms artifacts into digital surrogates with high-resolution scans and detailed photography, aiming not just to replicate but to capture the essence of each item. Metadata enriches these surrogates, providing context and facilitating access. These are then preserved in digital repositories, securing their longevity and accessibility for future engagement.

Incorporating VR and AR technologies, the process extends beyond preservation, offering interactive and evocative experiences of cultural content. This chapter emphasizes that digital heritage development is a strategic conservation effort, guarding against heritage loss and connecting future generations to the richness of human culture in the digital realm.

2.1 Importance of Collaborative Work

In digital heritage development, interdisciplinary collaboration is key. It's crucial for ensuring that digital representations are technically robust and culturally nuanced. Teams comprising technologists, historians, and community stakeholders blend their skills to meet the complex needs of heritage digitization.

Such co-work guarantees cultural artifacts are digitized with accuracy and depth. Technologists may produce high-quality 3D models, but it's the historians and cultural experts who infuse them with meaningful narratives. Community involvement further enriches authenticity, with locals contributing invaluable perspectives.

Additionally, co-working promotes project flexibility, essential in a field marked by constant evolution in cultural landscapes and digital tech. Through teamwork, the multifaceted essence of human heritage is preserved and shared globally. Collaborative efforts are not just preferred but essential in creating holistic and sustainable digital heritage content.

2.2 Importance of Historical Verification

Historical verification is paramount in the digitization of cultural heritage, ensuring authenticity and integrity. It grounds digital recreations of artifacts and cultural narratives in factual history, crucial for the trustworthiness of digital heritage resources. In a fast-moving digital space, it prevents the spread of inaccuracies that can skew historical perception globally.

This verification not only maintains visual accuracy but also respects cultural contexts, which is vital for artifacts with deep cultural significance. It involves multidisciplinary experts, fostering collaborations that enrich digital artifacts with accurate historical contexts and contemporary relevance.

Especially within AR and VR, historical accuracy is crucial for authentic experiences. Historical verification in digital heritage content ensures narratively true and reliable digital representations of our cultural past, safeguarding our digital cultural legacy.

3 Agile Methodology for Digital Heritage?

In the scholarly pursuit of digitizing cultural patrimony, the Agile process has emerged as a beacon for facilitating collaborative work and fortifying the veracity of historical content within digital heritage projects. This methodology, characterized by its adaptive and iterative nature, lends itself to the complex and nuanced demands of heritage digitization, ensuring both the integrity of collaborative endeavors and the meticulousness of historical verification.

3.1 Enhancing Collaboration Through Agile Methodology

Agile methodology champions iterative development, dividing projects into short sprints, which fosters an environment where historians, digital specialists, and curators coalesce their expertise toward a common goal. This segmentation of work allows for an interlaced approach where each sprint builds upon the insights and contributions of a diverse team, thus harmonizing different perspectives into a cohesive digital narrative.

Regular communication touchpoints, such as daily stand-ups and sprint retrospectives, are central to Agile's structure, providing a rhythm for continuous dialogue and exchange among team members. These meetings serve as conduits for seamless information flow, ensuring that all voices are heard and integrated into the developmental process.

The cross-functional nature of Agile teams encapsulates a range of expertise within a single project group. This amalgamation of skills ensures that projects are not siloed into disparate components but are instead the product of a concerted and unified effort, inherently elevating the quality and coherence of the digital content produced. User stories, a staple of the Agile framework, place the needs and perspectives of end-users at the forefront of project development. This user-centric approach ensures that the digital heritage content resonates with its intended audience, be they the general public, academic scholars, or cultural institutions (Fig. 1).

Fig. 1. Agile Process

3.2 Rigorous Historical Verification in an Agile Context

The Agile framework, with its built-in feedback loops at the conclusion of each sprint, offers a systematic avenue for continuous historical verification. This iterative review process allows for the ongoing appraisal and rectification of the content, ensuring its historical accuracy and authenticity.

Agile's inherent flexibility is particularly advantageous when historical inaccuracies are identified or when new insights surface. The methodology's adaptive planning capabilities permit teams to reassess and modify their course of action swiftly, incorporating these changes into subsequent sprints without substantial disruption.

Stakeholder engagement, a cornerstone of the Agile philosophy, augments the process of historical verification. By involving experts and community stakeholders throughout the project lifecycle, the digital heritage content is vetted for accuracy and cultural sensitivity, bolstering its credibility and depth.

Furthermore, the emphasis on documentation and transparency within Agile projects creates a robust framework for accountability. Detailed records of decisions, sources, and methodologies form a comprehensive trail that underpins the historical verification process, rendering it both transparent and defensible.

In essence, the Agile process, with its collaborative ethos and commitment to accuracy, emerges as an indispensable framework in the development of digital heritage content. It meticulously weaves together the collective expertise required to produce digital representations that are not only technologically refined but also historically accurate, ensuring that digital heritage content stands as a testament to our cultural legacy, authenticated by scholarly rigor and collaborative excellence.

4 Agile Methodology for Digital Heritage

In the evolving landscape of digital heritage content development, the Agile process emerges as a paradigmatic shift, promising a more dynamic, collaborative, and accurate approach to preserving our cultural legacies. This paper delineates the integration of Agile methodologies into digital heritage projects, underscoring its potential to revolutionize the creation, verification, and dissemination of digital heritage content.

Agile, characterized by its iterative cycles, stakeholder collaboration, and adaptability to change, offers a framework inherently suited to the multifaceted demands of digital heritage content development. Unlike traditional, linear project management approaches, Agile facilitates a more nuanced engagement with the project lifecycle, enabling teams to respond to new discoveries, technological advancements, and stakeholder feedback with unprecedented flexibility (Fig. 2).

4.1 Fostering Collaborative Synergies

A core tenet of Agile is its emphasis on cross-functional teamwork, which aligns seamlessly with the interdisciplinary nature of digital heritage projects. Through the organization of work into sprints, Agile promotes intense collaboration among historians, technologists, curators, and community stakeholders. This collaborative ethos is

Fig. 2. Rome Reborn

further enhanced by regular stand-up meetings and retrospectives, ensuring ongoing communication and the integration of diverse perspectives and expertise.

Moreover, Agile's user story approach centers projects around the end-user experience, ensuring that the digital content developed is not only historically accurate but also accessible and engaging to a broad audience. This focus on user engagement is particularly salient in the context of digital heritage, where the primary goal is to make cultural heritage more accessible and relatable to the public.

4.2 Ensuring Historical Accuracy Through Iterative Verification

The iterative nature of Agile, with its regular reviews and feedback loops, is particularly conducive to the rigorous historical verification required in digital heritage projects. By facilitating ongoing assessments of digital content's historical accuracy, Agile allows for the continuous refinement and correction of content throughout the development process. This iterative approach to verification ensures that digital heritage content remains true to its source material, upholding the integrity and authenticity of cultural representations.

Additionally, Agile's adaptive planning capability allows projects to quickly pivot in response to new historical insights or corrections. This flexibility is invaluable in the realm of digital heritage, where the discovery of new artifacts or information can necessitate adjustments to digital representations or narratives.

4.3 Stakeholder Engagement and Transparency

Critical to the success of digital heritage projects is the engagement of a wide array of stakeholders, including subject matter experts, community representatives, and the broader public. Agile methodologies prioritize this engagement, involving stakeholders throughout the project lifecycle to ensure that digital content is developed in a manner that is both accurate and respectful of the cultural heritage it represents.

Furthermore, Agile's commitment to documentation and transparency supports the meticulous tracking of decisions and sources, essential for historical verification and scholarly research. This documentation creates a transparent record of the project's development process, facilitating peer review and public scrutiny.

The Agile process, with its emphasis on collaboration, iterative development, and stakeholder engagement, holds significant promise for the field of digital heritage content development. By integrating Agile methodologies, heritage projects can achieve a balance between technical innovation, historical accuracy, and community involvement, ensuring the preservation and appreciation of our cultural heritage in the digital age.

5 Metaverse as Platform

Within the scholarly exploration of digital heritage content development, the emergence of the Metaverse has catalyzed a reevaluation of traditional methodologies in cultural preservation. This section of the paper scrutinizes the Metaverse's role as a vanguard platform, transforming the dissemination and interaction with cultural and historical artifacts.

The Metaverse, with its inherent immersive capabilities, presents an unprecedented opportunity for the virtualization of cultural heritage, allowing users to navigate through historical eras and cultural landscapes in a three-dimensional digital continuum. This immersive engagement not only enriches the educational experience but also bridges the gap between the past and the present, making history palpable and accessible to a global audience.

Crucially, the Metaverse facilitates a collaborative ecosystem where professionals from various disciplines—historians, technologists, and educators—can converge to co-create digital heritage content. This synergy amplifies the depth and authenticity of digital reproductions, ensuring that they are not only technically accurate but also culturally and historically resonant.

Moreover, the Metaverse stands as a bulwark for the preservation of intangible cultural heritage, offering a dynamic platform for the documentation and experience of practices that are ephemeral in nature. Through virtual simulations, the Metaverse captures the essence of these traditions, safeguarding them against the erosion of time (Fig. 3).

However, the adoption of the Metaverse in digital heritage content development is not without its challenges. The platform's reliance on cutting-edge technology raises questions regarding accessibility and the digital divide, potentially limiting the reach of these virtual heritage experiences. Furthermore, the immersive nature of the Metaverse necessitates rigorous historical verification to maintain the integrity and accuracy of cultural representations, a task that demands meticulous attention to detail and a profound understanding of the cultural nuances being portrayed.

In conclusion, the Metaverse represents a promising frontier in the field of cultural preservation, offering innovative solutions to age-old challenges in accessibility, engagement, and preservation. As we venture further into this digital realm, it becomes

Fig. 3. Guyuk Metaverse for Mongolian digital heritage

imperative to navigate its challenges with a strategic and thoughtful approach, ensuring that the digital heritage content developed within the Metaverse serves as a testament to our collective history, accessible to and preserved for future generations.

5.1 Agile and A.I.?

In the rapidly evolving landscape of digital heritage, the integration of artificial intelligence (AI) within agile methodologies presents a promising frontier for content development. As a researcher in digital heritage, the intersection of these technologies and methodologies heralds a paradigm shift towards more dynamic, responsive, and user-centered heritage content creation.

Agile methodologies, characterized by their iterative cycles, adaptability, and collaborative approach, align closely with the needs and challenges inherent in digital heritage projects. These projects often deal with complex, sensitive, and multifaceted data that require a nuanced approach to digitization, interpretation, and presentation. AI, with its capabilities for data analysis, pattern recognition, and automation, can significantly enhance these processes, offering tools to handle the vast and varied data involved in heritage work more efficiently and insightfully.

Moreover, the integration of AI into agile processes facilitates a more engaged and inclusive approach to digital heritage. Through techniques such as natural language processing and machine learning, AI can help decipher and translate intangible cultural nuances, making heritage content more accessible and relatable to a diverse audience. This, in turn, can foster a deeper connection with the heritage being presented, encouraging a broader discourse around its significance and preservation.

However, the application of AI in this context is not without challenges. Ethical considerations, such as data privacy, cultural sensitivity, and the potential for bias in AI algorithms, must be meticulously addressed. Furthermore, the successful integration of AI within agile methodologies necessitates a multidisciplinary approach, combining expertise in technology, heritage studies, and project management to ensure that the digital content developed is both technologically sound and culturally respectful.

In conclusion, the synergy between AI and agile processes in the development of digital heritage content offers a dynamic and innovative pathway for the preservation and dissemination of cultural heritage. This approach not only enhances the efficiency and depth of digital heritage projects but also promotes a more inclusive and engaging user experience. As we move forward, the careful balancing of technological advancement with ethical and cultural considerations will be paramount in realizing the full potential of this integration, ensuring that digital heritage serves as a bridge between the past, present, and future.

6 Conclusion

In conclusion, the integration of artificial intelligence (AI) and agile methodologies within the burgeoning sphere of the metaverse signifies a pivotal shift in digital heritage content development. This paradigm offers unparalleled opportunities for enhancing the accessibility, interactivity, and engagement of cultural heritage, bringing it closer to global audiences in ways previously unimaginable. However, this integration is not without its challenges. Ethical considerations surrounding the use of AI, the sustainability of digital preservation, and the inclusivity of the metaverse platform must be diligently addressed to ensure that the digital representation of heritage is both accurate and respectful of its intrinsic values.

Future research directions should focus on developing robust ethical guidelines for AI applications in heritage projects, exploring sustainable digital preservation techniques, and enhancing the inclusivity of the metaverse as a platform for heritage dissemination. Additionally, interdisciplinary collaboration will be essential in leveraging the full potential of these technologies, ensuring that digital heritage projects remain grounded in cultural sensitivity and historical accuracy.

As we look towards the future, the agile process, augmented by AI and realized within the metaverse, offers a promising avenue for the development of digital heritage content. This innovative approach not only facilitates a more dynamic and interactive engagement with heritage but also poses critical questions about the role of technology in preserving and interpreting our cultural past. By addressing these challenges and harnessing the opportunities presented by these technological advancements, we can ensure that digital heritage remains a vibrant and accessible resource for generations to come.

References

1. Park, J.: Seokguram Digital Contents Prospects for the Proposal of Artificial Intelligence Extended Reality. Content Plus **19**(4), 47–61 (2021)
2. Guidi, G., Frischer, B., Lucenti, I.: Rome reborn-virtualizing the ancient imperial Rome. In: Workshop on 3D Virtual Reconstruction and Visualization of Complex Architectures. Trento, Italy (2007)

3. Choi, H., A Study on the Construction and Utilization of Korean Prehistoric Remains Database, Mobile Information system, 7081286, (2022)
4. Park, J.h., Kim, S.: Digital Representation of Seokguram Temple UNESCO World Heritage Site. In: Jung, T., tom Dieck, M. (eds.) Augmented Reality and Virtual Reality. Progress in IS. Springer, Cham (2018)

Using the Metaverse to Teach Prehistoric History - Centered on Paleolithic Ruins in Korea -

Choi Hee-Soo and Ku Cha Yong(✉)

Sangmyung University, 20 Hongjimun2-gil, Jongno-gu, Seoul, Korea
koostar@smu.ac.kr

Abstract. This study examines the history education plan for prehistoric times in Korea using the metaverse. History education aims to cultivate historical thinking skills, and historical thinking skills are cultivated through imagination and vicarious experiences based on historical facts. In that respect, the metaverse becomes a useful tool for history education. Although a lot of information about the remains of the Paleolithic Age in Korea has been accumulated, the information is not comprehensively and systematically serviced. As a way to overcome this, the construction of an electronic map was conducted on a trial basis. After systematically organizing and sharing information on Paleolithic ruins through an electronic map, a metaverse construction plan was derived based on this. The core elements of the metaverse are the elements of experience and the contents of history education, and based on this, the history education plan for prehistoric times was summarized.

Keywords: Metaverse · History Education · Electronic map

1 Introduction

The current era is one in which digital technologies are evolving in a different direction and at a different pace than in the 1990s and 2000s. Artificial intelligence, superintelligence, chatGPT, and the arrival of the technological singularity are all being talked about at this point. Accordingly, social changes are also rapidly unfolding. Innovative changes are expected not only in traditional teaching and learning, but also in teaching methods and educational contents.

In the age of digital transformation, metaverses have recently gained prominence in education. The metaverse is a parent concept of virtual reality and augmented reality, a system that extends reality into a digitally based virtual world and allows all activities to be performed in virtual space [1]. The metaverse has spread rapidly since the COVID-19 pandemic hit the world in early 2020. In an era of hyperconnectivity, the coronavirus has created "disconnection" and "isolation," and the metaverse has gained traction in the transition from face-to-face to non-face-to-face based on the desire for connection. Almost all daily life, such as games, performances, festivals, events, and education, were implemented in the metaverse in a non-face-to-face situation [2].

The potential for metaverses in the field of history is very high. This is because the metaverse is basically composed of virtual or fictional worlds that cannot be experienced

G. Marreiros et al. (Eds.): DCAI 2024, LNNS 1151, pp. 100–108, 2025.
https://doi.org/10.1007/978-3-031-80946-0_10

in real life, and history is also about the past that cannot be experienced in real life. And of course, history education. This is because history education using metaverses can provide indirect experiences that allow learners to glimpse the thoughts of various cultures and historical figures of the time. I think it can help to realize the main purpose of history education, which is to cultivate historical thinking skills.

Utilizing the metaverse is especially important for prehistory because it is an era without historical records. However, while individual prehistoric sites have been well researched and studied, the overall distribution of Paleolithic sites, the process of change over time, and the distribution of characteristic indicator artifacts have not been summarized. Therefore, historical education about the prehistoric period is also limited to representative sites. This study aims to overcome this situation. To this end, we will examine, first, the current status of research on Paleolithic sites, second, the sharing of Paleolithic site discoveries, third, the creation of a nationwide distribution centered on electronic maps and detailed information input, and fourth, the construction of a metaverse and history education.

2 Theoretical Background

It is the cultivation of historical consciousness, the basic purpose of history education. Historical consciousness is to recognize the historical role of society in which one is currently located, to grasp the present status between the past and the future, and to conceive a desirable future based on this. To this end, in history education, historical thinking skills are formed through understanding the causal relationship of historical facts and learning to infer through logical thinking of those who study history. Historical thinking ability is the ability to interpret and judge objects according to historical perspectives based on historical knowledge. The sub-elements of historical thinking ability include chronological grasp, historical inquiry, historical imagination, and historical judgment [3].

Understanding history is the way to cultivate historical thinking skills. Understanding history is the understanding of an action or event in a certain context. Understanding history requires evidence of the period, event, and person. People who want to understand history based on evidence recognize that they may not discover something behind it through imagination or obtain it directly through their senses. In other words, imagination is an intellectual activity that reconstructs historical facts from understanding history. Understanding history is key to entering a strange world different from the present where people who want to understand history live. We must enter the time, place, and situation we have not experienced in order to understand the history of the time. Therefore, it is important in history education to make a habit of thinking so that we may enter a world we have not experienced. At this time, what is necessary is imagination, and the method used to imagine is vicarious experience [4].

Vicarious experience is the process of grasping other people and other societies. It is reenactment that enhances imagination as well as experience. Through reenactment, people can experience things like historical empathy. This empathy is accompanied by curiosity and imagination and is based on background knowledge about the past era. In other words, the richer the historical knowledge of the past society, the easier it is to empathize through ideological understanding [5].

In the end, history education must develop a historical imagination to properly understand the context through vicarious experiences, reenactments, and empathy for the past. Therefore, it is essential to expand the experience that can enter the world of the past. With the development of digital technology, this imaginary world can be unfolded unlike the past.

What can help in this process of historical education is the emergence of metaverses, other worlds that allow us to experience elements that we cannot experience in the real world. Experience-expanding historical content that utilizes digital technologies can help fulfill the essential purpose of history education. The reason why the metaverse platform has attracted attention for educational purposes is that various educational functions in the classroom could be accommodated in the metaverse platform. At first, it was used mainly for simple events, but as educational activities using platforms such as Gathertown and Naver JET were conducted, it began to be in earnest. In 2021, research on the guidelines for using education of the metaverse platform was also conducted [6].

Metaverse-based education consists of five types: lecture type, experience type, discussion type, problem solving type, and creative type [7]. Each type for each subject is used independently or fused. Metaverse-based education was also provided in history education. For example, in the social studies course of Dongyang Elementary School in Wanju-gun, Jeollabuk-do, prehistoric experiences were implemented through the Minecraft platform. This content, built in a game type, was positively evaluated by instilling students' perception of 'interesting history lesson'.

3 Paleolithic Research and Services in South Korea

3.1 Status of Paleolithic Research in South Korea

Excavations for prehistoric sites in Korea began during the Japanese occupation. Investigations by Japanese scholars heavily distorted the results and denied the rich history of the Korean peninsula (Archaeological Association of Korea, 2013). Since independence from the Japanese occupation, Korean excavations have been a way of correcting the history that was ignored or intentionally distorted by Japanese scholars. Excavations since liberation have revealed that the Korean peninsula, like other parts of the world, is rich in prehistory from the Paleolithic to the Bronze Age.

Initially, it was mainly national research organizations that were in charge of excavations, but as the discovery of sites progressed, local university research institutions and private research organizations became increasingly involved. However, it is difficult to obtain comprehensive information on prehistoric sites because excavation reports written by various research institutions are not systematically organized or serviced and are managed and serviced by each institution or commercially serviced by private companies. Therefore, it is urgent to establish a systematic information service for prehistoric sites.

However, there are several limitations. First, information related to prehistoric sites is not integrated online or offline but is scattered individually. There is a need for a unified platform to provide this scattered information. Secondly, information related to prehistoric sites is not systematically managed in various materials such as books, digitized files, photos, drawings, and videos. Most of the information is available in the excavation reports of excavation organizations, but digital files such as videos and 3D objects are not available in the reports. It is necessary to systematically manage digital information. Third, despite the fact that the core information of prehistoric sites is spatial information, prehistoric site information is not yet provided in the form of cultural maps. Therefore, it is urgent to build an electronic map of prehistoric sites.

3.2 Paleolithic Excavation Report Service in Korea

The Paleolithic excavation report service in Korea has the meaning of information sharing, but as already pointed out, the excavation report services established by the agencies involved in the research, such as the Cultural Heritage Administration, the National Institute of Cultural Properties, the National Museum of Korea, and the Korea Cultural Heritage Foundation, have focused on their own excavation report services, with no interconnection or reference functions [8].

There are several problems with these excavation report services. First, the variety of organizations and types of services that provide excavation reports makes it difficult to find the desired excavation report easily, i.e., there is no integrated service or integrated search service for excavation reports.

The second is that the content of the service is concentrated only in the form of books called excavation reports. In the case of the National Museum of Korea, the bibliographic information is well organized, but the contents of the important excavation reports are not available unless you look at the original texts. The same is true for the rest of the services. These problems highlight the urgent need to share excavation reports, or the results of research, in prehistory education and research.

4 Building an Electronic Map of Korea's Paleolithic Sites

4.1 Transforming Spatial Information for a Korean Paleolithic Site Inventory

One of the best ways to share research results and provide services to the public is to build an electronic map. Therefore, in this study, we converted the results of the Korean Paleolithic site cataloging project published by the Jeongok Prehistory Museum into a spatial map [9]. To do this, we extracted location information through geocoding based on the addresses in the list of Paleolithic ruins and combined the extracted location information with the list data as attribute information. Geocoding is the process of extracting coordinate information (longitude and latitude) of a point using the address in the list. First, we extract the field corresponding to "Address" from the list of Paleolithic sites and apply Google Maps' geocoding function (Fig. 1).

Spatial information was created in QGIS using the latitude and longitude values assigned through the geocoding process. Point objects were created using the latitude and longitude values and then combined with the attribute information of the Paleolithic Inventory. This process was used to create the spatial information of the Paleolithic Inventory. The result of geolocating the Paleolithic Inventory is shown in the figure. As shown in the figure, the location information of 489 Paleolithic sites was extracted (Fig. 2).

```
import pandas as pd
df = pd.read_csv('data.csv')

import googlemaps
my_key = "My_GoogleMap_API_Key"
maps = googlemaps.Client(key=my_key)
lat = []
lng = []
for address in df['address']:
    try:
        geo_location = maps.geocode(address)[0].get('geometry')
        lat.append(geo_location['location']['lat'])
        lng.append(geo_location['location']['lng'])
```

Fig. 1. Geocoding process using Google Maps

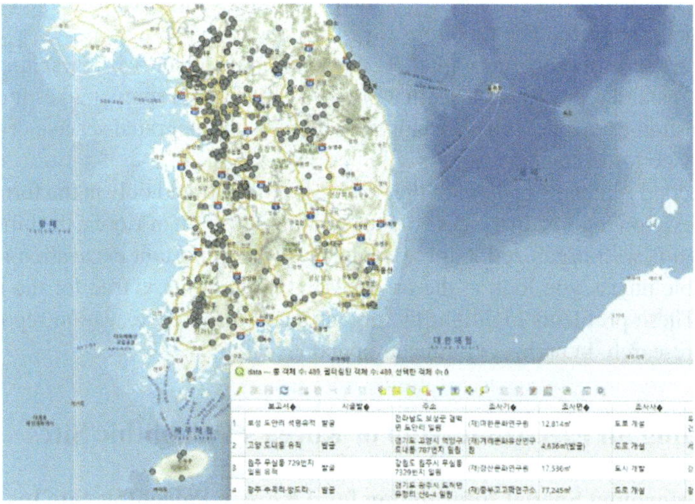

Fig. 2. Location information of Paleolithic sites in Korea

4.2 Problems in the Spatial Information Conversion Process of the Korean Paleolithic Site Inventory

There are several limitations in constructing an electronic map using the currently established list of Paleolithic ruins. The first is that it is difficult to extract comprehensive information on historical sites because the list of historical sites is provided annually based on excavation surveys. This means that the list should be reorganized based on the historical sites.

Second, the current address of the historical site has inaccurate address values and area information such as the exact number index, ~one-to-one, ~dong, and ~ area. In other words, information for location indication is not systematic on an electronic map.

Third, information on historical sites has not been extracted from the excavation report. There is little additional information on the condition of the site, the current status of the excavation of relics, the types of relics, and the nature of relics and relics.

Only when this is built into a database can electronic map information of the Paleolithic ruins be properly established.

The most important thing above is to ensure the integrity of location information [10]. An accurate electronic map of Paleolithic sites can be drawn only when spatial information is expressed in a facet type, that is, a polygon type, rather than a point type.

4.3 Proposed Electronic Map Service for Paleolithic Sites in Korea

If the inventory of Paleolithic sites in Korea is compiled in the form of a database, and if it is built into an electronic map, it will provide several services to the public and researchers on Paleolithic history. First, it can provide a distribution map of Paleolithic sites. The overall distribution of Paleolithic sites allows us to understand the geographical features of Paleolithic sites, their temporal distribution, and the movement of Paleolithic people. Second, the analysis of representative artifacts from a site can reveal the characteristics of Paleolithic regional life. Even the same lithic tools can be distinguished in time according to their processing forms and methods, and the interconnection between sites where the same type of stone tools were produced can be used to trace their migration routes. Through the analysis of animal bones and human bones, we can learn about the diet and skeleton of people at that time.

Simply looking at it can provide a variety of ways to obtain information about the Paleolithic Age. On an electronic map, it is possible to provide a distribution map by period, experience life in the Paleolithic Age through the movement of avatars, provide objects such as stone tools, dwellings and food, and provide 3D images of the time of excavation and restored or reproduced images through visits to historical sites by period.

5 Discussion

5.1 Building a Metaverse of Paleolithic Sites in Korea

There are several ways to build the remains of the Old Stone Age in the metaverse. Existing commercial platforms may be used, or new platforms may be built. It is safe to use Naver's jet platform, which is used the most in Korea, or foreign countries such as Gathertown, Roblox, and Minecraft.

To build a Korean Paleolithic site in the Metaverse, you must first create a map. The most important thing to consider is what aspects of the Korean Paleolithic sites you want people to experience. You can build all the sites on the electronic map, or you can build only the important ones. For the purpose of history education, it is better to select important sites rather than building all of them. This decision process is called planning. The process of building a metaverse of Paleolithic sites in Korea can be schematized as follows (Fig. 3).

The planning process involves two parallel processes. One is the process of building detailed information and content about the Paleolithic ruins, and the other is the process of building an electronic map of the Paleolithic ruins. The first step in the content building process is the excavation report, which contains detailed information about the ruins. We organize excavation reports for selected important sites, analyze the contents, and

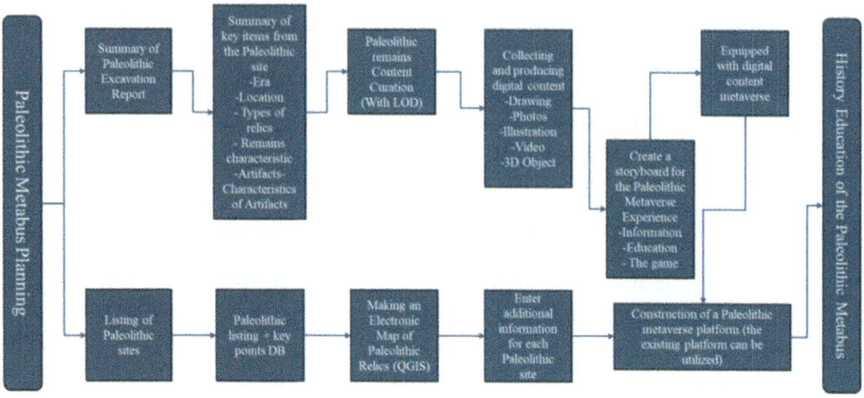

Fig. 3. Planning Process for Paleolithic Metaverse in Korea

organize key items. For example, the time period, location, types and features of the ruins, artifacts and features, and excavated items. The items are then curated to connect them to each other within a certain context. They also collect, create, and prepare digital content about the site and its artifacts [11].

The electronic mapping process begins with an inventory of Paleolithic sites and the creation of a database that itemizes the main features of each site. Based on this, an electronic map is created using the QGIS program. Each site is linked to provide detailed information. In addition, link the items created as digital content.

Once you've gotten to this point, you'll need to start building your metaverse, but first you'll need to create a storyboard. Create a storyboard by categorizing the content available in the Metaverse into information, education, and games, and organizing what you need to teach Paleolithic history. Then, build those items into the Metaverse platform so that the service can be delivered.

5.2 Teaching Korean Paleolithic History Through a Metaverse

For historical education about the Paleolithic, it is important to consider what experiences users can have within the Metaverse [12]. Since the Paleolithic is a prehistoric period without written language, we need to understand the historical context of the time through ruins and artifacts [13] For example, the following table shows (Table 1).

Based on the above, a simple example of the metaverse education scenario can be taken through the following process.

- Entry (Avatar selection) → Full view of the site → Entry into the cave → Conversation with Paleolithic people → Experience living in the Paleolithic Age (making stone, eating, making fires, etc.) [Providing information on related relics and remains] → Hunting experience (group hunting) → Funeral experience (burial customs) → Moving life (exploration of food sources) → Exit

Table 1. History Education of the Paleolithic Age

History education Topics	Ruins and artifacts	History education content	Remarks
Paleolithic diet	Animal Bones	Recognize what animals Paleolithic humans hunted for food from animal bones	Images of excavated ruins, animal bones, 3D objects, and bone and animal guessing puzzles
The hunting life of Paleolithic humans	Animal Bones, Chipped Stone tools (chopper, stemmed points, hand axes)	Estimating hunted animals from animal bones and recognizing hunting methods from stone tools	Images of excavation sites, images of animal bones, 3D objects, images of Paleolithic animals, and imagined hunting scenes
How Paleolithic people lived	Caves, Shacks (Late)	Understanding the creation and utilization of natural dwellings and the transformation of man-made dwellings and living environments through cave and later shack remains	Cave Ruins Images, Cluster Ruins Images, 3D Objects
Community life among the Paleolithic, Paleolithic Periodization, Paleolithic and Neolithic etc			

6 Conclusion

History education about Korea's prehistory relies on the results of archaeological excavations. Prehistoric history education through textbooks is based on fragmentary information, which makes it difficult to understand the overall history of the Paleolithic period. From this point of view, Metaverse is suitable for prehistoric history education. It is a great tool for graphically recreating past worlds that we can't visit, and for instantly traveling to sites that we can't visit and exploring related information.

Prehistoric history education using the metaverse can be effective when it combines information on historical sites and artifacts, contextualized historical knowledge, and game elements that make historical knowledge interesting and natural. Here, the four elements of the metaverse, such as virtual world, augmented reality, life logging, and mirror world, can be operated in combination. This will help to understand the history of the time in three dimensions. And this approach should be actively utilized in the education of the next generation of students who are accustomed to the digital environment of the digital age.

References

1. Hyunjae, J., Sungkyung, C., Eunshil, J.: Principles of learner-centered instructional design in education using metaverses. Korean J. Liter. Res. **13**(2), 17–18 (2022)
2. Jinchul, P.: Exploring the possibility of using Metaverse in Korean language education. J. Int. Netw. Korean Lang. Cult. **18**(3), 122 (2021)
3. Choi, S.-H.: Subcategories of historical thinking, history education and awareness of history, pp. 60–80 (2005)
4. Lee, Y.-h.: Discipline of history and history teaching, the theory of history education, p. 172 (2009)
5. Lee, Y.-h.: Discipline of history and history teaching, the theory of history education, p. 99 (2009)
6. Korea Education and Research Information Service: A Study on the Guidelines for Educational Use of Metaverse, KERIS, 43 (2022)
7. Choi, H.-S., Ku, C.Y.: Establishing the contents curation and digital map of humanities information: the case of prehistoric remains in Korea. Geograph. J. Korea **55**(4), 471–481 (2021)
8. Jeongok Prehistoric Museum Homepage. https://jgpm.ggcf.kr/boards/prehistoric%20remains/articles/529. Accessed 25 Mar 2024
9. Jinmu, C.: Geocoding method to record archaeological site information. Geo-graph. J. Korea **53**(1), 21–30 (2019)
10. Choi, H.-S.: A study on activation of Confucian people information service through digital curation. Confucian. Res. **50**, 401-425 (2020)
11. Choi, H.-S.: A study on content curation plan of local cultural resources archives: focusing on region N culture service. Global Cult. Cont. **42**, 85–103 (2020)
12. HeeSoo, C., Sangheon, K.: A research on metaverse content for history education. Global Cult. Cont. **26**, 209–226 (2017)
13. Hee-Soo, C.: Experiential content and the Korea ancient history - focusing on enhancing the historical imagination. J. Korean Anc. History **84**, 191–222 (2016)

DCAI - SS-GILL

Special Session on and for Female Technopreneurs and Early Career Researchers in Distributed Computing and AI Organised by the Gendered Innovation Living Labs (GILL) Project

Scope

The special session aims to support female technopreneurs and women in distributed computing and AI. It is organised by the Horizon Europe Gendered Innovation Living Labs (GILL) funded project. It aims to reduce gender inequalities in innovation and entrepreneurship in the EU by developing tools and methods to create more inclusive Innovation Ecosystems. Gender Responsive Smart Innovation and Entrepreneurship (GRSIE) has been described by Woodcock and Christensen (2023) as a reaction against one sided innovations that only consider the technological side and benefit only one group. GRSIE should be disruptive, reactive and challenging. It provides a means of responding to real-world challenges holistically and with new indicators that prioritise and address gender and other intersectional characteristics together with material innovations. 'Smartness' is implied not only in the use of technology but also in the mindsets and methods of those engaged in the innovation ecosystem. Products and systems are 'smarter' when they are integrated with and build on existing solutions, when they are inclusive, i.e., benefitting everyone, not creating worse problems by prioritising the needs of one group, or innovation at the expense of others. 'Gender responsive smartness' requires aspiring innovators and entrepreneurs to work in new ways, with new partners and cutting-edge technologies - to disrupt old ways of thinking and services, particularly in enabling women and those from previously excluded groups to have a voice and to encourage innovations and socio-material designs and services, which are based on their shared experiences. It also requires the development of a culture which values and uses these inputs. GILL activities address the need to fix cultures and ways of working which impede GRSIE.'

Topics

The session's main objective is to share the research and experiences of early career researchers and female technopreneurs who support Gender Responsive Smart Innovation and Entrepreneurship (GRSIE). The aims of the session are:

To demonstrate how the GILL project is addressing the known challenges of female technopreneurs
To share information on barriers and enablers to career progression
To showcase research designed, led and produced by early career researchers in AI and distributed computing
To provide a supportive and collegial environment to develop new relationships

Barriers and enablers for women working in Distributed Computing and AI
Gender disaggregated data
AI and gender equality and bias
Feminist considerations in AI
Gender stereotyping in AI systems
Gendered use of new technology
Cultural disparities
How AI and distributed computing can create a more equitable society
Indigenous entrepreneurship
Case studies and new research

Organizing Committee

Andree Woodcock	Coventry University, UK
Marta de los Rios White	European Network of Living Labs, Belgium
Javier Prieto Tejedor	University of Salamanca, Spain
Marta Plaza Hernández	University of Salamanca, Spain
Ana Belén Gil González	University of Salamanca, Spain

Addressing Gender Equality in AI and Robotics, with Special Reference to Pakistan and Developing Countries

Andree Woodock[1](✉), Alison Halford[1], James Brusey[1], Hilda Christenen[2], Nicola Marsden[3], and Riaz Uddin[4]

[1] Coventry University, Coventry, UK
a.woodcock@coventry.ac.uk
[2] University of Copenhagen, Copenhagen, Denmark
[3] Heilbronn University, Max-Planck-Str. 39, 74081 Heilbronn, Germany
nicola.marsden@hs-heilbronn.de
[4] NED University of Engineering and Technology, Karachi, Pakistan

Abstract. Existing issues in terms of female equality in AI and robotics are discussed in terms of strategies used in the EU funded Gender Innovation Living Lab (GILL) project to fix the number, the knowledge, institutions, and culture. These issues were investigated during a staff /student mobility exchange visit between Coventry University (UK) and 5 representative universities in Pakistan in 2023, with results confirming the literature. From this a set of recommendations is proposed to address acknowledged gender issues.

Keywords: Gender · STEM · Pakistan · AI and robotics

1 Introduction

The data science and Artificial Intelligence (AI) professions provide an opportunity to break the cycle of gender inequalities in science, technology, engineering, and mathematics (STEM) [1]. In 2018, WIRED and Element AI reviewed the AI research pages of leading technology companies and found that only 10%–15% of machine learning (ML) researchers were women [2], and in Google AI pages only 60 of the 641 people in machine intelligence were women. While other STEM fields have reduced the gender gap, 'white men remain overrepresented in data-related fields' [3].

The digital divide [4] is more gendered in developing countries where women have less access to information, technology, and communication (ITC), resources, and opportunities. Despite significant sectorial growth opportunities women are underrepresented particularly in high-value and skill jobs [5] due to gender bias and stereotypes, lack of a gender-friendly work environment and negative socio-cultural attitudes.

Women are more likely to occupy lower paid [6] and lower status jobs [7], e.g., in data preparation, quality assurance, less technical and customer support roles, despite their higher qualifications. Men occupy more positions in ML, big data, general purpose

G. Marreiros et al. (Eds.): DCAI 2024, LNNS 1151, pp. 113–123, 2025.
https://doi.org/10.1007/978-3-031-80946-0_11

computing and have computer science skills traditionally associated with more prestigious and higher paying careers. Worryingly as women enter technological subdomains, such as front-end development, these fields lose prestige and experience salary drops [8]. A similar picture has been found in academia with female authors accounting for 12% of contributions in leading ML conferences in 2017 [2, 10] and 13.8% in AI [11]. Men form the majority of AI department faculty and 80% of AI professors [12].

Proposing elements of a framework on gender equality and AI, [9] stressed that simply increasing the number of women was not enough; there was a need to address culture and power, ensuring that women are hired in core roles, such as in development and coding where they had the ability to exert influence.

The gender gap in AI is more than an equal opportunities issue, it is a matter of how the world is designed, for whom and by whom [13]. Technologies are socially shaped by gender power relations and cultural beliefs. These influence the design, technical content, and usage. Wajcman (2010) [14, p.149] commented that 'gender relations are materialized in technology, and masculinity and femininity in turn acquire their meaning and character through their enrolment and embeddedness in working machines'. If data science and AI are to solve societal problems, they need to be informed by a wide range of perspectives and experiences and accurate data. Research suggests that gender bias is already being built into and amplified by AI systems [15] and robotics. For example, [16] critiqued gender stereotypes and cultural norms in robot design, providing examples (from toy to assist robots) of the ways these are gendered and targeted towards female dominated occupations – thus devaluing women's contributions to society. Gendered robots, designed by men, reinforce existing norms.

1.1 Pakistan and Developing Countries

Pakistan was ranked 142nd in the 2023 Gender Global Gap [17] despite progress in economic participation. Gender inequality policy efforts often centre on encouraging more women to pursue STEM [18] without acknowledging the deeper cultural and structural barriers which inhibit their entrance and advancement. The persistence of the 'brogrammer' culture [20] deters inclusion and diversity. As such, women occupy only 14% of IT jobs, mostly at the lower levels. 'Invisible women' perform skilled work (e.g., in robotics) but are underpaid, undervalued, and unprotected by labour laws [21].

The metaverse provides potential for women to achieve equality in digitally enabled workplaces, political and economic life [19] if they can occupy this space. E-commerce provides the safety and security for women to set up their own businesses, access legal aid, secure banking, and new markets. Social media channels allow women to rise as role models and thought leaders less constrained by traditional cultural and social expectations and acquire new skills to facilitate personal and professional development. Women can adopt different identities, free from cultural and social restrictions, engage in different activities and acquire new skills which facilitate personal and professional development. Maqsood et al. (2022) [22] argued that by understanding technology, gender inclusive standards can be developed, along with the skills needed to implement, and that these can be translated back into society.

2 Strategies to Address the Gender Gap

The transatlantic Gendered Innovations project [23] argued that addressing societal challenges required supporting innovations that improved the lives of women as well as men – now taken as meaning all genders. This requires a number of transformative and far reaching 'fixes' related to the:

- Numbers, i.e., increasing participation of women and underrepresented groups.
- Institutions i.e., structural changes in organizations, institutions, and services to address women and family unfriendly structures.
- Knowledge or data, underpinning excellence in science and technology by integrating sex, gender, and intersectional analysis into research

In addition, the GILL project [24] introduced a 4th fix related to.

- Culture, defined as beliefs, norms and expectations within a society that materialize or become explicit in behaviors, practices, or artefacts like images.

Barriers to implementing the fixes [25] include *Soft ecosystem* components, such as education, training, mentoring, which create barriers to women's entering STEM when absent: *Hard ecosystem components* such as funding, finance, research and innovation, which tend to be skewed towards men – impeding women's growth, progression and ambition; *Compliance components*, such as regulatory frameworks, state run financial and support systems which create barriers when they favour sectors dominated by men, and health and welfare policies, such as lack of childcare, which create barriers for women entering the workforce; *Transversal components,* such as stereotypes, bias and lack of coordinated policy and interventions in all areas.

3 Knowledge Transfer Opportunities

The paper has developed through the opportunistic collision of two project – H2020 Horizon Europe funded project GILL (Gendered Innovation Living Lab – which aims to support gender responsive smart innovation and entrepreneurship) and the British Council (BC) funded Pakistan – UK Staff and Student Mobility Scheme focusing on AI and robotics. Staff and students from the mobility scheme have contributed their experiences and reflections on gender related problems in AI and robotics in Pakitsan (gathered through a small survey, lab visits and conversations with purpose), and GILL partners have interrogated their project to identify areas where there could be knowledge transfer. The result is a set of initial recommendations on how gender inequalities in AI and robotics in Pakistan universities could be addressed.

4 Barriers in Pakistan and Potential Solutions

On an exchange visit to five universities in Pakistan (National University of Science and Tehcnology (Islamabad), AIR University (Islamabad), Pak Austria Fachhochschule: Institute of Applied Sciences and Tehcnology (Haripur), NED University of Engineering and Technology and Institute of Business Administration (Karachi), the Coventry team (of approximately 18 people) found significant AI and robotics research addressing 'wicked' societal problems of Pakistan and developing countries which implicitly acknowledged the need for inclusivity and multidisciplinarity. Results from firsthand observations, surveys and email correspondence regarding the barriers faced by women in AI and robotics are shown Table 1.

There was a surprising number of female PhD students, postdocs, and women leading research centres. Such women were perceived as knowledgeable and resilient, displaying characteristics typical of those working in 'female - hostile environments' [26]. However, they were mostly missing from higher university offices. A student noted that in AI and robotics labs that around 90% of the staff were men, whereas '*almost 90% of the postgraduate students in linguistics were female. Over time, these patterns reinforce themselves, and more men tend to avoid such courses, viewing them as female dominated.*'

Staff and students recognised barriers related to the lack of role models, discriminatory employment practices, 'brogrammer' cultures, lack of family support and societal/cultural expectations on women's roles after marriage, and a lack of reliable, datasets especially in health-related fields.

When asked whether AI and robotics might reduce gender equalities, students were undecided. One student commented, '*Under this culture, as there are fewer female participants than male. The more technology advances, the gap will widen.*' Another suggested that the employment gap provided an opportunity for women to move into this space. '*As more women become visible in these areas, they can serve as role models and encourage others to follow in their footsteps.*' And that, at a cultural level as '*more women are aware of their rights and feel more confident about voicing out their opinions. This will lessen the inequalities as women are challenging the cultural barriers'*.

5 Suggested Gender-Inclusive Principles and Guidelines for AI and Robotics Research in Pakistan and Developing Countries

To mitigate gender disparities in AI and robotics within developing countries, an initial step must be the recognition of existing inequalities, a willingness to address these holistically and that gender equality is not simply a women's issue [27]. This involves quantification and fostering a commitment among all parties and organizations to address it. Subsequently, a suite of strategic interventions should be deployed based on a proactive willingness and ambition amongst all stakeholders and organisations to make changes.

A set of overarching principles emerged from the BC funded mobility programmes. In general, rather than imposing guidance when considering systems and approaches that encourage greater opportunities for all, we prefer:

- To place responsibility for avoiding gender bias on the institution rather than the individual affected by it.
- To value the longevity and well-being of the individual over that of the institution.
- To measure the effect of bias from whatever source (cultural, institutional, algorithmic) over blind attempts to correct for perceived bias.
- Algorithmic systems that affect the public that are open and transparent in their decision-making.

Table 1. Student perceptions of barrier and solutions to gender inequalities in AI and robotics in Pakistan.

Fix needed	Barriers	Solutions
Numbers	Cultural biases in Pakistan undervalue women's aptitude for STEM, diminishing their participation in these fields. This hinders enrolment in STEM education degrees	Implement outreach and recruitment programs that provide scholarships, internships, and mentorship opportunities specifically tailored to support women in AI and robotics fields
Institutions	Inflexible corporate hierarchies limit career growth equality	Family friendly work base policies e.g., flexible working (time and place), nurseries, financial allowance to support childcare
	The traditional family structure often does not support working women	Promote societal change to recognize women's career capabilities and encourage equal partnership at home
	Gender and experience biases at the recruitment stage led to assumptions that women will prioritize family over career. '*Recruiters assume young women will get married sit at home, wasting investment made on their training*'	Educating employers that women have the same abilities as a man
		Encourage women in the industry to positively showcase their professional capabilities
Data	Healthcare data is biased and unreliable. Gender-specific needs, e.g. around reproduction, '*are not widely known, or if known, they are discouraged from being openly discussed.*'	Prioritize the inclusion and discussion of gender-specific medical needs to improve healthcare data accuracy
	Bias and unreliability of data can lead to biased and unreliable results of AI algorithms	Invest in R&I to improve data collection methodologies to capture gender - diverse perspectives effectively
Culture	Ingrained family roles and societal expectations of childcare and homemaking. In Pakistan, women often abandon careers post-marriage due to the expectation of taking on traditional family roles. '*They have to let go of their careers as their responsibilities increase after marriage.*'	Increased societal awareness of rigid family roles and women's career aspirations

(*continued*)

Table 1. *(continued)*

Fix needed	Barriers	Solutions
		Women should understand their abilities and be empowered to use them. This should start from the home with equal sharing of responsibilities, '*where the family and partner is supportive. Instead of more responsibilities after marriage, the responsibilities should be lessened by a partner to share with*
	Women do not have the right to be financially independent	Encourage policies that support women to pursue financial independence and retain their career aspirations Develop and publicize recruitment and retention policies which enable women to return to their careers later in life
	'*It is believed that only the man of the house is responsible for earning, women do not need to work after marriage, leading a higher dropout rate among female students. Although women may have equal opportunities for education, they might lack of motivation. Career aspirations after marriage diminishes their appreciation for these opportunities*'	
	Marketplaces essential for technical fields like AI and robotics are male-dominated and can be unwelcoming to women. Women '*need male assistance to visit these places and buy the required equipment.*'	Tackle cultural stereotypes that lead to harassment to make technical marketplaces more accessible and welcoming to women, ensuring their independence in professional endeavours

5.1 Practical Measures

Following these overarching principles, based on the experiences of the two projects we suggest a set of practical measures aimed at enhancing diverse representation and ensuring substantive engagement. These are designed to affirm and elevate the contribution of women within the AI and robotics research ecosystem.

Fixing the Numbers. A starting point in terms of employment is getting and maintaining accurate intersectional employment data through demographic auditing and benchmarking segmented by gender, age, position, and other characteristics to show organisational and workforce trends. From this, gender and diversity action plans [28] can be co-designed to address inequalities, with key performance indicators to measure progress.

This category also refers to increasing the number of women entering and staying in the AI and robotics by understanding and addressing their (lifelong) needs, improving access to courses, providing role models, and stimulating interest. Targeted initiatives include mentorship programs, scholarships, and outreach activities. These should be complemented by training and professional development for staff employed in laboratories, research centres aimed at creating a nurturing and gender-inclusive atmosphere.

Fixing the Data. The application of data in AI contexts reveals deficiencies in legacy data sets. Accurate, reliable, trustworthy, and complete datasets underpin good science and innovation. This is problematic in developing counties where data on women and marginalized groups has not been routinely collected and data collection cannot be computerized. Training should be given to scientists and lab managers to improve the quality of data collection to ensure new data sets are bias free and can be segregated by gender and other characteristics. Research funding could be linked to the delivery of open-source gender disaggregated datasets and methods which diversify data extraction to ensure gender-representative sampling, and prioritization of inclusivity in data collection protocols.

Existing database biases should be acknowledged, measured, and addressed. Data sets with bias and missing data sets can lead to discrimination embedded in the system and reinforced or systematized by AI. For example, intersectional biases in decision making systems, have remained invisible to users and developers until discriminatory trends are evidenced in practice [29]. Shrestha and Das (2022) [30] analyzed 120 papers on gender biases in ML and AI algorithms to understand the research landscape and reviewed different mitigation and detection methods. Participatory approaches [31] ensure that AI development is informed by a broad range of perspectives, reducing the risk of ingrained biases and making discrimination more visible and addressable.

The 'Gender-Extended Research and Development model' [32] has been designed for use in the tech industry and related research fields to scrutinize biases in data, focusing on intersectionality and the representation of specifically disadvantaged groups. Such a nuanced approach helps avoid oversimplification and promotes a deeper understanding of discrimination's multifaceted nature.

Globally, regulatory frameworks are being developed to establish mandatory standards and control processes, prohibit the use of discriminatory systems, especially in critical areas and address liability issues. Also, initiatives such as those by UNESCO [33] aim to assist governments and corporations in ensuring balanced female representation in AI development, application, and governance and AlgorithmWatch [34], based in Berlin and Zurich appraises the extent to which automated decision-making systems are compatible with fundamental rights and democratic principles.

Fixing the Institution. Strategies could include promoting equitable workplace practices by establishing a more granular, contextual baseline so that progress in equality, diversity and inclusion (EDI) can be measured holistically. Discriminatory workplace practices can be identified and reduced, for example, by using the tools and methods developed and promoted by GILL, such as bias interrupters [35] and tweaks in everyday workplace practices [36]. Stereotypes can be addressed through institution wide anti-discrimination training and awareness and the appointment of EDI champions.

Fixing the Culture. This is hard to accomplish in STEM areas dominated by men and when embedded in cultures that do not value women. There must be a will to change, and a vision that is supported at all levels. Researchers and their managers should be fully accountable and responsible for project design, delivery outputs and outcomes that embed and respect the dignity, rights, safety, and well-being of all.

[37] developed a simple, general-purpose organisational change model, based on change agents, having the support of senior management who could implement and celebrate small changes in their workplace. In this, low-cost support mechanisms - allyships, research hubs, networks - have a critical role in problem sharing, and the contributions a diverse research ecosystem brings to the sector.

Crucial to this is the need to rethink the epistemology and ontology of AI and robotics to become more inclusive. Actions are needed that accelerate a paradigm shift in AI and robotics research away from the 'male gaze' and historical biases rooted in sexism, racism, homophobia, and classism. By re-evaluating the foundations of scientific inquiry, including the notions of subjectivity, objectivity, and the role of the scientist, we can redirect the focus towards a more inclusive and equitable scientific community that values diversity in both its practitioners and its research focus.

The metaverse provides direct opportunities for women and the possibility to shape the on-line culture. However to do this, we have already seen a worldwide failure to address training and safeguarding measures to ensure women's (and'vulnerable' groups) safety and privacy (e.g., in using AR and VR), prevent and respond to on-line harassment and cyberbullying [38].

6 Concluding Comments and Where to Start

The pervasive gender disparities in AI and robotics demand a multi-faceted approach that includes both top-down and bottom-up initiatives. This is a global problem with historical roots. Effective organizational change requires the active involvement of change agents who celebrate small victories and secure leadership buy-in to sustain momentum. It is essential not to assume that improvements will happen on their own; rather, every stakeholder must actively participate to create an inclusive environment that combats gender biases and promotes equality. This paper has outlined some strategic interventions in this rapidly developing field AI and robotics., aimed at mitigating gender disparities, particularly within Pakistan, offering a roadmap for initiating meaningful societal change.

Ethics. The survey was approved by Coventry University Ethics, P175150.

Acknowledgments. This work was partially funded by The British Council under the Mobility Partnership Grant for Faculty PAK-UK Education Gateway [award F137 and the UK Research and Innovation (UKRI) under the UK government's Horizon Europe funding guarantee [grant number 10049511] supporting the project "Gendered Innovation Living Labs (GILL)", Grant Agreement Number 101094812, funded by the European Union.

Disclosure of Interests. The authors have no competing interests to declare that are relevant to the content of this article. Views and opinions expressed are those of the authors.

References

1. Berman, F.D., Bourne, P.E.: Let's make gender diversity in data science a priority right from the start. PLoS Biol. **13**(7), e1002206 (2015). https://doi.org/10.1371/journal.pbio.1002206. Accessed 22 Apr 2024
2. Simonite, T.: AI is the future - but where are the women? WIRED, 17 August 2018. https://www.wired.com/story/artificial-intelligence-researchers-gender-imbalance/. Accessed 22 Apr 2024
3. D'Ignazio C., Bhargava, R.: Data visualization literacy: a feminist starting point. In: Engebretsen, M., Kennedy, H. (eds.) Data Visualisation in Society, pp. 208–223. Amsterdam University Press, Amsterdam (2020). https://doi.org/10.2307/j.ctvzgb8c7.19. Accessed 22 Apr 2024
4. World Development Report, Digital Dividends (2016). https://www.worldbank.org/en/publication/wdr2016. Accessed 22 Apr 2024
5. Powell, C., Chang, A.M.: Women in Tech as a driver for growth in emerging economics (2016). https://www.cfr.org/report/women-tech-driver-growth-emerging-economies. Accessed 22 Apr 2024
6. Loukides, M.: 2021 data/AI salary survey. O'Reilly Media (2021). https://www.oreilly.com/radar/2021-data-ai-salary-survey/. Accessed 22 Apr 2024
7. Howcroft, D., Rubery, J.: Bias in, bias out': gender equality and the future of work debate. Labour Ind. **29**(2), 213–227 (2019). https://doi.org/10.1080/10301763.2019.1619986.Accessed22Apr.2024
8. Broad, E.: Computer says no. Inside Story (2019). https://insidestory.org.au/computer-says-no/. Accessed 22 Apr 2024
9. UNESCO: Artificial intelligence and gender equality: key findings of UNESCO's global dialogue. Paris: UNESCO Publishing, p. 27 (2020). https://unesdoc.unesco.org/ark:/48223/pf0000374174.locale=en. Accessed 22 Apr 2024
10. Mantha, Y., Hudson, S.: Estimating the gender ratio of AI researchers around the world (2018). https://medium.com/element-ai-research-lab/estimating-the-gender-ratio-of-ai-researchers-around-the-world-81d2b8dbe9c3. Accessed 22 Apr 2024
11. Stathoulopoulos, K., Mateos-Garcia, J.: Gender diversity in AI research, Nesta, July(2019). https://media.nesta.org.uk/documents/Gender_Diversity_in_AI_Research.pdf. Accessed 22 Apr 2024
12. Perrault, R., Shoham, Y., Brynjolfsson, E., Clark, J., Etchemendy, J., Grosz, B., et al.: Artificial Intelligence Index 2019 annual report. Human-Centered AI Institute Stanford University, vol. 291 (2019). https://hai.stanford.edu/ai-index-2019. Accessed 22 Apr 2024
13. Wagman, K.B., Parks, L.: Beyond the command: feminist STS research and critical issues for the design of social machines. In: Proceedings of the ACM on Human-Computer Interaction, 5(CSCW1), vol. 101 (2021). https://arxiv.org/pdf/2102.00464.pdf. Accessed 22 Apr 2024
14. Wajcman, J.: Feminist theories of technology. Cambridge J. Econ. **34**(1), 143–152 (2010). https://www.jstor.org/stable/pdf/24232027.pdf. Accessed 22 Apr 2024
15. Young, E., Wajcman, J., Sprejer, L.: Mind the gender gap: inequalities in the emergent professions of artificial intelligence (AI) and data science. New Technol. Work Employ. **38**, 391–414 (2023). https://doi.org/10.1111/ntwe.12278. Accessed 22 Apr 2024
16. .Marchetti-Bowick, M.: Is you Roomba male or female. The Role of Gender Stereotypes and Cultural Norms in Robot Design. Intersect 2, 1 (2009). https://ojs.stanford.edu/ojs/index.php/intersect/article/download/171/629/2793. Accessed 22 Apr 2024
17. World Economic Forum: The Global Gender Gap Report (2023). http://reports.weforum.org/globalgender-gap-report-2023. Accessed 22 Apr 2024
18. Ashraf, M., Cagliesi, G., Hawkes, D., Rab, M.: Tackling the gender biases in higher education careers in Pakistan: potential online opportunities post COVID-19. Emerald Open Res. **1**, 12 (2023). https://doi.org/10.1108/EOR-12-2023-0005/full/html. Accessed 22 Apr 2024

19. Tran-Nguyen, A., Zampett, A.B.: Trade and Gender: opportunities and challenges for developing countries (2004). https://unctad.org/publication/trade-and-gender-opportunities-and-challenges-developing-countries. Accessed 22 Apr 2024
20. Morris, R.: Beyond 'Brogrammers': Can AI create a meritocracy? (2017). https://www.bbc.co.uk/news/business-38393802. Accessed 22 Apr 2024
21. Gray, M.L., Suri, S.: Ghost work: how to stop Silicon Valley from building a new global underclass. Eamon Dolan Books (2019). https://doi.org/10.1177/0019793919864564.Accessed22Apr.2024
22. Maqsood, M., Bokhari, S.F., Bukhari, S.F.: Facebook to Metaverse: a philosophical insight into monetization, practices and preferences in Pakistan. Global Digital Print Media Rev. **1**, 154–163 (2022). https://doi.org/10.31703/gdpmr.2022(v-i).1
23. Gendered Innovations in Science, Health, and Medicine, Engineering and Environment. https://genderedinnovations.stanford.edu. Accessed 22 Apr 2024
24. GILL - Gendered Innovation Living Labs. https://gi-ll.eu/. Accessed 22 Apr 2024
25. Christensen, H., Breengard, M.H., Stensgaard, S., Křížková, A., Volejníčková, R.M, Pospíšilová, M., Vohlídalová M.: D1.1 Roadmap for GRSIE, Unpublished deliverable of GILL project (2022). Available from authors
26. Woodcock, A., Rosenau, S., Constantin, I., Dambiano, C.: Barriers to women's employment in the transport sector in Europe (under review) (2024) Draft available from authors
27. Balka, E., Wagner, I., Weibert, A., Wulf, V.: Gender and Technology at Work - From Workplace Studies to Social Justice in Design: Cambridge University Press (2024)
28. Christensen, H., Levin, L., Woodcock, A: Gender and Diversity Action Planning (GaDAP) in the transport sector: gender smart mobility for smart cities. In: 14th ITS European Congress, Toulouse, France (2022). https://pure.coventry.ac.uk/ws/portalfiles/portal/54089953/toulouse_workshop_29.5.2022.pdf. Accessed 22 Apr 2024
29. Mehrabi, N., Morstatter, F., Saxena, N., Lerman, K., Galstyan, A.: A survey on bias and fairness in machine learning. ACM Comput. Surv. (CSUR) **54**(6), 1–35 (2021). https://doi.org/10.48550/arXiv.1908.09635
30. Shrestha, S., Das, S.: Exploring gender biases in ML and AI academic research through systematic literature review. Front Artif. Intell. **11**, 5:976838, October 2022. https://doi.org/10.3389/frai.2022.976838
31. Bødker, S., Dindler, C., Iversen, O.S., Smith, R.C.: Participatory design. Synthe. Lect. Hum.-Cent. Inform. **14**(5), i–143 (2021). https://doi.org/10.2200/S01136ED1V01Y202110HCI052
32. Draude, C., Maaß: Making IT work: integrating gender research in computing through a process model. In: Proceedings of the 4th Conference on Gender and IT, pp. 43–50 (2018). https://doi.org/10.1145/3196839.3196846
33. Women4Ethical AI. https://f5vip1.unesco.org/en/artificial-intelligence/women4ethical-ai. Accessed 22 Apr 2024
34. AlgorithmWatch. https://algorithmwatch.org/en/. Accessed 06 May 2024
35. Williams, J.C., Mihaylo, S.: How the best bosses interrupt bias on their teams. Harvard Bus. Rev. (2019). https://hbr.org/2019/11/how-the-best-bosses-interrupt-bias-on-their-teams. Accessed 22 Apr 2024
36. Holtzblatt, K., Marsden, N.: Retaining Women in Tech - Shifting the Paradigm. Springer, Cham (2022). https://doi.org/10.2200/S01153ED1V01Y202111PRO006

37. Nienaber, A.M., Spundflasch, S., Soares, A.E.: Behavioural change in local authorities to increase organisational capacity. In: Woodcock, A., Saunders, J., Fadden-Hopper, K., O'Connell, E. (eds.) Capacity Building in Local Authorities for Sustainable Transport Planning. Smart Innovation, Systems and Technologies, vol. 319. Springer, Singapore (2023). https://doi.org/10.1007/978-981-19-6962-1_6
38. Sabir, R.I., Majid, M.B.: Factors affecting women empowerment: a micro financing perspective. J. Entrepreneurship Bus. Ventur. **3**(1), 7–15 (2023). https://doi.org/10.56536/jebv.v3i1.19

The GILL Platform: A Pan-European, Virtual, Learning, Sustainable and Collaborative Hub Built on the Living Lab Principles

Marta Plaza-Hernández[1] (✉), Fotis K. Liotopoulos[2], Carlos Tello-Alvarado[1], Enrique Mesonero-Ronco[1], Javier Merchán-Benito[1], Diego Benito-Manzanas[1], Mario Rodrigo-Marcos[1], Alberto Montero-Fernández[1], David Sánchez-Sánchez[1], Álvaro Rosa-González[1], Carolina Villoria-Torres[1], and Javier Prieto-Tejedor[1]

[1] BISITE Research Group, University of Salamanca, Edificio Multiusos I+D+I, Calle Espejo 2, 37007 Salamanca, Spain
{martaplaza,carlostello,id00779874,javierm07,diego.benito, mariodrg,albertomonfdez,davidsansan,alrogo,carolvt, javierp}@usal.es

[2] SBOING, K. Karamanli 124, Diavata, 57008 Thessaloniki, Greece
liotop@sboing.net

Abstract. The Gendered Innovation Living Labs (GILL) project consortium is developing a new IT platform that will be iteratively developed through a series of co-creation approaches to meet the needs of all gender innovation stakeholders. The platform will provide access to training, case studies, tools and methods, supporting gender-responsive innovation related to European challenges. It will facilitate collaborative work, knowledge sharing and creation between scientists, formal organisations and citizens engaged in open innovation cycles or gender innovation careers. Ultimately, it strives to become a sustainable open innovation resource driving cultural and systemic change through inclusive practices. Once the platform is populated, it will integrate advanced digital technologies to support and enable GILL to reach an interested community of organisations, projects, and individuals across Europe.

Keywords: IT platform · Living Labs · learning hub · collaborative hub · cloud

1 Introduction

Information and Communication Technologies (ICT) are increasingly influencing our daily lives. This highlights the importance of considering sociocultural factors in the development of software systems and attributes a significant impact of software engineering on society. Gender and cultural diversity play a crucial role in software development, its sustainability, and the society that employs it. Incorporating these diversity dimensions not only enriches the relevance and accessibility of the software but also promotes inclusive design, which is fundamental for creating systems that are truly universal. Recognising and valuing gender and cultural diversity in the early stages of

G. Marreiros et al. (Eds.): DCAI 2024, LNNS 1151, pp. 124–131, 2025.
https://doi.org/10.1007/978-3-031-80946-0_12

software development can lead to more innovative and accessible products that effectively meet the needs of a broader and more varied user base. Therefore, it is imperative that software engineering professionals adopt a holistic approach that includes these aspects of diversity, ensuring that software systems are not only technically sound but also socially responsible and culturally aware.

The Gendered Innovation Living Labs (GILL) project [1] aims to address gender and diversity blindness in the Living Labs (LL) discourse and practices for: enabling organisational and cultural changes; enhancing professional development; increasing the integration of gender and diversity into product design, technologies, and innovation; allowing gendered educational practices.

The project consortium is developing a pan-European collaboration and learning hub based on the LL principles [2, 3] to become the open innovation framework for all European actors for open-gendered innovation. Partners are designing sustainable mechanisms to increase Gender Responsive Smart Innovation and Entrepreneurship (GRSIE) by fixing practices and culture across the European ecosystems [4]. In parallel, GILL actions act along a matrix of horizontal spheres of innovation - individual & team, tools & methods, innovation & products, and societal processes - and vertical priorities - health & resilience, green transition, and digital transformation.

The project is implemented through an iterative co-creation approach structured on a four-phase cycle -understand, co-design, implement, and evaluate- which will be repeated twice during the project lifetime to incorporate the feedback and evaluation results in fine-tuned and validated results. Through this cycle, GILL will deliver a systemic review of barriers to GRSIE, leading to the co-design of proven methodologies, services, and tools. In line with the LL methodology, co-creation and open innovation will be fostered among the main actors of the quadruple helix model (citizens, government, industry, and academia), who will be engaged all along the project through different participatory activities [5].

This conference paper presents the GILL platform, also called the GILL hub, describing its primary modules, key elements, and functionalities. The pilot hub was launched in December 2023, and it will evolve during the remaining two years of the project, adding instruments and functionalities as it is populated with content and visual materials. The rest of the paper is organised as follows: Sect. 2 presents the GILL hub, and Sect. 3 draws the main conclusions and describes future work.

2 The GILL Hub

The GILL project aims to sustainably operate as a recognised pan-European collaborative hub comprising multiple LLs. The hub will work to drive significant cross-sector progress on gender innovation, partner with sector stakeholders to establish principles for greater gender equality and develop a portfolio of gender-focused tools and methods that can increase and accelerate innovations with positive gender impact at a large scale.

The platform is being developed based on LL principles, iteratively improved to meet the needs of all gender innovation stakeholders [6], including existing innovation centres, LLs, female entrepreneurs, startups, technology transfer offices and more. The platform's specifications have been co-created with stakeholders to match the consortium's experience in implementing platforms, and it will gradually develop into a pan-European cloud-based learning and collaboration platform with distributed ledger technologies, adaptable interfaces, and modules.

The platform is currently hosted at **gill.bisite.usal.es**, and it is designed to be openly accessible to all, requiring only basic signup details to become a registered user. Key information is surfaced in an engaging, visually oriented layout that motivates navigation into the core modules - learning hub, collaborative hub, and GILL cloud -. A brief description orients users to the hub's purpose and what they can gain from exploring it further. A self-assessment learning tool will soon be available, providing comprehensive support in identifying users' knowledge gaps, enabling them to access the necessary information and choose the appropriate learning package to boost their proficiency in specific subject areas (Fig. 1).

Fig. 1. The GILL hub landing page.

2.1 Learning Hub

The *learning hub* module, also known as the GILL capacity-building repository, is it currently under development. It will host a variety of project materials, such as guidelines, reports, deliverables, training resources, best practices, lessons learned, and results from collaborative activities. This module will act as a one-stop shop for finding full support for building community capacities and integrating gender-responsive approaches into innovation initiatives. The diversity of content caters to all backgrounds and learning preferences.

It will have a hierarchical structure organised around a central knowledge base. Its homepage will provide users with a wealth of resources organised into four main sections. The first is a search engine, allowing one to find relevant materials across the repository easily. The second section features learning modules centred around key themes: webinars, workshops, innovation & entrepreneurship. Each module will contain educational content and best practices related to its topic area. The third section covers two help topics of high importance for GILL: Living Labs and gender & cross-cutting. Living Labs will explore the nuts and bolts of establishing and managing these platforms, and gender & crosscutting will investigate how to integrate gender-responsive practices. The fourth section is a robust training offer with several modules to suit different needs and learning techniques, such as best practices, canvases, policy briefs, and guidelines.

2.2 Collaborative Hub

A collaborative ecosystem in the context of an IT platform refers to a technology-enabled environment that facilitates cooperative relationships and collective value creation among organisations and individuals through digital tools and instruments. It encourages all participants to act interdependently, leveraging a shared technology fabric for mutual benefit through cooperation, co-innovation, and co-creation. This module is already available in the platform, including the following sub-modules:

The Polling System. This functionality will be used to co-create and test a polling platform for LLs to design, distribute, and collect custom questionnaires efficiently and quickly for feedback or evaluation of gender and innovation issues, ideas, challenges, etc. Multilingual campaigns will be designed and run concurrently, exploitation of crowdsourcing and incentivisation practices to increase the survey yields, custom survey templates to evaluate gender innovation, and growing a crowd to support LLs surveys.

The Toolbox. A set of good practices, methods and tools were identified during four co-creative workshops conducted with consortium partners and external experts. These tools were iteratively developed to support GRSIE within the local quadruple helix ecosystem. An initial list of fifty tools was compiled. They are already available in the hub and can be categorised by *objective* (organisational and cultural Change, professional development, product design-technologies-innovation, educational practices), *focus* (doing, reflecting), *sphere of innovation* (innovation & products, societal processes, tools & methods, individual & team), and *phase* (evaluate, understand, implement, co-design) (Fig. 2).

The Network. This tool fosters collaboration and synergies among aligned initiatives and organisations. It allows projects and partners to connect by sharing critical details about their work and goals. Users can input a brief description of their initiative showcasing what problems they address and how their approach or solutions contribute to the shared vision. They can also provide helpful website links and contact information so interested parties can quickly learn more about their efforts. Sharing appropriate information and linking builds visibility and makes it effortless for like-minded groups to find potential areas for partnership or complementary approaches. The network also helps avoid duplicating efforts by bringing together all related work underway. It facilitates coordination, enables initiatives to share lessons and resources, and even combines strengths where synergies exist.

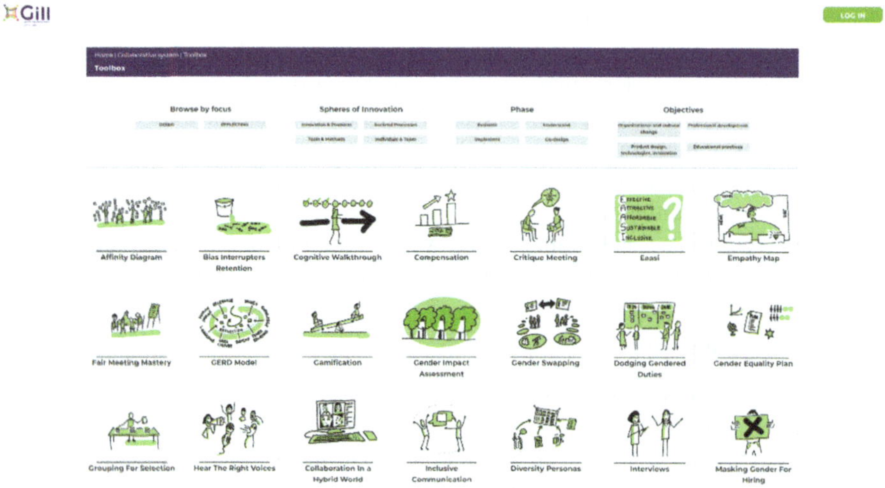

Fig. 2. GILL hub toolbox, a submodule within the collaborative ecosystem.

The Forum. This instrument facilitates interaction and collaboration between users. It is designed to stimulate synergies and knowledge exchange. Fixed overarching macro topics were established to organise initial discussions on the first release of the hub. Each topic includes a short description outlining its focus area. The goal is to nurture conversations around key subjects relevant to building community resilience and inclusion. Users can initiate new discussion threads or contribute to existing conversations under the relevant macro topic headings. This structure provides a starting point and framework around which debates and sharing of perspectives can unfold productively (Fig. 3).

Fig. 3. GILL hub forum, a submodule within the collaborative ecosystem.

The Observatories. This dashboard instrument, already available, allows data analysis and visualisations by ingesting external data [7]. Users gain insight through four key graph iypes: bar, treemap, line and radar charts. Together, these visualisations clarify complex information into understandable formats (Fig. 4).

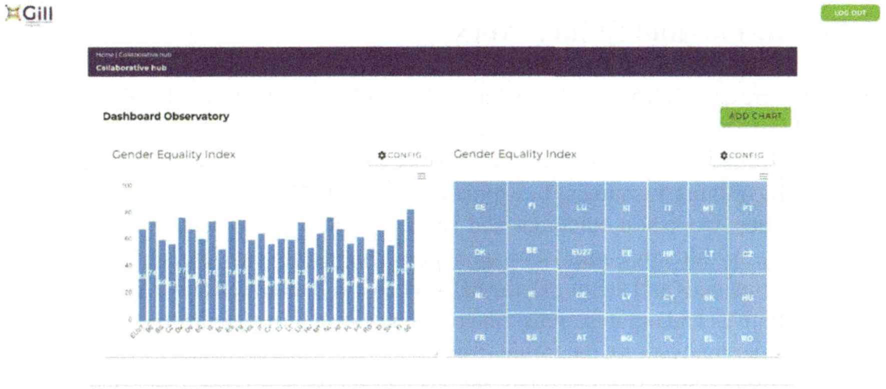

Fig. 4. GILL hub observatories, a submodule within the collaborative ecosystem.

2.3 GILL Cloud

The *GILL cloud* module will be a data exchange and collaboration environment using the open-source platform OwnCloud [8] and is it currently under development. Its purpose is multifold: to provide a data-sharing and collaboration environment supporting the LL hubs; to serve as a public repository for open data sharing with the broad public; and to become a free, private space as incentivisation for the LL crowd. It will have the following key features:

- File synchronisation and sharing. It will allow users to synchronise files and folders seamlessly across various devices. This ensures that users have access to the latest version of their files, promoting collaboration and productivity. The platform supports file sharing within the LLs, with external users and public access, making it a versatile solution for diverse collaboration needs.
- Security and encryption. To safeguard sensitive data, it will employ robust encryption measures, both in transit and at rest. Users can encrypt files before uploading them, ensuring only authorised individuals can access the decrypted content.
- Collaboration tools. It will enhance collaboration with features such as document editing and versioning. Users can collaborate in real-time on documents, spreadsheets, and presentations without the need for third-party applications. It will support integration with various third-party tools and services, offering an open API that enables developers to create custom applications and integrations. This flexibility allows organisations to tailor it to their specific workflows and requirements.
- Deployment Options. The GILL cloud will provide flexibility in deployment, catering to diverse user preferences: on-premises hosting (on their own servers, giving LLs complete control over data storage and security); cloud hosting (on popular cloud

platforms like Azure or Google Cloud); and hybrid deployment, allowing the LLs to combine on-premises and cloud hosting to meet specific performance, security, and compliance needs.

3 Conclusions and Future Work

This paper presented, from a non-technical point of view, the GILL IT platform (also called the GILL hub), detailing its primary modules, key elements, and functionalities.

The *learning hub* will host a variety of project materials, such as guidelines, reports, deliverables, training resources, best practices, lessons learned, results from collaborative activities, workshop outcomes and action-oriented experimentation studies. The *collaborative hub* already comprises a set of powerful instruments, including the toolbox (set of good practices and methods), the network (for collaboration and synergies among aligned initiatives and organisations), the polling system (customed survey templates to evaluate gender innovation), the forum (for collaboration and knowledge exchange), and the GILL observatories (data analysis and visualisation). The *GILL cloud* will provide a data exchange and collaboration environment.

The GILL hub, launched in December 2023, will evolve during the remaining two years of the project. The initial structure, key modules and functionalities have been designed based on the feedback provided by project partners and external experts. During the project's second year, the platform will be tested by a selected group of users, including project partners and stakeholders, to improve its usability. Additionally, a first batch of content and materials will be added. Once the hub is populated, it will integrate advanced digital technologies, aiming to empower the GILL project as a vibrant centre supporting gender innovators across Europe. Continuous evaluation and iterative improvements will ensure it effectively meets stakeholders' needs over time. Ultimately, the platform strives to become a sustainable open innovation resource driving cultural and systemic change for the better through inclusive practices.

Acknowledgments. This research has been supported by the project "Gendered Innovation Living Labs (GILL)", Grant Agreement Number 101094812. Funded by the European Union, views and opinions expressed are, however, those of the authors only and do not necessarily reflect those of the European Union or the European Research Executive Agency (REA). Neither the European Union nor the granting authority can be held responsible for them.

Disclosure of Interests. The authors have no competing interests to declare that are relevant to the content of this article.

References

1. GILL - Gendered Innovation Living Labs. https://gi-ll.eu/. Accessed 04 May 2024
2. Ståhlbröst, A., Holst, M.: The Living Lab Methodology Handbook. Centre for Distance-spanning Technology, Sweden (2012)
3. Bergvall-Kåreborn, B., Ståhlbröst, A.: Living Lab: an open and citizen-centric approach for innovation. Int. J. Innov. Reg. Dev. **1**(4), 356–370 (2009)

4. Gendered Innovations in Science, Health & Medicine, Engineering and the Environment. https://genderedinnovations.stanford.edu/what-is-gendered-innovations.html. Accessed 29 Apr 2024
5. Schuurman, D., Baccarne, B., De Marez, L., Veeckman, C., Ballon, P.: Living Labs as open innovation systems for knowledge exchange: solutions for sustainable innovation development. Int. J. Bus. Innov. Res. **10**(2/3), 322 (2016)
6. Mastelic, J.: Stakeholders' engagement in the co-design of energy conservation interventions: the case of the Energy Living Lab. ArODES (2019)
7. Corchado, J.M., et al.: Deepint.net: a rapid deployment platform for smart territories. Sensors **21**(1), 236 (2021)
8. OwnCloud - infinite scale. https://owncloud.com. Accessed 15 Mar 2024

Measuring Citizen's Perceptions of Gender Equality Measures

Margarita Cea-López[1(✉)], Andrés Langoyo-Martín[1], and Ana Belén Gil-González[2]

[1] AIR Institute, IoT Digital Innovation Hub, Av. Santiago Madrigal, 39, 37003 Salamanca, Spain
{mcea,alangoyo}@air-institute.com

[2] BISITE Research Group, University of Salamanca, Edificio Multiusos I+D+i, Calle Espejo 2, 37007 Salamanca, Spain
abg@usal.es

Abstract. Social media platforms serve as dynamic arenas where individuals freely express their opinions, furnishing a rich repository of data conducive to analyzing the efficacy of gender initiatives. Within the ambit of the GILL (Gender Innovation Living Labs) project, this study posits the extraction and analysis of Twitter and Reddit data employing advanced Artificial Intelligence (AI) and Natural Language Processing (NLP) techniques, including automatic topic identification, term frequencies, and sentiment analysis.

Keywords: Social media analysis · Citizen perception · Gender Equality · Artificial Intelligence · Natural Language Processing

1 Introduction

Social media platforms are now vital for connecting people, exposing personal lives, and enabling public discussions on various topics, offering unprecedented opportunities to analyze public opinion on a large scale and in real-time.

Our involvement in the Gender Innovation Living Labs (GILL) project focuses on evaluating public perceptions of gender and equality initiatives through social media. This method enhances our understanding of gender equality policies and public attitudes, using advanced Artificial Intelligence (AI) techniques. AI includes machine learning and image processing, enabling machines to perform tasks that typically require human intelligence.

AI's ability to reason, learn, and interpret data unveils deep insights, crucial for scientific research and technological advancements across various fields. In our study, Natural Language Processing (NLP), a branch of AI, will be used to analyze human language from various sources, combining computational linguistics and AI to process and understand natural language. [1].

Integrating AI and NLP into our project improves our analysis of social media discussions on gender equality, helping us uncover patterns and sentiments with high precision [2, 3]. This advanced approach aids in making informed public policy decisions

G. Marreiros et al. (Eds.): DCAI 2024, LNNS 1151, pp. 132–141, 2025.
https://doi.org/10.1007/978-3-031-80946-0_13

and enriches the understanding of gender equality debates, contributing to academic research and policy formulation [4–6].

The GILL project adopts an iterative co-creation methodology rooted in Living Labs principles, spanning two cycles with four phases each: understand, co-design, implement, evaluate. This approach fosters open innovation and active user involvement over the project's two-year duration, facilitating adaptation to evolving user needs. Through repeated cycles, feedback is integrated to refine project deliverables continuously. Additionally, a matrix structure ensures alignment with Gender-Responsive Smart Innovation and Entrepreneurship goals, addressing cultural and organizational change, professional development, product design, and educational practices. The methodology begins with a comprehensive study, literature review, and social media data collection, followed by analysis techniques leading to an interactive dashboard for data visualization and detailed analysis (Fig. 1).

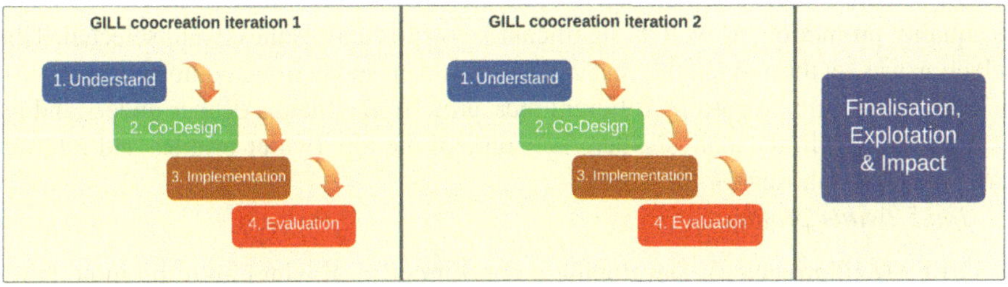

Fig. 1. Schematic graphic of the GILL methodology approach (iterative cycles).

This conference paper analyzes users' reactions to gender-related measures on social media platforms, an action-oriented experimentation in the framework of the GILL project. The rest of the paper is organized as follows: Sect. 2 introduces the research methodologies employed, Sect. 3 presents the preliminary results of the analysis, and Sect. 4 draws the main conclusions and describes future work.

2 Work Performed

During the first iteration cycle efforts were made on the understanding, co-design and implementation phases. This involved the literature review, social media research, data extraction from the platforms and selection of analysis methodologies.

2.1 Understanding and Co-Design Phase

The work carried out so far has focused on the understanding and co-design phase. Initially, an exhaustive investigation was conducted to understand how to measure public opinion on social media and to identify AI techniques that could yield significant results in this context [4–6]. This preliminary stage was crucial for establishing the technological foundations of the project.

Subsequently, efforts were focused on researching and defining the characteristics the data set should have to ensure its adequacy and relevance for achieving the desired conclusions, aiming for it to reflect public opinion on equality accurately.

The acquisition of the dataset posed a significant challenge. During this phase, the feasibility of Application Programmer Interfaces (APIs) from different social networks for downloading the necessary data was evaluated. Here, several limitations arose, most of them did not allow proper information download or had download restrictions associated with costs. In particular, Twitter required a paid license for the correct download of its content.

Despite the limitations encountered, access to the necessary resources from Twitter [10] and Reddit [11] was finally secured, as they offer the most suitable APIs for the project's needs. Moreover, these platforms are widely recognized as online debate spaces on a variety of topics, making them the epicenter and a valuable resource.

Another crucial step was choosing which information to include in the dataset. With the collaboration of several partners, relevant profiles of experts and organizations related to equality promotion, as well as significant keywords and themes, were selected. This selection was set throughout the year 2023 to obtain an updated and comprehensive view of public perception regarding different measures. To facilitate a deeper understanding of the data used in the analysis, here are some of the top Twitter profiles and relevant hashtags/topics chosen for this study:

Top 5 Twitter profiles:

- GILL (@gillprojecteu): Developing a pan-European #LivingLab to promote fairer open #gendered innovation across Europe and beyond!
- EQUALS-EU (@Equals_EU): #H2020 project promoting #genderequity through #socialinnovation and #entrepreneurship.
- Eige (@eige_eu): The European Institute for Gender Equality, advancing gender equality through research and policy support.
- ATHENA (@ATHENA_Equality): Committed to gender equality, addressing gender imbalances in decision-making processes.
- CALIPER EU(@calipereu): Supports RPOs and RFOs in Europe to develop Gender Equality Plans and engage with local Innovation Hubs.

Selected Twitter hashtags and topics. #GenderedInnovation, #GILLprojectEU, #GenderGap, #GenderEquality, #WomensRights, #WomenEmpowerment, #diversity, #inclusion.

2.2 Implementation Phase

Currently, the project is in the implementation phase, focused on downloading and constructing the complete data set. This set includes profiles from Reddit and Twitter, accessed through their respective APIs. Reddit data were obtained without issues, extracting the 1000 most relevant posts from 15 communities along with their comments. However, with Twitter data, we faced certain limitations.

Initially free, the Twitter API now offers various priced access tiers. For this study, we have access to the basic tier, which also introduces a new limitation encountered during data download. The extracting replies to tweets are limited to only the past seven days. This factor is limiting as it does not allow for the collection of all user opinions on the selected tweets and may compromise some of the conclusions we had planned to draw. Future phases will require adjusting NLP techniques to minimize this limitation's impact on the study's outcomes.

Simultaneously, some of the selected AI and NLP techniques began to be applied and evaluated to determine their suitability and feasibility, as well as to identify any necessary adjustments in data collection and for adapting our analysis strategies to the dynamic and sometimes ambiguous nature of language used on social platforms. Moreover, this process allows us to start developing the models' scripts, facilitating a more efficient implementation once the full data set is available. Through an iterative process of testing and adjustments, these algorithms will be optimized to improve their ability to identify, classify, and analyze opinion and sentiment patterns in the data.

3 AI Techniques and Preliminary Results

Within the scope of our project, we have selected and proposed the application of various advanced NLP techniques to analyze perceptions and opinions on social media regarding gender equality measures. The proposed techniques are detailed below. It is important to note that the data observed in Figs. 2, 3 and 4 are simple examples just for some clarifications, taken from studies carried out with other data; the results of the current project may be different.

Sentiment Analysis. This technique allows for the categorization of user messages in various ways, from a simple distinction between positive or negative sentiments to the identification of specific emotions such as aggression, hatred, fear, sadness, joy, or anger [12].

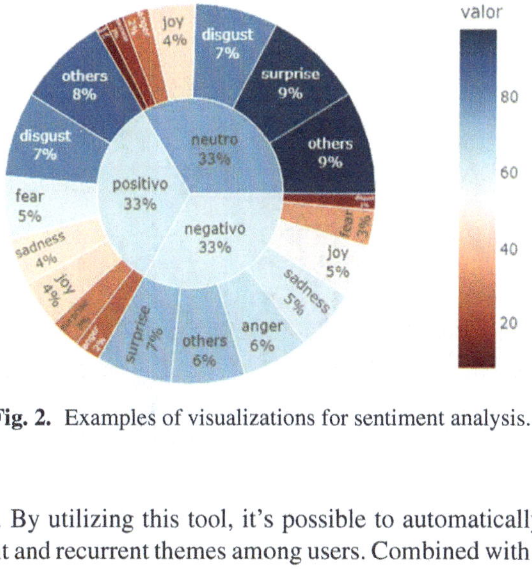

Fig. 2. Examples of visualizations for sentiment analysis.

Main Topic Study. By utilizing this tool, it's possible to automatically identify words, phrases, or relevant and recurrent themes among users. Combined with different types of

Fig. 3. Example of topic analysis visualization.

filtering, this technique unveils differences between users or communities and interesting patterns of discussion.

Frequency Studies. This methodology enables the observation of posting frequencies and their correlation with various types of emerging themes at specific moments.

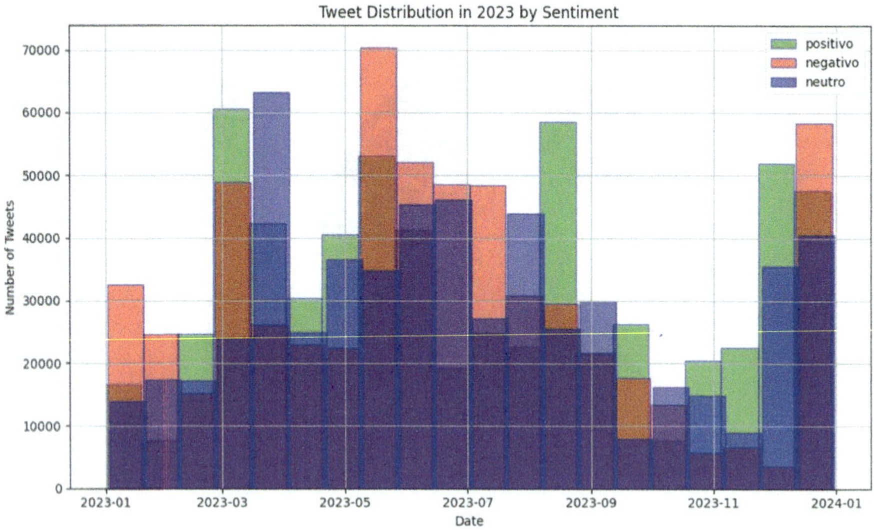

Fig. 4. Example of frequency analysis.

3.1 Pilot Results

To start preparing and adjusting the analyses to be carried out, the implementation and adjustment of these techniques to the currently available Reddit social network dataset have begun. Initial applications have yielded insightful visualizations and data aiding our model development. Below are several of the results obtained through these analyses. For instance, Fig. 5 shows the *temporal distribution* of posts in the 'Feminism' community, highlighting peaks in activity. These increases could correspond to moments when specific feminism topics gained relevance in public discourse, possibly in response to social events, awareness campaigns, or commemorative dates. The marked rises could be starting points for investigating correlations between the most discussed topics in the community and external events during those periods. *Sentiment analysis* applied to the posts of various communities allows for a comparison of what percentage of them exhibit different emotions, as shown in Fig. 6. This visualization reveals the range of sentiments in gender equality discussions. For instance, observing these graphs, most communities display a predominance of positive emotions, though feelings like disgust and joy are also significant, highlighting opinion polarization. Communities centered on LGBTQIA + and gender themes tend to show more positive and joyful sentiments, suggesting supportive attitudes.

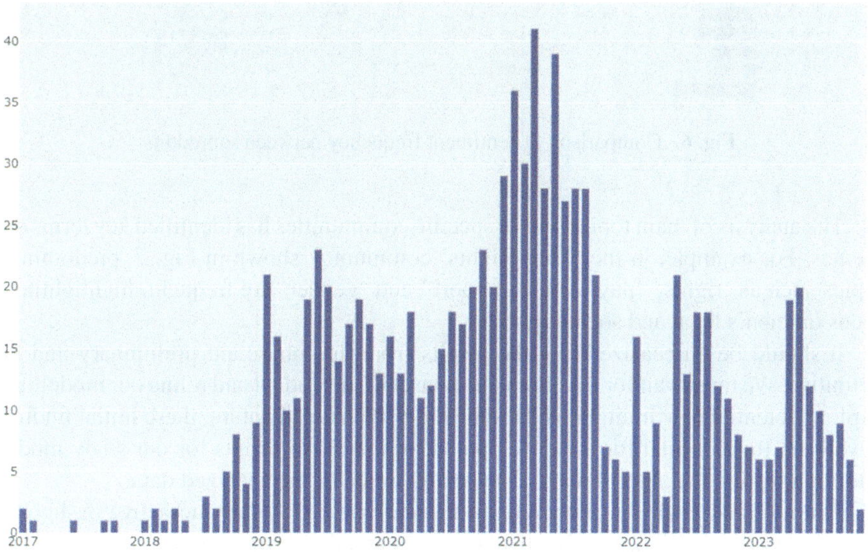

Fig. 5. Topic frequency in the "Feminism" community.

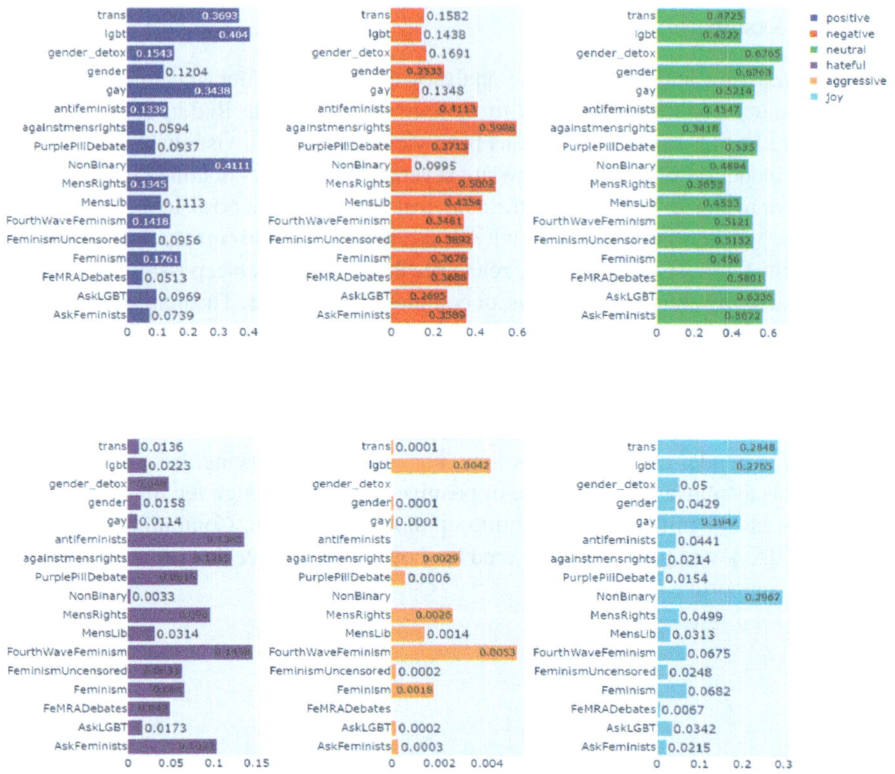

Fig. 6. Comparison of sentiment frequency between subreddits.

The analysis of main topics within specific communities has identified key terms and themes. For example, in the 'MensRights' community, shown in Fig. 7, predominant topics such as 'rights,' 'pay,' 'child support,' and 'gender' are frequent, highlighting a focus on men's legal and social rights.

It should be emphasized that the results from this phase are preliminary and not definitive. We must wait for the complete dataset to fully adapt and refine our models and explore potential new limitations. Despite their provisional nature, these initial findings have been fundamental for the development of definitive scripts for our study models and for adapting the algorithms to the specific needs of the analyzed data.

The ultimate goal is to integrate all these analyses into an interactive dashboard, allowing users to explore the visualizations and draw their own conclusions about public perception of gender equality measures on social media.

Fig. 7. Topic frequency in subreddit "MensRights".

4 Conclusions and Future Work

This study provides valuable insights into public perception on social media regarding gender equality and the rights of various minorities. The use of advanced sentiment analysis techniques and Natural Language Processing (NLP) has unveiled the complexity and diversity of opinions present in these digital communities. The results highlight the existence of polarized discussions, reflecting the dynamics and social climate surrounding these crucial issues.

AI and NLP techniques proved to be powerful tools for classifying large volumes of text and revealing significant trends and patterns in the data. This methodological approach has been efficient in extracting qualitative information from quantitative data, facilitating a deeper understanding of the attitudes and sentiments expressed online. Furthermore, the development of an interactive dashboard for data visualization is expected to be highly beneficial for policymakers, providing an accessible means to interpret the gathered information and generate additional applicable knowledge.

Given the evolving nature of social media and its impact on public opinion formation, future research might focus on longitudinal analysis to observe how perceptions and attitudes change over time. Additionally, it would be worthwhile to explore in depth the correlation between activity peaks on social media platforms and key global events to better understand how the global context influences discussions on gender equality.

This study marks a starting point towards a more nuanced understanding of public opinion in the digital space and opens the door to future research efforts that encompass an even broader spectrum of communities and discourses.

The evaluation phase will play a crucial role in the project, setting specific criteria designed to quantify the impact and relevance of our findings. This phase will facilitate both technical refinement of results and a critical assessment of our organizational structure and methodology. We will implement a review strategy involving internal evaluations and consultations with external experts in gender equality, AI, and social network analysis. This multidisciplinary input is vital for enhancing our understanding and approach to the project. The feedback collected during this stage will be decisive for making meticulous adjustments to our methods and in the interpretation of the data, ensuring that our research conclusions are robust and aligned with our initial questions. This reflective process will support continuous improvement and strategic adaptations for the subsequent work cycle, considering the lessons learned and the challenges identified during the first year.

In the project's second cycle, the identified improvements will be implemented to optimize both the way of working and the collection and analysis of data. This might involve expanding the data set to include additional profiles or extending our analysis to the year 2024, thus providing a broader and more updated view of public perception on social media regarding equality measures. Furthermore, the application and final adjustment of Natural Language Processing (NLP) techniques will be completed, with the goal of refining our analysis of public opinion on equality measures on social media.

Acknowledgments. This research has been supported by the project "Gendered Innovation Living Labs (GILL)", Grant Agreement Number 101094812. Funded by the European Union, views and opinions expressed are, however, those of the authors only and do not necessarily reflect those of the European Union or the European Research Executive Agency (REA). Neither the European Union nor the granting authority can be held responsible for them.

Disclosure of Interests. The authors have no competing interests to declare that are relevant to the content of this article.

References

1. Daniel, J., James, H.M.: Speech and Language Processing: An Introduction to Natural Language Processing, Computational Linguistics, and Speech Recognition. Prentice hall (2007)
2. Habib, M.B., Keulen, M.V.: Information extraction for social media. In: Proceedings of the Third Workshop on Semantic Web and Information Extraction (SWAIE 2014), Dublin, Ireland, vol. W14-62, pp. 9–16. Association for Computational Linguistics (2014). https://doi.org/10.3115/V1/W14-6202
3. Messaoudi, C., Guessoum, Z., Ben Romdhane, L.: Opinion mining in online social media: a survey. Soc. Netw. Anal. Min. **12**, 25 (2022). https://doi.org/10.1007/s13278-021-00855-8
4. Scarborough, W.J.: Feminist Twitter and gender attitudes: opportunities and limitations to using Twitter in the study of public opinion. Socius Sociol. Res. Dyn. World **4**, 237802311878076, January 2018. https://doi.org/10.1177/2378023118780760
5. Burns, N., Gallagher, K.: Public opinion on gender issues: the politics of equity and roles. Annu. Rev. Polit. Sci. **13**(1), 425–443 (2010). https://doi.org/10.1146/annurev.polisci.12.040507.142213

6. Schrading, J.: Analyzing Domestic Abuse using Natural Language Processing on Analyzing Domestic Abuse using Natural Language Processing on Social Media Data Social Media Data. https://repository.rit.edu/cgi/viewcontent.cgi?article=9949&context=theses
7. "Gendered Innovation Living Labs | GILL Project | Fact Sheet | HORIZON," CORDIS | European Commission. https://cordis.europa.eu/project/id/101094812. Accessed 08 April 2024
8. Ståhlbröst, A., Holst, M.: The Living Lab Methodology Handbook. Centre for Distance-spanning Technology, Sweden (2012)
9. Bergvall-Kåreborn, B., Ståhlbröst, A.: Living lab: an open and citizen-centric approach for innovation. Int. J. Innov. Region. Dev. **1**(4), 356–370 (2009)
10. Twitter, "Developer," Twitter.com (2017). https://developer.twitter.com/en
11. "PRAW: The Python Reddit API Wrapper — PRAW 7.4.0 documentation," praw.readthedocs.io. https://praw.readthedocs.io/en/stable/
12. Pérez, J.M., Giudici, J.C., Luque, F.: Pysentimiento: a python toolkit for sentiment analysis and SocialNLP tasks. arXiv:2106.09462 [cs], June 2021. https://arxiv.org/abs/2106.09462

DCAI - SS-TECTONIC

Special Session on New Perspectives and Solutions in Cultural Heritage (TECTONIC)

Scope

The special session entitled "New perspectives and solutions in Cultural Heritage" is a forum that will share ideas, projects, research results, models, experiences, applications, etc., focusing on Preventive Conservation of marine and aerial Cultural Heritage. New technological proposals based on Artificial Intelligence are currently being applied as measures and processes for conserving Cultural Heritage. This special session has its main objectives:

- To maximise the value of research outcomes by promoting knowledge exchange, interactions, partnerships and inclusive engagement between cultural heritage researchers, individuals and organisations outside the immediate research community.
- To encourage the implementation and transmission of research outcomes and to communicate them and the knowledge acquired among researchers and stakeholder sectors.

The best papers of this Special Session will be invited to submit an extended version to the Special Issue "Advanced Technologies for Maritime and Underwater Archaeology—2nd Edition" of the Journal of Marine Science and Engineering.

Topics

Preventive Conservation of Cultural Heritage (CH), particularly Underwater Cultural Heritage (UCH), is crucial in safeguarding tangible testimonies of past human life preserved on the seabed. Solutions to preventive conservation in the UCH sector are still costly and sometimes very difficult to develop and implement. This is because research activities in marine environments imply more significant difficulties regarding methods and the use of advanced technologies than in the aerial environment. Underwater sites are generally difficult to access and more dangerous compared with working on dry land. Also, working in the marine environment and using specific equipment requires a lot of experience, highly skilled operators and the coordination of different "technical actors", thus keeping such solutions out of the reach of "common operators".

In this context, the whole management of an underwater site, starting from evaluating its state of conservation until the restoration works, is often carried out by "external experts," implying enormous costs, only sometimes manageable and applicable.

In this framework, the main objective of the session is to share research solutions tailored to the technical needs of all the actors and operators working in the Preventive Conservation of Cultural Heritage, and in particular, but not limited to this, in the UCH field (i.e., underwater site manager, managing institution, managing authority, etc.).

The session tries to join experts and researchers who recently focused their work on intelligent approaches for more sustainable management of aerial and UCHs, facilitating monitoring operations, encouraging conservation in situ, and providing accurate guidelines for correct maintenance.

The main topics of the session (although not limited to these) are:

Monitoring systems provide awareness of the artefact conservation status and the surrounding environment.
AI-based decision support system (DSS) that helps to collect data from the aerial and marine environment.
Data processing using AI techniques is used to infer subjacent relations among the parameters registered and to perform a comparative data analysis able to define which environmental aspects have a deeper impact on the preservation of the CH.
Remotely Operated Vehicles (ROVs) to perform and support the monitoring and maintenance activities based on an innovative modular and configurable architecture.
Systems management and protocols for the maintenance and enhancement of CHs.
Underwater conservation and restoration strategies.
Methodologies and digital tools for underwater and aerial archaeology.
Sensors and innovative monitoring for the aerial and marine environment.
Green Conservation strategies applied to the Cultural Heritage.
Sustainability and Cultural Heritage: new perspectives.

So, the overall objective is to share ideas to promote and visualise different technological and digital solutions for safeguarding CHs and to understand better threats and solutions for more successful and sustainable management of CH conservation practices.

Organizing Committee

Mauro Francesco La Russa	University of Calabria, Italy
Michela Ricca	University of Calabria, Italy
Javier Prieto Tejedor	University of Salamanca, Spain
Antonio Lagudi	University of Calabria, Italy
Marta Plaza Hernández	University of Salamanca, Spain

Program Committee

Natalia Rovella	National Research Council of Italy - Institute on Membrane Technology (CNR - ITM), Italy
Silvestro Antonio Ruffolo	University of Calabria, Italy
Fabio Bruno	University of Calabria, Italy
Vladimir Djapic	H2O Robotics, Croatia
Alessandra Bonazza	National Research Council - Institute of Atmospheric Sciences and Climate (CNR-ISAC), Italy

Alessandro Sardella	National Research Council - Institute of Atmospheric Sciences and Climate (CNR-ISAC), Italy
Yeray Mezquita	University of Salamanca, Spain
Sergio Márquez	AIR Institute, Spain
Jorge Herrera	University of Salamanca, Spain
Dolores Elkin	CONICET, Argentina

Forensic Investigations for Discriminating Decontextualized Artifacts: Preliminary Data from a Research Project in Cultural Heritage

M. P. Albanese[1(✉)], M. Ricca[1], M. F. Alberghina[1,2], N. Rovella[3], S. Schiavone[2], and M. F. La Russa[1]

[1] Department of Biology, Ecology and Earth Sciences (DiBEST), University of Calabria, 87036 Arcavacata di Rende (CS), Italy
mariapia.albanese@unical.it

[2] S.T.Art-Test, Via Stovigliai, 88, 93015 Niscemi, CL, Italy

[3] Institute on Membrane Technology, ITM-CNR, Via P. Bucci, Cubo 17/C, 87036 Arcavacata di Rende (CS), Italy

Abstract. Clandestine excavations, widespread throughout the Italian territory, represent a significant threat to cultural heritage. As the cradle of Magna Graecia, Southern Italy is an undisputed witness to this destructive phenomenon. These illicit activities strip artifacts of their original context, compromise history, and hinder research. The profit-driven counterfeit industry thrives due to the ample supply of raw materials essential for the counterfeiting processes within the region. In addressing these challenges, this study adopts a multidisciplinary approach, merging forensic investigations with archaeometric analyses, to authenticate ceramic artifacts recovered from clandestine excavations by the Carabinieri Unit for the Protection of Cultural Heritage and Anti-Counterfeiting in Cosenza, Calabria, Italy. Despite some documented stylistic- iconographic incoherent features compared to the same typology finds certainly authentic, the archaeometric results achieved so far demonstrate the compatibility of the majority of the examined samples. Additional studies with further analytical investigations using microscopic, geochemical, and physical methods on micro-samples will be essential to confirm the preliminary data.

Keywords: Illicit trafficking · multidisciplinary analytical approach · decontextualized finds

1 Introduction

Cultural heritage serves as a tangible bridge to our past, enhancing our comprehension of ancient civilizations and their legacies [1]. When conducting archaeological research, it is critical to understand that artifacts are puzzle pieces including fragments of history. Questions related to an object, such as where and how it was made and the raw materials it consists of, cannot be answered by archaeology alone. Archaeometry plays a fundamental role in this context, where a multidisciplinary approach is essential. Therefore, a

G. Marreiros et al. (Eds.): DCAI 2024, LNNS 1151, pp. 147–153, 2025.
https://doi.org/10.1007/978-3-031-80946-0_14

key issue in illicit trafficking is determining the authenticity of artifacts because archaeological falsification can harm academic research, distort historical understanding, and compromise research. In the global market for ancient artifacts, counterfeit items represent an ongoing challenge, especially as demand for authentic pieces continues to increase.

The widespread occurrence of clandestine excavations represents a significant threat to cultural heritage, especially in southern Italy, an area rich in Magna Graecia testimonies. This work focuses on the harmful effects that illicit excavations have on the history of archaeological finds and how counterfeiting spreads within these criminal contexts. Through an in-depth examination of carefully selected confiscated artifacts, this study aims to shed light on the complexities of distinguishing real artifacts from counterfeit ones [3–5]. In pursuing this goal, research is committed to protecting our cultural heritage for the benefit of future generations. The specimens examined in this study constitute a subset of a larger collection of 28 artifacts confiscated by the Carabinieri Unit for the Protection of Cultural Heritage and Anti-Counterfeiting in Cosenza (Calabria, southern Italy), as part of the investigation codenamed "Brettii". This study will focus just on 4 finds, providing preliminary data into the investigations conducted thus far (Fig. 1). These 4 items belong to two distinct ceramic categories: overpainted ceramics, known as "Gnathia" and "Daunia" ceramics.

Fig. 1. Selected Artifacts investigated.

2 Materials and Methods

The paper reports a preliminary study about the application of a multidisciplinary approach to discern the authenticity of confiscated ceramic finds. Scientific analyses allow for the assessment of compositional and microstructural properties, as well as the provision of typological-stylistic information, which aids in enhancing our knowledge of the materials and techniques utilized in their production. Furthermore, the archaeological-stylistic examination, coupled with comparison to existing literature, supports the identification of discrepancies and potential counterfeiting. Starting from a stylistic-iconographic investigation aimed at recognizing the shapes and ceramic classes to which the selected artifacts belong, non-invasive analyses were carried out and then micro-samples of ceramic material were then collected for the preliminary diagnostic investigations. Specifically: X-ray fluorescence spectroscopy (XRF), digital optical microscopy (digital OM), and X-ray diffraction (XRD) (Fig. 2).

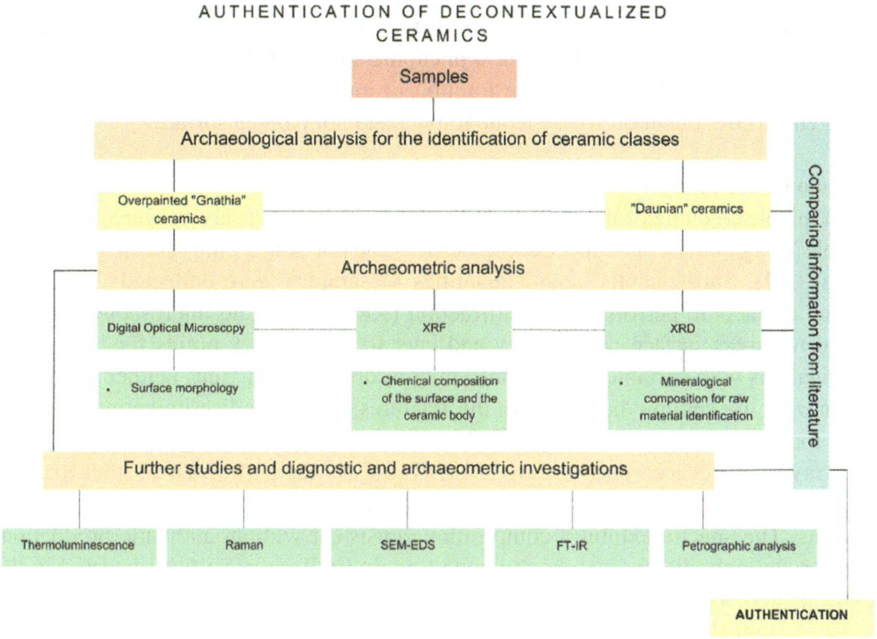

Fig. 2. Schematic diagram of the methodology

3 Results and Discussion

Archaeological analysis

The archaeological-stylistic analysis allowed for the identification of two distinct ceramic categories: overpainted "Gnathia" ceramics and "Daunian" ceramics. Furthermore, comparisons have led to hypothesizing their placement within a potential production area of the ancient Apulia.

"Daunian" ceramic emerges at the beginning of the 8th century BC, made of refined clay. The decoration is monochrome, with a matte brown-black color applied in precise geometric patterns on the light surface of the finds.

Regarding the ceramic class "Gnathia," it developed from the 5th century BC in Greek colonies, where the major production centers were located in the areas of Metaponto and Taranto. It is characterized by a simple technique, where the surface is coated with black paint upon which scenes and figurative details are painted in white, with touches of yellow and crimson [6, 7]. The decorative repertoire consists of two fundamental categories: geometric and vegetal, sometimes integrated with figurative scenes. Furthermore, from a stylistic-archaeological perspective, the examined samples present discrepancies with authentic ceramic material, both morphologically and iconographically. Specifically, discontinuities are evident in the two *skyphoi*, where Greek stylistic precision is lacking in the discontinuous rendering of laurel leaves (MdF-34). Another useful element for the preliminary discrimination of the studied specimens has been the encrustations (Fig. 3, first picture) [8, 9]. They are absent, and if present but not visible, they never cover the iconographic elements (MdF-40), a phenomenon that naturally occurs in authentic ceramics. A possible previous restoration of an original find generally provide a homogeneous treatment on whole iconographic and not painted surface.

Archaeometric analysis

For an initial screening of the ceramic artefacts, non-invasive analyses were conducted: portable X-ray fluorescence spectroscopy (XRF) and portable digital optical microscopy (digital OM). Subsequently, small quantities of samples were collected from the 4 selected artifacts to perform X-ray diffraction (XRD). DinoLite microscopy was performed to assess surface morphology and later to identify key points for conducting investigations with portable XRF. From a microscopic point of view, neither on the surface, nor on the ceramic body, nor in the fractures, were any discriminating elements found for counterfeit production (MdF-26 and Md-40; Figs. 3 and 4).

X-ray fluorescence spectroscopy (XRF) [10–12] was performed on both painted surface areas and ceramic body to evaluate their composition in terms of major and trace elements. The spectra exhibit a composition consistent with an authentic production; however, the only discriminating compound might be Barium Sulfate, detected in the yellow paint of sample MdF-40 (Fig. 5). Although comparisons in the literature suggest its usage from the 19th century onwards, further studies are underway to justify its presence as modern white pigment added in mixture or as impurities of earth pigments used for yellow pictorial layers.

X-ray diffraction (XRD) [13–16] revealed the mineralogical composition of the material at a qualitative and semi-quantitative level. Through the identification of newly formed minerals, hypotheses regarding firing temperature and the type of raw materials used were formulated. From the diffractograms (Fig. 6), Quartz (Qz) is the dominant mineral phase, followed by smaller quantities of clay minerals (Illite/Muscovite) as well as Plagioclase (Pl) and Feldspars (Fs). The mineralogical composition of the ceramic paste is compatible with authentic ceramic production. This result also suggests that the forger may have used raw materials from the Apulia production area.

Fig. 3. MdF-26: Digital Optical Microscopy (digital OM; magnification x200) on the surface and ceramic body.

Fig. 4. MdF-40: Digital Optical Microscopy (digital OM; magnification x200) on the surface and ceramic body.

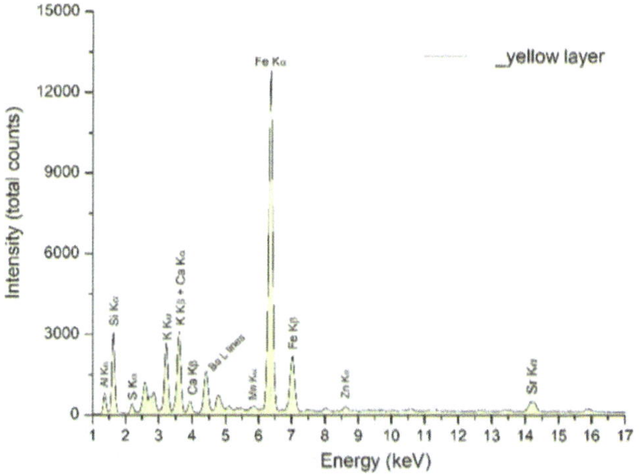

Fig. 5. XRF spectrum showing the presence attributable to barium sulfate on the yellow layer of sample MdF-40.

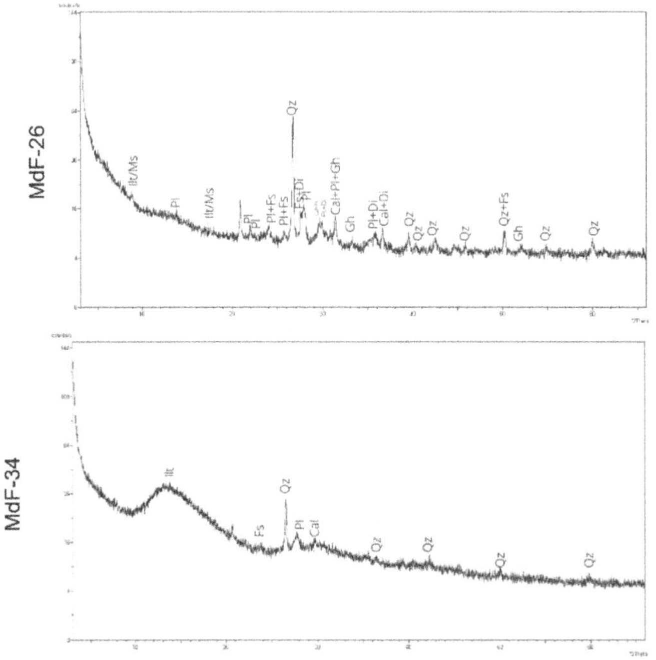

Fig. 6. Diffractograms of two analyzed samples: MdF-26 and Mdf-34.

4 Conclusion

The preliminary research has achieved some important data:

- After comparing the discovered objects with authentic pieces, in-depth humanistic investigations revealed stylistic and iconographic differences. These variations are evident in details such as decorative motifs or iconographic representations, which are not present in well-documented authentic artefacts. Consequently, these observations raise doubts about the authenticity of the discovered objects.
- Archaeometric analyses have revealed that, despite some stylistic inconsistencies, the materials used in the findings are compatible with those of authentic archaeological productions. This means that the artifacts do not present significant compositional differences compared to authentic materials. These results suggest that the same raw materials, likely from Apulian production, were used for the production of the artifacts.
- The presence of barium sulfate detected by XRF in the yellow surface decoration raises doubts about the authenticity of the artifacts. Barium sulfate is a material used in modern times, and its significant presence may indicate a modern intervention or forgery. This result contrasts with the expected characteristics of an authentic artifact, suggesting that the specimen in question may be a fake.
- To resolve doubts and authenticate the finds, further studies and archaeometric investigations will be necessary. Thermoluminescence investigations are currently underway which could provide useful dating information. Future investigations will

include mineral-petrographic analysis, scanning electron microscopy combined with microanalysis, and Raman spectroscopy.

Disclosure of Interests. The authors have no competing interests to declare that are relevant to the content of this article.

References

1. Matthes, E.H.: The Ethics of Cultural Heritage, The Stanford Encyclopedia of Philosophy (Fall 2023 Edition). In: Edward, N., Zalta, Nodelman, U. (eds.). https://plato.stanford.edu/archives/fall2023/entries/ethics-cultural-heritage/
2. Brodie, N., Kersel, M.M., Mackenzie, S., Sabrine, I., Smith, E., Yates, D.: Why there is still an illicit trade in cultural objects and what we can do about it. J. Field Archaeol. **47**, 117–130 (2001)
3. Rovella, N., Comite, V., Ricca, M.: The methodology of investigation on red-and-black figured pottery of unknown provenance. Int. J. Conserv. Sci. **7**, 954–964 (2016)
4. Randazzo, L., Gliozzo, E., Ricca, M., Rovella, N., Berikashvili, D., La Russa, M.F.: Ceramics from samshvilde (Georgia): a pilot archaeometric study. J. Archaeol. Sci. Rep. **34**, 102581 (2020)
5. Montana, G., Tsantini, E., Randazzo, L., Burgio, A.: SEM-EDS analysis as a rapid tool for distinguishing Campanian a ware and Sicilian imitations. Archaeometry **55**, 591–608 (2013)
6. Taliano Grasso, A.: Patrimonio recuperato e falsi archeologici tra forma, iconografia e materia in un sequestro operato dai Carabinieri del Nucleo TPC di Cosenza. In: Malacrino C., Taglietti B. (a cura di), Salvati dall'oblio. Tesori d'archeologia recuperati dai Carabinieri per la Tutela del Patrimonio Culturale., Reggio Calabria, p. 107 (2022)
7. Cuno, J.: Museums and the acquisition of antiquities. Cardozo Arts Entertain. Law J. **19**(1), 83–96 (2001)
8. Thorn, J., Glascock, M.: New evidence for Apulian red-figure production centers. Archaeometry **52**(5), 777–795 (2010)
9. Giacobello, F.; Dall'argilla al vaso. Produzioni e tecniche in (a cura di) Salvadori M., Baggio, M., Zamparo, L.: Argilla. Storie di vasi, pp. 23–33. Padova University Press (2021)
10. Giacobello, F.: Aspetti produttivi delle officine del pittore dell'ilioupersis e del pittore della patera in (a cura di) Giacobello F.: Savoir-Antichi e moderni pittori e officine cermaiche nell'Apulia del V e IV secolo A.C.; Atti della giornata di studi Vicenza, Gallerie d'Italia-Palazzo Leoni Montanari, pp. 115–130 (2015)
11. Le imitazioni di ceramica greca in (a cura di) Ghiotto A.R.; A Decoro Di Schio Intellettuale e Studiosa". Le Collezioni Archeologiche Nel Museo Civico Di Schio (2023). https://hdl.handle.net/11577/3496349
12. Ricca, M., et al.: A combined non-destructive and micro-destructive approach to solving the forensic problems in the field of cultural heritage: two case studies. Appl. Sci. **11** (6951) (2021). https://doi.org/10.3390/app11156951
13. Giannossa, L.C., Muntoni, I.M., Laviano, R., Mangone, A.: Building a step by step result in archaeometry. Raw materials, provenance and production technology of Apulian Red Figure pottery, J. Cult. Herit. **43**, 242–248 (2020)
14. Janssens, K., et al.: Use of microscopic XRF for non-destructive analysis in art and Archaeometry. X-Ray Spectrom. **29**, 73–91 (2000)
15. Rendle, D.F.: X-ray diffraction in forensic science. Rigaku J. **19**, 11–22 (2003)
16. Giannossa, L.C., Forleo, T., Mangone, A.: The distinctive role of chemical composition in archaeometry. the case of apulian red figure pottery. Appl. Sci. **11**, 3073 (2021)

Optimizing Underwater Acoustic Survey Strategies with Autonomous Underwater Vehicles: A Simulation-Based Framework

Antonio Lagudi(✉), Umberto Severino, and Fabio Bruno

Department of Mechanical, Energetic and Management Engineering, University of Calabria, 87036 Arcavacata di Rende, Italy
{antonio.lagudi,umberto.severino,fabio.bruno}@unical.it

Abstract. Ocean exploration and inspection missions involving underwater robotic systems have gained increasing popularity in various domains such as marine science, archaeology, and defense in recent years. However, deploying vessels and robotic solutions for such research activities entails significant costs for acquisition and maintenance. This is primarily due to the logistical challenges involved in transporting equipment between locations, which necessitates a substantial financial investment, in addition to expenses associated with crew and research personnel. Effective planning of survey campaigns is crucial to avoid costly missteps arising from poor organization. It ensures that expenditures are justified and that objectives are achieved without unnecessary setbacks or errors. In this context, the utilization of simulation methodologies adds invaluable value in mitigating costs and exploring various solutions before the mission is carried out in the real world. Their utility extends beyond the examination of specific sensors or setups; they are also a valuable tool for planning different campaigns with different mission objectives. In this paper, a simulation framework that integrates an Autonomous Underwater Vehicle with a Multibeam Echo Sounder as the perception sensor is presented. The objective is to streamline acoustic survey mission planning and generate synthetic acoustic image data incorporating the modeling of typical error sources, simplifying the evaluation of output data for real applications. This system can also be utilized as a Software-in-the-loop system for testing and verification of control systems and algorithms.

Keywords: Simulation · AUV · Underwater Robotics · Acoustic Imaging · Unity3D · Robot Operating System

1 Introduction

The deployment of robotic systems in ocean exploration and inspection, such as Autonomous Underwater Vehicles (AUVs) or Remotely Operated Vehicles (ROVs), has experienced significant expansion in recent years, emphasizing their role across diverse domains such as marine science, archaeology, and defense [1].

However, the utilization of vessels and robotic systems for such research endeavors imposes a significant financial burden. Procuring and sustaining these assets involves

G. Marreiros et al. (Eds.): DCAI 2024, LNNS 1151, pp. 154–163, 2025.
https://doi.org/10.1007/978-3-031-80946-0_15

substantial expenses, further compounded by the logistical intricacies of equipment transport and the compensation of crew and research personnel. In light of these challenges, meticulous planning of survey campaigns is paramount. A methodical approach reduces the likelihood of costly mistakes or delays, guaranteeing the prudent allocation of resources and the efficient attainment of objectives.

In this context, the application of simulation methodologies emerges as indispensable. These tools not only facilitate the development of algorithms for robots but also enable the examination of specific vehicle setups, serving as invaluable aids in planning scientific campaigns.

Over the years, many underwater simulators have been developed. The UUV Simulator [2] is a Multi-Robot simulator for underwater interventions implemented in Gazebo [3], which is a simulation suit for the ROS framework [4]. It extends the functionalities of the underwater environment in Gazebo, by integrating the capability to simulate robots performing various tasks, including robotic manipulators. It implements underwater hydrostatic and hydrodynamic phenomena, thrusters, sensors, and disturbance effects. While employing different simulation tools, UWSim [5] serves as another example of a simulator that integrates the capabilities of multiple software platforms. Its purpose is to provide a tool capable of testing new control logic for underwater missions. The simulator is built using OSG (Open Scene Graph) [6], a 3D graphics application also utilized for game development and virtual reality implementation. Dynamic models, on the other hand, are the result of programming in Matlab [7]. The scenario can be configured by adding vehicles and sensors, thereby integrating the simulator with existing control architectures and enabling Hardware-in-the-Loop (HIL) simulations. UNav-Sim [8] aims to provide a simulator with a high-detail rendering in order to transfer algorithms trained in the simulation environment to real-world applications.

An underwater simulator, which offers a virtual environment for simulating vehicles equipped with an acoustic sensor such as a Multibeam Echo Sounder (MBES), is outlined in [9]. In this study, the simulation environment is utilized for testing typical lawnmower-like patterns and exploring the interaction between the sensor and the environment. However, it should be noted that the customization options for the instrument are limited, and the simulator does not fully incorporate typical sources of noise encountered during data acquisition.

The current study aims to enhance the physical behavior of MBES sensors by integrating models of typical error sources, commonly visible in output data. This expanded simulator allows for more faithful evaluation of output data. Furthermore, the software tools support mission management and MBES sensor interfacing, adopting the communication protocol of real sensors. Illustrative examples of achievable outcomes through typical area surveys are provided in this paper, emphasizing the practical effectiveness of simulation in advancing marine research endeavors.

2 Framework Architecture

Within this simulation framework, the principal components or actors comprise the environment itself, the underwater vehicle, and the MBES sensor. Each of these elements plays a pivotal role in the simulation ecosystem, contributing to the comprehensive understanding and effective planning of underwater exploration activities.

Following the modeling of the simulator, users can plan missions using the mission planner module developed in Unity3D. The development environment considers the interaction between two frameworks: Unity3D and ROS. The following table (Table 1) shows the functions deputed to each software:

Table 1. Unity3D and ROS tasks.

Software	Functions
Unity3D	· World creation
	· Vehicle importation
	· Creation of motor thrust curves
	· Navigation sensor implementation
	· MBES sensor implementation
	· Creation of a graphical interface for mission commands and data visualization
ROS	· Sensor data processing
	· Vehicle pose estimation
	· PID tuning for position and velocity control
	· Propulsion system modeling
	· Navigation algorithm
	· Time reference

The seamless communication between the two modules is facilitated by the *ROS-Bridge* communication protocol. This protocol serves as an additional abstraction layer within the ROS framework, acting as a bridge between ROS and web applications. Its primary purpose is to empower developers by furnishing them with a robust technology tailored for the development of web-based applications that seamlessly integrate with ROS.

Hence, considering the capabilities of each framework and how they are related, the following architecture has been defined (Fig. 1):

3 Development of the Simulation Environment

A typical modeling approach has been followed for the definition of the simulator, consisting of defining: the scope of the model, the level of detail, assumptions, and simplifications, as depicted in Fig. 2.

As an initial phase, a simulation scenario mirroring a marine environment has been crafted using Unity3D. Specifically, the Suimono asset [10] has been employed for this purpose. Suimono facilitates the rendering of water surfaces, wave behavior, and other environmental aspects, both above and below the waterline. After selecting a model of an actual AUV, it has been used as model within the scene. Through the development of custom scripts, efforts have been made to simulate the effects of propulsion from the

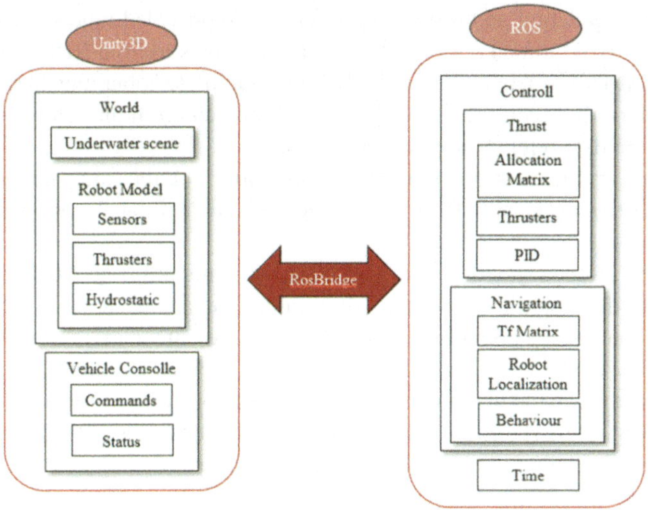

Fig. 1. Architecture of the Unity3D-ROS simulation

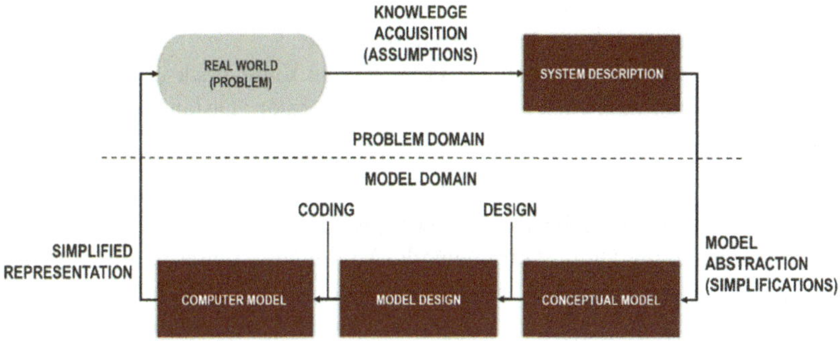

Fig. 2. Simulator Modelling Approach

native motors, buoyancy in underwater conditions, and the operation of a mass variation mechanism.

The model of the vehicle is provided following the guidelines reported in [11]. In terms of propulsion, five thrusters can be identified, which, individually or in combination, enable the vehicle's navigation. Thrusters do not facilitate roll control. However, this degree of freedom is managed by the vehicle's structural design, with masses positioned to stabilize rotation around the longitudinal axis. The control of the thrusters is handled in the ROS framework. The same framework is used to simulate the navigation sensors: Inertial Measurement Unit (IMU), Doppler Velocity Logger (DVL), Global Positioning System (GPS), and the altimeter.

It should be noted that this study did not encompass the development of the underlying physical phenomenon governing the operation of the navigation sensors; rather, it focused on extrapolating and processing information provided inherently by the environment in Unity3D to make it readable by the packages used in ROS. Each sensor is associated with a frame having a specific ID, which facilitated obtaining the correct transformation matrix between the sensor frame and the vehicle frame, thereby converting sensor data into data referenced to the AUV frame.

Regarding vehicle navigation and perception, data are utilized by the *robot_localization* package in ROS to estimate the vehicle's state by applying a Kalman filter.

In the presented work, the main focus lies in simulating the MBES sensor. The objective is to develop a tool that provides realistic results to the user, while offering flexibility in configuring both the sensor and the desired outcomes. Among the variables of this type of sensor, users can adjust parameters such as the number of beams, the maximum slant range (acquisition range), the angular beam aperture, and the operating frequency (ping rate). The chosen configuration for the beams is of the constant angular increment type, meaning that based on the angular aperture and the number of beams, there will be a certain angular discretization. The MBES model operates through the following steps:

- Creation of the acoustic signal reception beam;
- Reading of the acquired data;
- Processing of the result with the addition of errors;
- Creation and saving of images (Polar View, Wedge View, Backscatter);
- Saving of detected data for each ping in CSV format (Distance, Intensity, Pings, Time);
- Geolocation of acquired points.

The sensor's beams have been modeled using the *Raycast* function of the Physics class, integrated into the Unity3D framework. A schematic representation of the beam generation is illustrated in Fig. 3. The virtual sensor operates at a frequency equivalent to the ping rate, employing the same logic as other sensors within the simulator (Coroutine). An analysis of the code's performance led to the conclusion that a simple *for loop* would slow down the sensor's operation, as all calculation operations are executed on the main thread alone. Therefore, the utilization of parallel computing has been sought. For each *Raycast* of the swath, factors considered include the intensity, described by Lambert's law [12], the dependency of the reflected signal on the material of the object struck, and the phenomenon of the recorded distance, which is influenced by the angle of the beam. To establish the dependency on the angle of incidence, a term linked to Lambert's law has been incorporated into the model. This is achieved by utilizing the vector product between a vector oriented along the radius and a vector relative to the normal of the struck surface (i.e., the direction of the normal of the hit mesh element). The result is a maximum intensity (equal to 1) when the surface is perpendicular to the listening direction and approaches zero as the angle of incidence approaches zero.

The focus of the simulation is to generate typical output acoustic images of the sensor rather than perfectly emulate its physics. Data generation commences with populating an array of *Unity3D Color32* objects (representing RGBA colors in byte format) to be associated with a Texture2D, which will subsequently be saved and/or displayed at

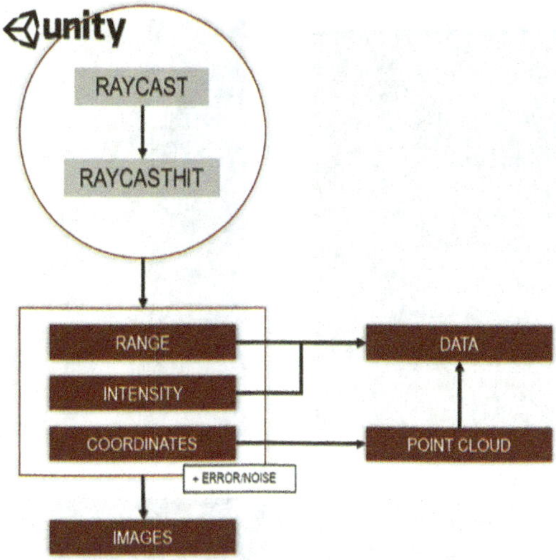

Fig. 3. Schematic representation of the beam generated in Unity3D to simulate the MBES

runtime. For each image, a texture is initialized for display, and another one for saving, thus segregating the two processes and offering flexibility of choice.

In the image, one dimension corresponds to the number of pixels as there are sensor rays, while the other dimension is arbitrary. To represent the distance of the points for each beam, a scale factor between the number of pixels and the distance observation window is required. This ensures that all points within the image fall within the maximum slant range and beyond the minimum detectable range. The Backscatter Image follows a different construction. It still maintains a number of pixels in one dimension equal to the number of beams. However, for each ping, the image is updated by adding the main lobe backscatter effect at the last acquisition moment and consequently removing the tail (the texture dimensions remain constant throughout the code execution, Fig. 4).

In the *wedge* representation (Fig. 4c), the reconstruction logic is similar to that of polar visualization (Fig. 4a). It involves converting coordinates from the sensor frame to coordinates relative to the image resolution, while considering the cone configuration (opening angle and maximum distance).

Furthermore, to simulate real-world data more accurately, artifacts and noise typical of MBES images have been incorporated. Therefore, effects such as concentric circles and errors around the ideal point have been introduced into the image. For an enhanced user experience, parameters that modify the result have been made adjustable, allowing for testing phases to find the image most closely resembling reality in terms of amplitude and intensity of the concentric circles and disturbances around the ideal points (Fig. 5).

4 Simulation of a Survey Mission

The mission planner integrates both Unity3D and ROS frameworks, interconnected via RosBridge. The interaction between them follows a server-client communication logic. Data exchange occurs via topics and messages, in an asynchronous and bidirectional

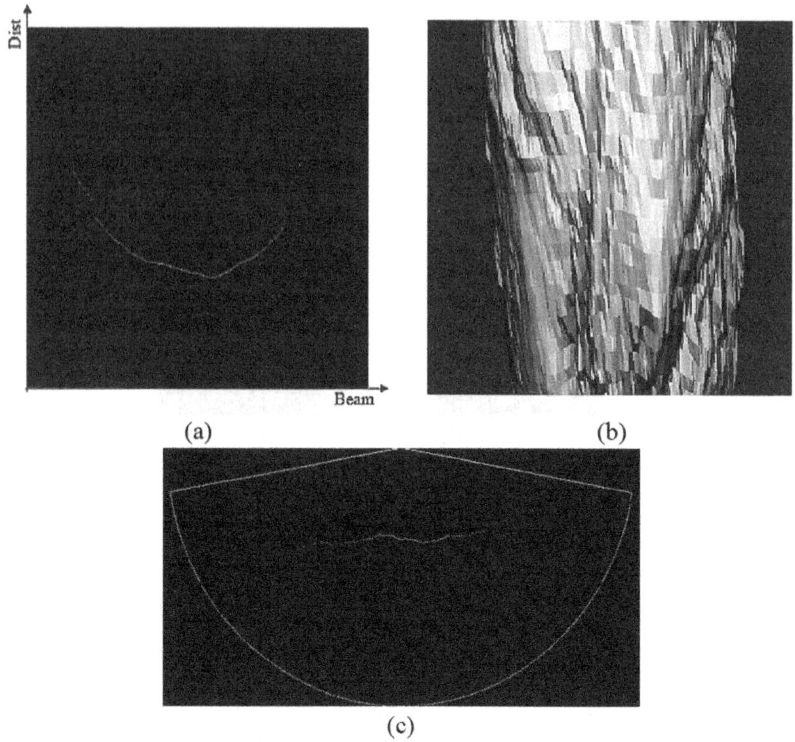

Fig. 4. MBES images representing the acoustic data: (a) polar representation of the water column, (b) backscatter image, (c) wedge view.

manner between the two entities: the client requests the server to perform operations by sending the goal, while the server executes operations through a callback and sends feedback to the client during execution, which can be processed by the client. At the end of the callback algorithm, the server returns the task result to the client.

In this specific scenario, the goal encompasses details regarding the initiation of vehicle control type (automatic or manual), the target's position, the desired distance from the seabed, and the relative orientation of the target in relation to the AUV's starting frame. This information is crucial for directing the vehicle towards the target. Regarding feedback, it has been determined that relaying the vehicle's positional data to the client during automatic mission execution would facilitate visualization within the implemented GUI. Users can configure all aspects of the mission and sensor settings via the GUI. Specifically, they have the option to manually control the vehicle or initiate an autonomous mission after defining the transects within the inspection area. As for the MBES setup, the following configurations can be adjusted (Table 2):

An example of the simulator GUI while the AUV is surveying is depicted in (Fig. 6).

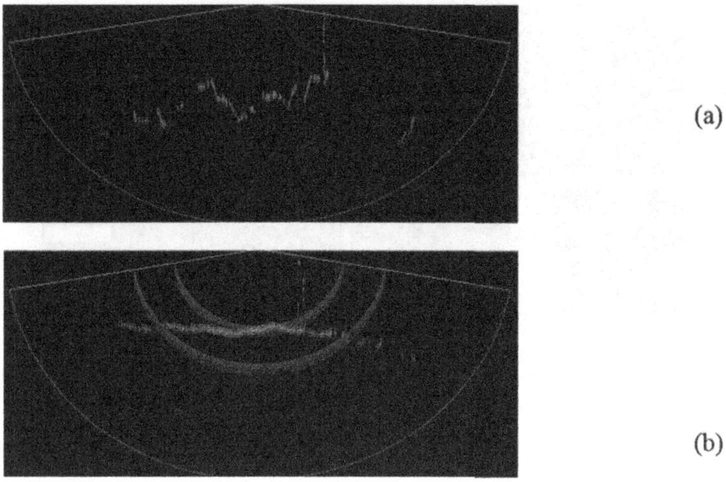

Fig. 5. Wedge views of the water column with noises and artifacts: (a) error around the ideal point; (b) concentric sidelobe artifacts.

Table 2. List of setting parameters in the mission planner.

Parameter	Description
Beam Number	The number of Raycasts
Slant Range	The maximum acquisition distance
Min Distance	The minimum acquisition distance
Fan Open	The angular aperture of the sensor beam, expressed in sexagesimal degrees
Ping Rate	The operating frequency of the sensor provided as input to the EnoughTimePassed function to manage the moment when the RaycastReader (acquisition) is called
Diffuse	A boolean variable that allows enabling material dependency to test the effects due to geometric factors alone or considering the composition of the seafloor
Send Msg	A boolean variable to activate the SendPointCloudMsg function and receive the geolocation of points from ROS

5 Conclusions

In this paper, a simulation-based framework has been developed to optimize acoustic underwater surveys carried out by AUVs, capable of virtually recreating different operational scenarios for a broad range of applications. Both ROS and Unity3D frameworks have been integrated to offer sophisticated simulation experiences, boasting advanced graphical capabilities in Unity3D and robust software performance. Leveraging ROS's process communication model facilitates the creation of a user-friendly system, easing development stages even with varied programming languages within ROS nodes.

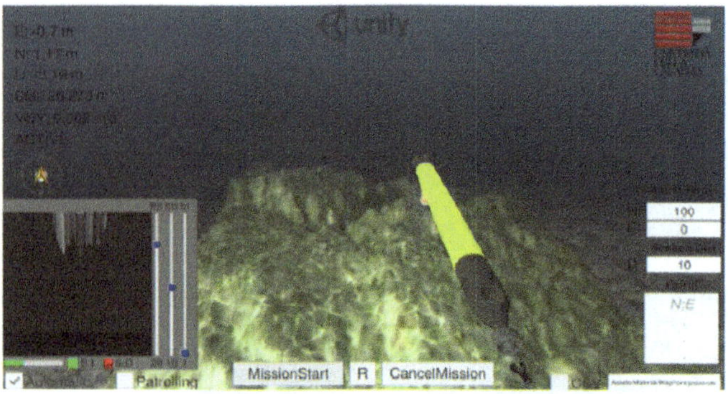

Fig. 6. Simulator user interface during a surveying mission of the AUV. Backscatter data are depicted in the bottom right of the image.

In the context of this paper, simulating the mission allows for the testing of a particular setup before embarking on an actual survey mission. This is particularly pertinent in sensitive archaeological surveys focused on acquiring underwater cultural heritage sites.

Looking ahead, there is potential to enhance sensor implementation by more accurately modeling their behavior through the addition of additional error sources. Implementing additional simulation scenarios for testing and developing obstacle-avoidance algorithms and/or modeling further features of the multibeam sensor equipped on the vehicle could provide added value. For instance, introducing a multi-detection function for generating a water-column image with variable detail levels based on factors such as the acoustic receiver's acquisition frequency, sound velocity in the medium, and the considered distance domain could be beneficial. Additionally, the developed system enables the generation of synthetic data to train artificial intelligence networks, offering a preliminary dataset closely mirroring real-world conditions.

References

1. Yuh, J., Giacomo, M., Richard, B.D.: Applications of marine robotic vehicles. Intel. Serv. Robot. **4**, 221–231 (2011)
2. Manhães, R.T., et al.: UUV simulator: a gazebo-based package for underwater intervention and multi-robot simulation. In: Oceans 2016 MTS/IEEE Monterey, pp. 1–8 (2016)
3. GAZEBO, "Gazebo Homepage,". https://gazebosim.org/home. Accessed April 2024
4. ROS, "ROS Homepage." https://www.ros.org/. Accessed April 2024
5. Prats, M., Perez, J., Fernández, J.J., Sanz, P.J.: An open source tool for simulation and supervision of underwater intervention missions. In: 2012 IEEE/RSJ International Conference on Intelligent Robots and Systems, pp. 2577–2582 (2012)
6. Graph, O.S.: "OSG Homepage." http://www.openscenegraph.org. Accessed April 2024
7. Mathworks, "MATLAB Homepage." https://it.mathworks.com/?s_tid=gn_logo. Accessed April 2024
8. Amer, A., et al.: UNav-Sim: a visually realistic underwater robotics simulator and synthetic data-generation framework. In: 2023 21st International Conference on Advanced Robotics (ICAR), pp. 570–576 (2023)

9. Chaudhary, A., et al.: Development of an underwater simulator using unity3d and robot operating system. In: OCEANS 2021: San Diego, Porto, pp. 1–7 (2021)
10. Store, U.A.: "Suimono Water System." Tanuki Digital. https://assetstore.unity.com/packages/vfx/shaders/suimono-water-system-4387. Accessed April 2024
11. Fossen, T.I.: Handbook of Marine Craft Hydrodynamics and Motion Control. John Wiley & Sons (2011)
12. Jung, B.G., et al.: Numerical modeling and simulation technique in time-domain for multibeam echo sounder. Int. J. Naval Arch. Ocean Eng. 225–234 (2018)

Climate Change and Underwater Cultural Heritage: The Case Study of Emma Wreck in Argentina

Virna L. Meccia[1(✉)], Alessandro Sardella[1,2], Linda Canesi[1], Julieta Frère[3,4], Mónica Grosso[4], Cristian Murray[4], Dolores Elkin[3,4], and Alessandra Bonazza[1,5]

[1] National Research Council of Italy, Institute of Atmospheric Sciences and Climate (CNR-ISAC), Via Gobetti 101, 40129 Bologna, Italy
{v.meccia,a.sardella,l.canesi,a.bonazza}@isac.cnr.it
[2] Department of Environmental and Prevention Sciences, University of Ferrara, Corso Ercole I d'Este 32, 44121 Ferrara, Italy
[3] National Scientific and Technical Research Council of Argentina (CONICET), Godoy Cruz, 2290, C1425FQB Buenos Aires, Argentina
[4] Latin American Thought (INAPL), Ministry of Culture of Argentina, National Institute of Anthropology, 3 de Febrero 1370, C1426BJN Buenos Aires, Argentina
[5] Istituto Superiore per la Protezione e la Ricerca Ambientale, Via Brancati 48, 00144 Roma, Italy

Abstract. One of the main objectives of the H2020-MSCA-RISE-2019 TECTONIC project is the development of a tool to support managers and decision makers for the risk assessment of Underwater Cultural Heritage (UCH) in a changing environment. The completion of this tool foresees a series of activities to be implemented at three pilot sites including the Emma wreck, located in the Nuevo Gulf, Chubut Province, Argentina. We here present an overview of the current environmental conditions of the site and discuss what is expected in the future according to climate projections, focusing mainly on those parameters which could have a priority on the deterioration processes of the site. To this end, we identified the environmental variables that should be considered for the development of a tool to decision makers for the safeguarding of UCH. In addition, we evaluate how the state-of-the-art climate models represent the current climate conditions by investigating the available datasets to better evaluate the future climate projections in the area where Emma pilot site is located. The present work is necessary to constrain the climate models to include in the "Risk Mapping Tool for Cultural Heritage Protection", an online tool geared towards policy and decision makers in support of the identification of risk areas and vulnerabilities for cultural heritage exposed to climate change.

Keywords: Risk Assessment · Shipwreck · Climate Modelling · Copernicus Marine Service · Sea Surface Temperature and Salinity · Northern Patagonia

G. Marreiros et al. (Eds.): DCAI 2024, LNNS 1151, pp. 164–174, 2025.
https://doi.org/10.1007/978-3-031-80946-0_16

1 Introduction

The evaluation of the impact of climate change on cultural heritage is increasingly gaining attention and several initiatives are currently in progress at international level for fostering the development of scenarios in the near and far future, which could be of support to policy and decision makers in the management of heritage sites at risk [1–4]. Within this framework, the "Risk Mapping Tool for Cultural Heritage Protection", developed in the framework of the Interreg Central EU projects ProteCHt2save and STRENCH, is an online tool conceived with the aim of interactively visualizing risk maps at territorial level (with a spatial resolution of ~ 12 km) for cultural heritage exposed to climate change [5, 6]. The maps are intended to show the changes of specific climate parameters and extreme standardized indices (related to events of heavy rain, flooding, drought, and extreme heating) for 2 future 30-year periods (2021–2050 and 2071–2100) and under the RCP4.5 and RCP8.5 future scenarios proposed by the International Panel on Climate Change (IPCC), with respect to the reference historical period (1976–2005). This way, the maps, developed with outputs from regional and global climate models, describe how and where the possible changes related to climate change are projected to affect the area where selected heritage case studies are located. So far, the "Risk Mapping tool for Cultural Heritage Protection" includes aerial case studies. However, in the context of the "Technological Consortium TO develop sustaiNabIlity of underwater Cultural heritage" (TECTONIC) project, we aim at including evaluation and analysis of the effects of climate change on underwater cultural heritage (UCH) into the tool. The selected TECTONIC Pilot Sites are: the Puerto Madryn "Goleta Emma Underwater Park" (Patagonia, Argentina), which includes the Emma wreck; the Marine Protected Area "Capo Rizzuto" (Calabria, Italy), which includes the Bengala wreck and the Capo Cimiti wreck; and the Saronic Gulf area (Greece), which includes the ancient harbor of Aegina, Saronic Islands (Attica) and the Villa of Dolia in Epidaurus (Argolis). The inclusion of UCH sites implies the evaluation of the marine environment and the consideration of the ocean variables that, eventually, are more difficult to both monitor and represent in climate models, than the atmospheric ones. In general, the UCH might be vulnerable to climate change for several reasons. Some examples include: seabed erosion and mechanical abrasion caused by changes in ocean currents and sediment movements; corrosion due to chemical changes affecting water pH (acidification) and salinity; biological overgrow caused by rising surface water temperature and dissolved oxygen; submersion of sites currently not underwater due to sea level rise; increased intensity and frequency of extreme weather events including sea storms [7, 8].

The aim of this paper is to present a study of the environment in which the Emma shipwreck is located and to evaluate how the state-of-the-art global climate models perform in the area in order to allow for the inclusion of this pilot site of TECTONIC into the "Risk Mapping Tool for Cultural Heritage Protection".

2 The Emma Pilot Site

2.1 The Shipwreck

Emma was a two-masted schooner with wooden hull, launched at the Deering & Donnell shipyard in Maine, USA in 1883. It was 25.35 m long, 6.17 m wide, and had a registered tonnage of 73 tonnes. It sailed the coasts of Argentinean and Chilean Patagonia, including the ports of Punta Arenas, Ushuaia, Río Grande, Río Gallegos, Puerto Deseado, Comodoro Rivadavia, and the Malvinas Islands. It was used for various purposes such as cargo and passenger transport, sealing, and fishing [9]. The vessel was lost in 1947 due to an accidental fire while anchored in front of Puerto Madryn. The ship is valued by local diving pioneers because it was one of the first frequented diving sites, where underwater freediving contests were held. This wreck site was one of the first dive sites in Patagonia in the 1950s [9].

The remains of Emma shipwreck consist mainly of the lower part of the hull structure (parts of the keel and keelson, frames, and internal and external planking). Preservation has been affected by the impact of anchors, and by the removal of parts of the ship by scuba divers in the past decades. The bow of the Emma, where the windlass is located, points south-west. In the central part of the wreck, there is an accumulation of pebbles which are interpreted as the ballast of the vessel. The engine and part of the propeller shaft are located next to the stern.

2.2 The Sedimentation Conditions

The Emma shipwreck is located at 42°45.932'S – 65°01.288'W, 1000 m from the coast and 16 m deep at mean high tide (Fig. 1). The range of visibility is from 2 to 6 m, and the bottom is composed of sand and other sediment fine grain size. The environment in which the site is located is of low energy and currents are not strong.

Two sediment samples associated with the Emma pilot site were analyzed [9]. A sample in the central area of the wreck is characterized by the dominance of psephytes (74%), which do not correspond to the local sediment, but to the ballast of the vessel. A sample outside the wreck presented predominance of psamites (80%) and fine to very fine sands corresponding to the natural sediments of the rest of the bay. Analyzing the granulometric parameters, all of them are poorly selected. The organic matter of both samples is relatively low and ranges between 2% and 4%. Bioclasts in both samples are very poorly represented with a predominance of pelecypods. In one of the core samples, mollusk shells of *Lyrodus pedicellatus* were found, confirming the action of wood-boring organisms.

2.3 The Atmospheric Conditions

The climate in Patagonia is characterized mainly by the westerly flow and the effect of the Andes mountains. The westerlies, defined as the prevailing winds between 30°S and 60°S, are the main climatic feature in the midlatitudes of the Southern Hemisphere [10].

In coastal areas, wind circulation patterns are strongly affected by the local phenomenon of sea breezes. This breeze occurs mostly during the afternoon in the spring

Fig. 1. The location of the Emma shipwreck.

and summer months, when the differential heating is higher. In Nuevo Gulf, northern Patagonia (Fig. 1), diurnal sea water temperature fluctuations were caused by sea breezes which forced consecutive downwelling-relaxation periods. The breeze also could be relevant as a transport mechanism for coastal organisms at larval stages.

Regarding precipitation, the Patagonian coast is characterized by extreme dryness conditions due to scarce rainfall and high potential evapotranspiration as a consequence of intense winds and suffers prolonged and frequent droughts.

2.4 The Oceanographic Conditions

The coast of northern Patagonia is located between 40°S and 43°S and includes the Nuevo Gulf (NG), San José Gulf (SJG), and the Valdés Peninsula (VP, Fig. 1). Regarding the water circulation, in situ observations are scarce and the average circulation in the area is mainly inferred from hydrographic data, biological indicators, and numerical models that show that the region is dominated by high-frequency, inertial fluctuations and tides [11].

Tides have a large amplitude, reaching more than 6 m in the Nuevo Gulf [12]. The bottom topography and the complex coastline are such that there is a phase shift of nearly a half period on both sides of the 7-km-long isthmus that separates NG from SJG. This peculiarity has promoted the interest for the generation of tidal energy [13]. The interaction of the large tidal currents with the complex bottom topography produces one of the highest energy dissipation rates of the South Atlantic Ocean. The vertical mixing induced by the tidal energy in combination with the increased heat fluxes in spring and summer allows the homogenization of the water column in shallow areas and the occurrence of thermal fronts that separate these well-mixed waters from the stratified waters. Associated with the dynamics of these frontal systems, there is a high rate of primary and secondary productivity [14, 15].

The surface heat flux changes direction throughout the year, but the freshwater flow is always toward the atmosphere, because evaporation exceeds precipitation. Therefore, there is a large annual thermal amplitude and relative maximum of salinity within the gulfs [16, 17].

2.5 The Observed Trends

Significant positive observed trends for the maximum temperature in the coastal areas of Patagonia were found [18] but no significant observed trends for minimum temperature were reported. A poleward shift of the precipitation belt over extratropical latitudes has been observed in recent years, increasing the wetter conditions over Patagonia, especially in the cold season.

Changes in the climate system have impacts on coastal areas. On short timescales, the physical impacts come from floods, coastal erosion and coastal flooding, severe storms, and intense rainfall runoff. On longer timescales, the impacts come from the effects of climate change on wind and wave patterns.

The sea surface temperature (SST) shows a significant positive trend up to 0.04 °C per year north of ~50°S, while negative values up to − 0.02 °C per year are observed south of 50°S [11]. The spring chlorophyll a (Chl a) shows a linear trend of 2 mg m^{-3} per decade in the outer portion of the continental shelf between 50°S and 40°S [11]. Satellite altimetry measurements show a trend of the SSH of up to 3 mm yr^{-1} in Nuevo Gulf [11]. The global sea-level rise impacts on coasts and exacerbates coastal erosion and flooding in coastal areas of Argentina.

Given the southward migration of the semipermanent South Atlantic high-pressure system [19, 20], a higher frequency of northerly winds can be expected over this region of the Patagonian coast. This scenario could favor the occurrence of coastal upwelling events [21]. On the other hand, a decrease in the mixed-layer depth (5 m per decade) was estimated by a global numerical model in the SMG and NG accordingly between 2003 and 2017 [22]. This change should have strongly affected the summer circulation, the transport of the main gyres, and the local ecosystem. Yet, evidence of such changes based on observations has not been reported so far.

3 Available Datasets to Evaluate Present and Future Climates Around Emma Pilot Site

3.1 Ocean Reanalysis

Global ocean reanalyses are homogeneous 3D gridded descriptions of the physical state of the ocean covering several decades, produced with a numerical ocean model constrained with data assimilation of satellite and in situ observations.

Here we use the Copernicus Marine Environment Monitoring Service (CMEMS) Global Ocean Ensemble Reanalysis product [23]. This product uses four reanalyses to create the ensemble, covering the altimetric era period (starting from 1st of January

1993) during which altimeter altimetry data observations are available. The four reanalyses used by the Global Ocean Ensemble Reanalysis are: GLORYS2V4 from Mercator Ocean (France), ORAS5 from the European Centre for Medium-Range Weather Forecasts (ECMWF; UK), GloSea5 from Met Office (UK), and C-GLORSv7 from Euro-Mediterranean Center on Climate Change (CMCC; Italy). These four products provide four different time series of global ocean simulations 3D monthly estimates. The Global Ocean Ensemble Reanalysis provides the mean averages describing the ocean at 0.25° x 0.25° horizontal resolution. In this work we use the monthly means of sea surface temperature (SST) and salinity (SSS) which are vulnerable environmental parameters to climate change. These affect the benthic communities with an impact on biofouling, for example controlling the activity of wood boring bivalves, and have an effect on metal corrosion. The change in weather patterns (shifts in precipitation, wind, and currents) may influence sediment movement, which is a significant variable in the preservation of UCH sites. We start the period in January 1993 because the reanalyses are more reliable since it is the starting date of altimetry satellite data. We finished the comparison period in December 2014 to cover the historical simulations of the climate models according to the CMIP6 protocol (see Sect. 3.2 below).

3.2 CMIP6 Climate Models

The Coupled Model Intercomparison Project Phase 6 (CMIP6), is a collaborative international effort aimed at coordinating climate model simulations from research institutions around the world. These climate models are crucial tools used by scientists to better understand Earth's climate system and to make projections about future climate change under different scenarios. CMIP6 builds upon the previous phases of the project (CMIP5 and CMIP3), with improvements in model resolution, representation of physical processes, and incorporation of new data and scientific understanding. The models simulate various components of the Earth system, including the atmosphere, oceans, land surface, and ice, and their interactions.

On the one hand, by comparing model simulations of the historical period (1950–2014 defined for CMIP6) with historical observations, scientists can evaluate the ability of the models to reproduce past climate variability and change. This helps to build confidence in their ability to simulate future climate. On the other hand, by making future scenarios of greenhouse gas emissions and other factors, the model simulations of the future period (2015–2100 defined for CMIP6) are used to project future changes in temperature, precipitation, sea level, and other climate variables and how those changes may affect ecosystems, economies, and societies. This information helps decision-makers to identify and prioritize adaptation and mitigation strategies.

We here present an evaluation of some CMIP6 model simulations (27 for SST and 10 for SSS) in performing the SST and SSS in the grid point which is closest to the latitude and longitude of the Emma site pilot when compared to the ocean reanalysis. This step is fundamental for the selection of the climate models that will be included in the "Risk Mapping Tool for Cultural Heritage Protection" to provide information about the expected changes in the near and far future.

4 Results and Discussion

We here provide a comparison of the SST and SSS simulated by the CMIP6 climate models and the reanalysis during the period 1993–2014 at the Emma site. The aim of this analysis is to identify the models that perform better during the historical period to be used by the "Risk Mapping Tool for Cultural Heritage Protection". The time series of the annual mean of SST and SSS at the Emma pilot site are plotted in Fig. 2 and Fig. 3, respectively.

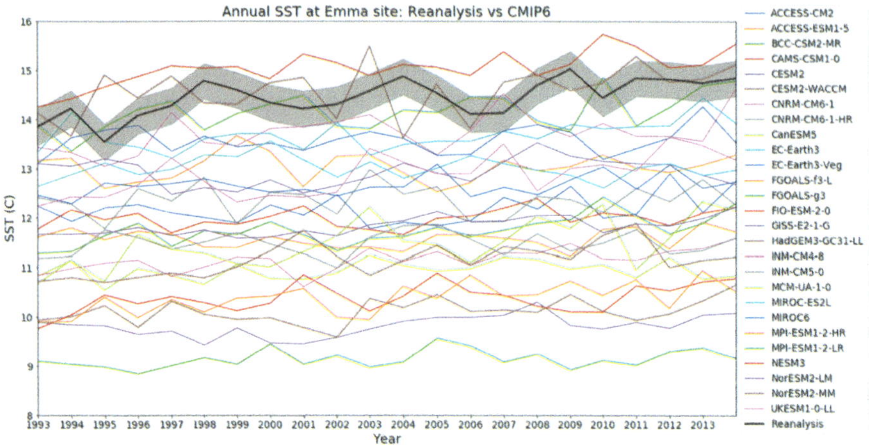

Fig. 2. Time Series of annual SST from 1993 to 2014 simulated by CMIP6 climate models (colors) and the ocean reanalysis (black line). The gray shaded plotted area indicates one standard deviation above and below the mean.

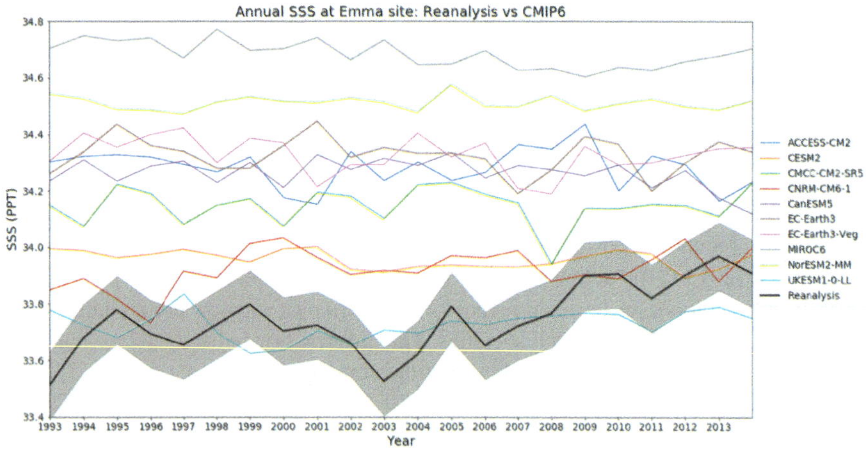

Fig. 3. Time Series of annual SSS from 1993 to 2014 simulated by CMIP6 climate models (colors) and the ocean reanalysis (black line). The gray shaded plotted area indicates one standard deviation above and below the mean.

In general, the CMIP6 models tend to underestimate the SST values and overestimate the SSS values at the Emma site in comparison with the reanalysis.

To quantify the differences between the simulated time series and the reanalysis, Table 1 and Table 2 show the correlation, the bias and the root mean square (RMS) of the monthly time series for SST and SSS, respectively. The models that better perform are highlighted in red. So far, from this preliminary analysis, the models that are selected for the inclusion in the "Risk Mapping Tool for Cultural Heritage Protection" are: UKESM1-0-LL, CNRM-CM6-1, BCC-CSM2-MR, HadGEM3-GC31-LL, NESM3, CESM2 and CMCC-CM2-SR5.

However, this is a preliminary analysis, and more variables will be evaluated in the future such as the seabed temperature and salinity, ocean currents and pH.

Table 1. Correlation, Bias (°C) and RMS (°C) between the monthly time series of SST (°C) simulated by CMIP6 and the Reanalysis. The models with the lowest Bias (less than 1 °C) and lowest RMS (less than 2 °C) are highlighted in red.

MODEL	Correlation	Bias	RMS
ACCESS-CM2	0.94	-2.26	2.8
ACCESS-ESM1-5	0.95	-1.4	2.09
BCC-CSM2-MR	0.97	-0.28	1.03
CAMS-CSM1-0	0.93	-2.45	2.94
CanESM5	0.92	-3.49	4.15
CESM2	0.94	-4.62	5.08
CESM2-WACCM	0.94	-4.33	4.81
CNRM-CM6-1	0.97	-0.7	1.59
CNRM-CM6-1-HR	0.97	-2.08	2.24
EC-Earth3	0.92	-0.75	1.88
EC-Earth3-Veg	0.92	-0.86	1.96
FGOALS-f3-L	0.94	-2.86	3.29
FGOALS-g3	0.92	-2.69	3.44
FIO-ESM-2-0	0.94	-4.05	4.46
GISS-E2-1-G	0.91	-1.51	2.5
HadGEM3-GC31-LL	0.94	0.14	1.67
INM-CM4-8	0.94	-3.28	3.88
INM-CM5-0	0.92	-2.96	3.65
MCM-UA-1-0	0.94	-2.94	3.34
MIROC6	0.9	-1.78	2.86
MIROC-ES2L	0.81	-1.42	2.98
MPI-ESM1-2-HR	0.94	-4.06	4.55
MPI-ESM1-2-LR	0.94	-5.3	5.59
NESM3	0.97	0.61	1.85
NorESM2-LM	0.77	-2.61	3.93
NorESM2-MM	0.69	-3.29	4.38
UKESM1-0-LL	0.97	-1.57	1.77

Table 2. Correlation, Bias (ppm) and RMS (ppm) between the monthly time series of SSS (ppm) simulated by CMIP6 and the Reanalysis. The models with the lowest Bias (less than 0.5 ppm) and lowest RMS (less than 0.5 ppm) are highlighted in red.

MODEL	Correlation	Bias	RMS
ACCESS-CM2	-0.13	0.54	0.57
CanESM5	-0.36	0.52	0.54
CESM2	-0.13	0.21	0.26
CMCC-CM2-SR5	-0.01	0.4	0.43
CNRM-CM6-1	0.23	0.18	0.24
EC-Earth3	0.06	0.58	0.6
EC-Earth3-Veg	-0.08	0.58	0.6
MIROC6	-0.28	0.94	0.95
NorESM2-MM	-0.09	0.77	0.78
UKESM1-0-LL	0.17	-0.02	0.15

5 Conclusions and Final Remarks

Overall, there are three main factors that determine the way in which a UCH site, and in particular a shipwreck, will be preserved. These are: a) the way in which the shipwreck occurred; b) the materials with which it has been built; c) and the environment (including anthropogenic factors) in which their remains were deposited [24]. Therefore, the evaluation of the impact of climate change on UCH and its future projections need a detailed study of the environment where heritage sites are located and of the changes of specific climate and oceanic variables in the future, such as sea surface temperature and salinity. Accordingly, there is the need to use climate modelling.

In this context, we have here presented preliminary results of the comparison between some of the Coupled Model Intercomparison Project Phase 6 (CMIP6) model simulations with the Copernicus Marine Environment Monitoring Service (CMEMS) Global Ocean Ensemble Reanalysis product in performing the sea surface temperature and salinity at the Emma pilot site, in order to select the most appropriate climate models to be included in the "Risk Mapping Tool for Cultural Heritage Protection" with the aim of providing information about the expected changes in the near and far future.

Future work will include the extension of this approach to the other two pilot sites of TECTONIC (Marine Protected Area "Capo Rizzuto" and Saronic Gulf area) and the integration in the Risk Mapping Tool of maps and time series in relation to further climate and marine parameters (such as dissolved oxygen and acidification) with priority in the deterioration processes of UCH.

Acknowledgments. This research has been supported by the project "Technological Consortium TO develop sustainability of underwater Cultural heritage (TECTONIC)", financed by the European Union (Horizon 2020 research and innovation programme under the Marie Skłodowska-Curie grant agreement No 873132).

Disclosure of Interests. The authors have no competing interests to declare that are relevant to the content of this article.

References

1. European Commission: Directorate-General for Education, Youth, Sport and Culture, Maxwell, I., Drdácký, M., Vintzileou, E., Bonazza, A., et al.: Safeguarding cultural heritage from natural and man-made disasters – a comparative analysis of risk management in the EU, Publications Office (2018). https://data.europa.eu/doi/10.2766/224310
2. Sesana, E.G., A.S., Ciantelli, C., Cassar, J., Hughes, J.J.: Climate change impacts on cultural heritage; A literature review. WIREs Clim Change 2021, 1–29 (2021). https://doi.org/10.1002/wcc.710
3. European Commission, Directorate-General for Education, Youth, Sport and Culture, Strengthening cultural heritage resilience for climate change: where the European Green Deal meets cultural heritage, Publications Office of the European Union. (2022)https://doi.org/10.2766/44688
4. Bonazza, A., Sardella, A.: Climate change and cultural heritage: methods and approaches for damage and risk assessment addressed to a practical application. Heritage **2023**(6), 3578–3589 (2023). https://doi.org/10.3390/heritage6040190

5. Risk Mapping Tool for Cultural Heritage Protection. https://www.protecht2save-wgt.eu. Accessed 09 Apr 2024
6. Sardella, A., Natali, S., Del Grande, C., Cacciotti, R., Bonazza, A.: The risk mapping tool for cultural heritage protection in Europe and mediterranean basin. In: 2022 IMEKO TC-4 International Conference on Metrology for Archaeology and Cultural Heritage, MetroArchaeo 2022, pp. 137–142 (2022)
7. Gregory, D., Dawson, T., Elkin, D., Tilburg, H., Underwood, C., Richards, V., Hollesen, J.: Of time and tide: the complex impacts of climate change on coastal and underwater cultural heritage. Antiquity **96**(390), 1396–1411 (2022). https://doi.org/10.15184/aqy.2022.115
8. Plaza-Hernández, M., Abbasi, M., Djapic, V., Prieto-Tejedor, J., Corchado-Rodríguez, E.: Environmental parameters that negatively impact Underwater Cultural Heritage. In: Distributed Computing and Artificial Intelligence, Special Sessions, 19th International Conference. DCAI 2022. Lecture Notes in Networks and Systems, vol. 585. Springer, Cham (2023)
9. Elkin, D., Murray, C., Grosso, M., Gutiérrez, G., Trassens, M., Bastida, R.: Investigaciones interdisciplinarias en sitios arqueológicos de naufragio en el Golfo Nuevo (Provincia del Chubut): Primeros resultados. Cuadernos del Instituto Nacional de Antropología y Pensamiento Latinoamericano **24**(1), 2039 (2015)
10. Pessacg, N., Blázquez, J., Lancelotti, J., Solman, S.: Climate changes in coastal areas of Patagonia: observed trends and future projections. In: Helbling, E.W., Narvarte, M.A., González, R.A., Villafañe, V.E. (eds.) Global change in Atlantic coastal Patagonian ecosystems. A journey through time. Springer, Cham (2021). https://doi.org/10.1007/978-3-030-86676-1_2
11. Saraceno, M., Martin, J. Moreira, D., Pisoni, J.P., Tonini, M.H.: Physical changes in the Patagonian shelf. In: Helbling, E.W., Narvarte, M.A., González, R.A., Villafañe, V.E. (eds.) Global Change in Atlantic Coastal Patagonian Ecosystems. A Journey Through Time. Springer, Cham (2021). https://doi.org/10.1007/978-3-030-86676-1_3
12. Palma, E.D., Matano, R.P., Piola, A.R.: A numerical study of the southwestern Atlantic shelf circulation: barotropic response to tidal and wind forcing. J. Geophys. Res. C Ocean **109**, C8 (2004). https://doi.org/10.1029/2004JC002315
13. Palma, E.D.: Tides and tidal energy in Valdés Península (Argentina). Revista geofísica **56**, 31 (2002)
14. Sabatini, M., Martos, P.: Mesozooplankton features in a frontal area off northern Patagonia (Argentina) during spring 1995 and 1998. Sci. Mar **66**, 215–232 (2002). https://doi.org/10.3989/scimar.2002.66n3215
15. Acha, E.M., Mianzan, H.W., Guerrero, R.A., Favero, M., Bava, J.: Marine fronts at the continental shelves of austral South America: physical and ecological processes. J. Mar. Syst. **44**, 83–105 (2004). https://doi.org/10.1016/j.jmarsys.2003.09.005
16. Lucas, A.J., Guerrero, R.A., Mianzán, H.W., Acha, E.M., Lasta, C.A.: Coastal oceanographic regimes of the Northern Argentine Continental shelf (34–43°S). Estuar. Coast. Shelf Sci. **65**, 405–420 (2005). https://doi.org/10.1016/j.ecss.2005.06.015
17. Tonini, M.H., Palma, E.D., Piola, A.R.: A numerical study of gyres, thermal fronts and seasonal circulation in austral semi-enclosed gulfs. Cont. Shelf Res. **65**, 97–110 (2013). https://doi.org/10.1016/j.csr.2013.06.011
18. Rusticucci, M., Barrucand, M.: Observed trends and changes in temperature extremes in Argentina. J. Clim. **17**, 4099–4107 (2004)
19. Barros, V.R., Doyle, M.E., Camilloni, I.A.: Precipitation trends in Southeastern South America: relationship with ENSO phases and with low-level circulation. Theor. Appl. Climatol. **93**, 19–33 (2008). https://doi.org/10.1007/s00704-007-0329-x
20. Blázquez, J., Nuñez, M., Kusunoki, S.: Climate projections and uncertainties over South America from MRI/JMA Global Model Experiments. Atmos. Clim. Sci. **02**, 381–400 (2012). https://doi.org/10.4236/acs.2012.24034

21. Pisoni, J.P., Rivas, A.L., Piola, A.R.: Satellite remote sensing reveals coastal upwelling events in the San Matías Gulf-Northern Patagonia. Remote Sens. Environ. **152**, 270–278 (2014). https://doi.org/10.1016/j.rse.2014.06.019
22. Franco, B.C., Palma, E.D., Combes, V., Acha, E.M., and Saraceno, M. Modeling the offshore export of Subantarctic shelf waters from the Patagonian shelf. J. Geophys. Res. Ocean **123** (2018). https://doi.org/10.1029/2018JC013824
23. https://data.marine.copernicus.eu/product/GLOBAL_MULTIYEAR_PHY_ENS_001_031/description. Accessed 08 Apr 2024
24. Grosso, M.: Estudios de cultura material en sitios históricos de naufragio en el litoral patagónico. El uso de la madera en artefactos del barco británico HMS Swift (siglo XVIII). Tesis Doctoral. Universidad de Buenos Aires (2011)

Biodeterioration and New Sustainable Conservation Strategies: The Case of Iblean Calcarenites in the Sicilian Baroque. Preliminary Results

M. Cirone[1,2](✉), N. Rovella[1,2], C. M. Belfiore[3], A. Figoli[2], B. Gabriele[4], F. Galiano[2], M. F. La Russa[1], A. Macchia[1], R. Mancuso[4], M. Ricca[1], S. A. Ruffolo[1], and M. Taverniti[5]

[1] Department of Biology, Ecology and Earth Sciences, University of Calabria, Via P. Bucci Cubo 12/B, Arcavacata di Rende (CS), 87036 Rende, Italy
martina.cirone@unical.it
[2] Institute on Membrane Technology, CNR-ITM, Via P. Bucci Cubo 17/C, Arcavacata di Rende (CS), 87036 Rende, Italy
[3] Department of Biological Geological and Environmental Sciences, University of Catania, Corso Italia 57, 95129 Catania, Italy
[4] Laboratory of Industrial and Synthetic Organic Chemistry (LISOC), Department of Chemistry and Chemical Technologies, University of Calabria, Via P. Bucci Cubo 12/C, Arcavacata di Rende (CS), 87036 Rende, Italy
[5] Institute of Informatics and Telematics, CNR-IIT, Via P. Bucci 17/C, Arcavacata di Rende (CS), 87036 Rome, Italy

Abstract. The rich architectural Baroque heritage of the cities in the Iblean area (eastern Sicily, Southern Italy) is characterized by the presence of important monuments built using different varieties of local calcarenite such as Ragusa, Comiso, and Modica stones. Unfortunately, these materials are significantly affected by degradation processes, such as biodeterioration, favored by surrounding environmental conditions and their intrinsic characteristics. In response to these challenges, in recent years there has been growing attention on the formulation of more sustainable alternatives to traditional chemical biocides.

The mineral-petrographic and physical characterization of the three stones provided important information to understand their responses to degradation processes and to define the most suitable conservation strategies. In this regard, biopolymeric membranes as controlled-release systems for natural biocides have been synthesized and tested.

In this first phase of investigation, the results obtained from petrographic and ultrasonic analyses reveal similarities and differences between the three lithotypes, suggesting that their heterogeneity may influence degradation phenomena. Further mineralogical, physical, and mechanical characterization tests will be carried out in the future. FTIR results showed the presence of limonene in the biopolymer membrane. In the next steps of the research, NMR analysis will also be conducted to confirm or disprove FTIR results. In addition, new membranes will be synthesized with other molecules with antimicrobic features.

G. Marreiros et al. (Eds.): DCAI 2024, LNNS 1151, pp. 175–183, 2025.
https://doi.org/10.1007/978-3-031-80946-0_17

Keywords: Biodeterioration · Biopolymeric Membranes · Calcarenites

1 Introduction

The preservation of stone artifacts from degradation phenomena in cultural heritage constitutes a paramount challenge.

Environmental conditions related to exposure to sunlight, fluctuations in temperature and humidity, atmospheric pollution resulting from human activities, and intrinsic characteristics of building materials undoubtedly contribute to the acceleration of stone deterioration.

The biological component also plays a significant role in the degradation processes of stone materials. It includes the direct action of the micro and macro-organisms and the effects of their metabolic processes. Cleaning procedures are often used for reducing or eliminating biological colonizers through the use of biocidal products. For a long time, chemicals such as acetone, ethanol, toluene, ethyl acetate, quaternary ammonium salts, etc., have been applied [1]. However, these substances can cause harm to human health and the environment, as well as potential damage to the stone surfaces themselves. Therefore, there is a current trend towards the exploration of innovative, ecological, and non-harmful approaches, such as novel biocides based on natural compounds, capable of inhibiting biofilm formation [1, 2].

The Baroque monuments of Noto Valley in eastern Sicily, constructed with local limestones, represent a clear example where visible signs of physical, chemical, and biological deterioration suggest the importance of adopting more efficient conservation strategies [3].

This research work will be articulated in two phases. On one hand, mineral-petrographic and physicomechanical analyses will be conducted on representative samples of three lithotypes selected because widely utilized in northeastern Sicily: Modica, Ragusa, and Comiso stones [4, 5]. These analyses are crucial for characterizing the materials and defining also the main predisposing factors for their degradation (e.g. porosity, roughness, secondary minerals). Moreover, the data collected will be useful to formulate the most suitable products in the subsequent phase. In this regard, on the other hand, an innovative approach will be pursued, focusing on the synthesis of biopolymeric membranes as controlled-release systems for natural biocides, able to be both stable and effective versus biodeterioration but also suistainable for the environment and economically [6]. In this regard, products will be tested on Modica, Ragusa, and Comiso samples that underwent biocolonization in a controlled laboratory environment.

2 Biodeterioration and Sustainable Conservative Strategies

Biological colonization represents one of the most hazardous deterioration processes for cultural heritage, seriously jeopardizing the integrity of materials of artistic and historical significance. A wide array of microorganisms, including algae, bacteria, lichens, and molds, can colonize such materials, posing a significant threat to their preservation [6].

Rocks, present both in natural geological outcrops and in stone monuments, constitute a common habitat for a wide range of microorganisms. Due to their ubiquitous presence and metabolic activity, these microorganisms contribute to the deterioration process of stone [7].

About biodeterioration, the traditional method typically involves the use of chemical biocides followed by the subsequent mechanical removal of the treated biological patina [6]. However, some of these products have been banned over time due to their toxicity and high costs, necessitating a transition towards more sustainable alternatives, among which is the synthesis of biopolymeric membranes. These ones serve as carriers for natural biocides, facilitating their controlled release over time. By utilizing a biopolymer as the matrix base, along with a naturally sourced active molecule known for its antimicrobial properties, the aim is to achieve controlled release and targeted action against biodeterioration agents.

Biopolymers are macromolecules composed of structural units sourced from biological origins, such as carbohydrates, proteins, and nucleic acids. These natural polymers are characterized by their renewable and biodegradable nature, making them pivotal for the development of sustainable solutions across various sectors, including medicine, agriculture, and advanced materials production. Poly (lactic acid) (PLA), obtained from natural sources like corn and sugarcane, is a biopolymer extensively employed for the production of biodegradable and biocompatible plastic materials [8].

Biopolymers, being macromolecules derived from biological sources, provide an excellent platform for the incorporation of natural active molecules, such as biocides, into membranes.

Currently, some scholars in the field of cultural heritage conservation are examining the potential use of naturally occurring active molecules found in essential oils. It is known that essential oils are a mixture of numerous organic compounds, but only some of them possess biocidal properties [6] such as limonene.

By incorporating natural active molecules with antimicrobial properties into these membranes, it is possible to achieve a gradual and targeted release of biocides, ensuring effective protection against the harmful action of microorganisms without the use of harmful chemical biocides. This approach combines principles of sustainability and innovation, offering advanced solutions for the conservation of cultural heritage and other application sectors [9].

3 Materials and Methods

The stones examined in this experimental study were collected from various areas within the Iblean region of Sicily. Three lithotypes (Fig. 1) were selected from Ragusa formation, a geological succession constituted by late Oligocene-Miocene limestones, often accompanied by marly limestones that, in certain areas, are impregnated with bitumen, lending them a dark hue [10]: Comiso Limestone, well-lithified with a colour ranging from yellow to grey to cream-white and a middle fine-grained; Modica Limestone, characterized by middle fine-grained, cream-brown colour and fossils visible to the naked eye; and Ragusa Limestone, displaying a middle-fine grained and a colour ranging from white grey to reddish-brown. Occasionally, the Ragusa stone may be impregnated with bitumen, resulting in a dark colour, and in such cases, it is locally referred to as Pece Limestone.

Samples of Comiso Stone will be directly extracted from quarries, while those of Modica and Ragusa stone will be obtained from natural outcrops [11].

Fig. 1. General view of the study areas in Sicily map (A) where find Comiso Stone (B), Modica Stone (C), Ragusa Stone(D).

The study of the susceptibility to degradation of building stones requires a comprehensive understanding of their composition, structure, and consistency.

This type of research can be initially tackled with traditional methods. Although superficially similar in colour and macroscopic characteristics, Ragusa, Comiso, and Modica stones exhibit subtle variations in their petrographic and physical attributes [12, 13]. Differences play a crucial role in defining the extent and nature of deterioration observed in each type of stone, also providing valuable data on the material's response to conservation practices. Factors such as mineralogical composition, pore size distribution, and mechanical properties exert significant influences on degradation mechanisms. To determine them, mineral-petrographic and physico-mechanical analyses on representative samples of each limestone variety will be conducted [4, 5].

The foreseen analyses include: Polarized Optical microscopy (POM) provides insights into the microstructure of the stone, while X-ray diffraction (XRD) elucidates its mineralogical composition. Mercury intrusion porosimetry (MIP) enables precise characterization of pore networks, facilitating a deeper understanding of moisture penetration mechanisms. Salt crystallization tests simulate environmental stresses, shedding light on the resilience of stones to external forces. Additionally, the determination of ultrasonic velocity, uniaxial compression strength, flexural strength, and elastic modulus provide crucial data regarding mechanical properties. At this stage of the research, just POM and ultrasonic velocity measurements were carried out.

Data collected will be useful to understand the interaction between these substrates and the product under evaluation for optimizing conservation strategies and providing valuable information on the treatment's effectiveness and compatibility with various limestone types.

As regards the biopolymer membranes, the first approach was the synthesis of a membrane based on Polylactic acid (PLA) and combined with limonene.

The membrane is characterized using analytical techniques [14] such as Fourier-transform infrared spectroscopy (FTIR) and, in the next experimental phase, also nuclear magnetic resonance spectroscopy (NMR). FTIR analysis enables the identification and interpretation of molecular interactions between limonene and the PLA matrix through vibrational band analysis, while NMR highlights the presence of limonene in the membrane.

In the future, in addition to preparing biopolymer membranes with other natural active molecules, will be conducted biocolonization tests in the laboratory-controlled environment on limestone samples representative of Modica, Ragusa e Comiso Stones, to evaluate the biocidal efficiency of membranes.

4 Results and Discussions

Microscopic observations allowed to identify the main mineral-petrographic characteristics of the three lithotypes (Fig. 2). The determination of the propagation speed of the ultrasound waves in μs (was carried out preliminarily on three $7 \times 7 \times 7$ cm cubic samples for all lithotypes), in each of the three orthogonal directions (A, B, and C) to evaluate possible anisotropies in the stones.

Modica and Ragusa samples are quite similar. They are classified as packstone or biomicrite, according to Dunham (1962) and Folk (1959) [15, 16] classifications, respectively. These stones are characterized by micrite (40%). The allochemical component (40%) consists of fragments of echinoderms, coral algae, bryozoa, and bivalves, other than foraminiferous debris. The porosity (20%) is both intragranular and intergranular types.

The Comiso samples show organogenous texture with microsparitic and sparitic cement (45%). The allochemical component (40%) is characterized by foraminifera and echinoderms fragments. The porosity (15%) is mainly intergranular, subordinately of the moldic type. According to Dunham (1962) and Folk (1959) [15, 16] it is classified as packstone or biosparite, respectively.

POM highlighted the minero-petrographic differences and analogies between the three lithotypes suggesting how the porosity and the high heterogeneity of the components (e.g. typology and percentages of allochems, micrite, sparite) could represent predisposing factors for the development of degradation phenomena.

The determination of the ultrasonic speed [17] was carried out preliminarily on one sample representative for every lithotype to verify the presence of probable variations related to the anisotropies inside the materials. Considering it, in the next phases the number of samples to analyse will be increased: at least three samples for lithotype.

The data reported in the histogram (Fig. 3) indicate the values of the time for propagation in directions A B and C referring to the three different types of calcarenite. The calcarenite of Ragusa appears to have less anisotropic behaviour in the three different directions, compared to those of Comiso and Modica. The time for propagation measured alongside A of the Comiso stone is lower than that recorded for sides B and C. While the time for propagation measured alongside A and side B of the Modica stone is lower than that recorded alongside C.

Fig. 2. Thin section microphotographs of 1) Modica 2) Ragusa and 3) Comiso stones

The different values measured in the three directions suggested the presence, especially in Comiso and Modica samples, of anisotropies potentially responsible for the different behaviour of the stones in the degradation processes.

Anisotropies found through ultrasonic velocities indicate that the stones have a non-uniform internal structure, with properties that vary in different directions. These differences in structure can influence the vulnerability of stones to degradation processes. Some directions may be more sensitive to weather or mechanical stress than others.

The synthesized membrane was analysed by FTIR to assess how the presence of limonene within the membrane varies over time.

From the interpretation of the peaks compared with the IR spectrum of PLA (Fig. 4), it emerged that the limonene molecule is not present. Limonene is a volatile compound so it may have gone away, or the FTIR may not be sensitive enough to be able to identify it. Another type of analysis, nuclear magnetic resonance spectroscopy, will have to be carried out for further information.

NMR analysis will be carried out in the later stages of the experimentation to provide a more precise evaluation of the presence of the limonene molecule within the membrane

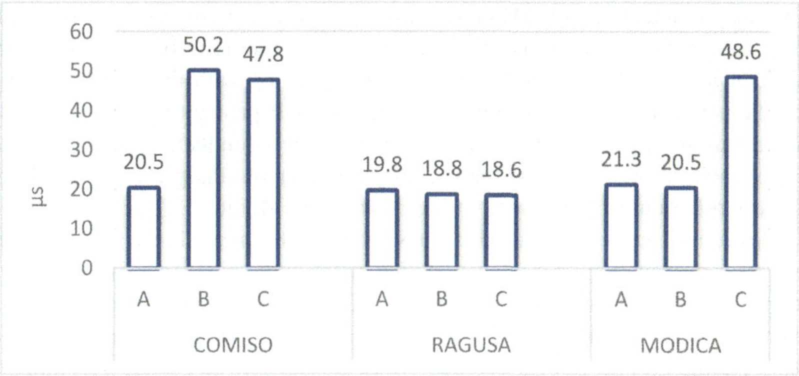

Fig. 3. Histogram with values of ultrasound propagation in the directions A B and C referred to three limestones.

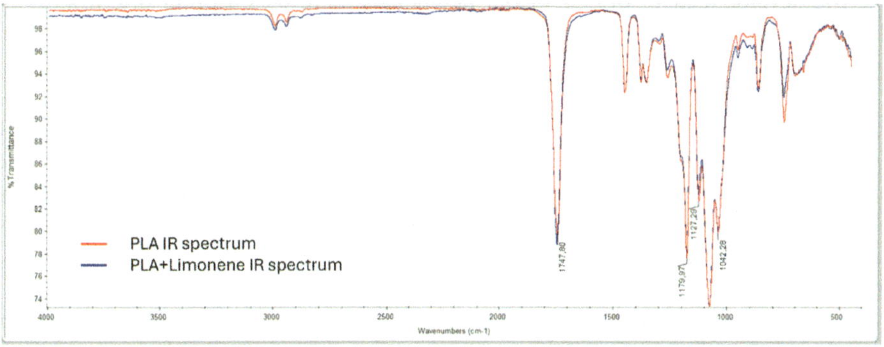

Fig. 4. IR spectra of PLA and PLA + Limonene compared.

networks. By comparing the FTIR results and those will be obtained by the NMR, it will be possible to understand if the limonene molecule was released from the membrane meshes or maintained. In this way, considering that the goal of these membranes is the release of the biocidal product, different concentrations of limonene could be tested to evaluate their efficiency.

5 Conclusions

The analysis of physical and chemical characteristics, coupled with a detailed understanding of deterioration processes, are crucial for the development of highly effective conservation strategies in the stone materials in built cultural heritage.

The present paper deals with two complementary topics: a first characterization of three Sicilian lithotypes used in the monuments of the Sicilian Baroque in Noto Valley: Comiso, Ragusa, and Modica limestones; and a first approach to the synthesis of biopolymer membranes based on PLA and limonene to be tested on these stones subject to biological deterioration.

The study conducted by POM emphasized the contrasting mineralogical and petrographic characteristics among the three lithotypes, highlighting both their similarities and differences. POM suggested that the porosity, along with the variability in constituent elements, might serve as influential factors predisposing these lithotypes to various degradation forms over time. The sonic velocity test conducted on a representative sample of each lithotype provided initial information about the presence of anisotropy particularly pronounced in the samples of Comiso and Modica.

These preliminary results provided clues about the differential behaviour of the three stones versus degradation processes, to define better in the next steps of the research.

In the future to get more information about anisotropies, the ultrasonic velocity will be determined in additional cubic samples of the three different types of stone. As the next steps, samples of the three lithotypes will be subjected to further tests for mineralogical, physical, and mechanical characterization, such as x-ray diffraction (XRD), water absorption by capillarity and immersion tests, mercury intrusion porosimetry (MIP), salt crystallization test, uniaxial compressive strength, flexural strength, and elastic modulus.

The development of biopolymeric membranes represents a promising frontier in the field of cultural heritage preservation by the exploitation of natural biocides within controlled-release systems.

The synthesized membrane underwent FTIR analysis to investigate the temporal evolution of limonene presence within the membrane. Observing the peaks of the IR spectrum, it was noted that the limonene molecule was absent. Considering the volatile nature of limonene, it is plausible that it may have evaporated over time, or the sensitivity of FTIR might not suffice for its detection. Consequently, additional analytical techniques such as nuclear magnetic resonance spectroscopy will be useful to discover further insights.

The next stages of the work involve the preparation of membranes containing various concentrations of limonene and the preparation of membranes with other active natural molecules. UV aging tests will be conducted to assess the strength and stability of these membranes when exposed to sunlight and ultraviolet radiation over time. Biological tests will be carried out in a controlled laboratory environment to examine the biocidal efficacy of the membranes. These tests will be performed both on the single membrane and on the membrane applied to the samples.

This approach will allow an accurate evaluation of the potential of membranes treated with active molecules against microbial growth on stone materials under conditions similar to those in reality.

Finally, after laboratory tests, the membranes will be tested directly on the sacrificial surfaces of the buildings to evaluate their biocidal capability.

Disclosure of Interests. The authors have no competing interests to declare that are relevant to the content of this article.

References

1. Cirone, M., et al.: Innovative methodologies for the conservation of cultural heritage against biodeterioration: a review. Coatings **13**, 1986 (2023)

2. Scrano, L., et al.: Natural biocides and bio-calcite: innovative tools for cultural heritage. IOP Conference Series: Materials Science and Engineering **949**, 012096 (2020)
3. Cardiano, P., Sergi, S., Triscari, M., Piraino, P.: Conservation studies on ornamental and building stones of north-eastern Sicily. Geomineralogical and porosimetric investigations. Annali di chimica **91**, 41–50 (2001)
4. Belfiore, C.M., Calabrò, C., Ruffolo, S.A., Ricca, M., Török, Á., Pezzino, A., La Russa, M.F.: The susceptibility to degradation of stone materials used in the built heritage of the Ortygia Island (Italy): a laboratory study. Int. J. Rock Mech. Mining Sci. **146**, 104877 (2021)
5. La Russa, M.F., Barone, G., Belfiore, C.M., Mazzoleni, P., Pezzino, A.: Application of protective products to "calcarenite (south-eastern Sicily): a case study for the conservation of stone materials. Environ. Earth Sci. **62**, 1263–1272 (2011)
6. Veneranda, M., Blanco-Zubiaguirre, L., Roselli, G., Di Girolami, G., Castro, K., Madariaga, J.M.: Evaluating the exploitability of several essential oils constituents as a novel biological treatment against cultural heritage biocolonization. Microchem. J. **138**, 1–6 (2018)
7. Sasso, S., Miller, S., Rogerio-Cadelera, M., Cubero, B., Coutinho, M., Scrano, L.: Potential of natural biocides for biocontrolling phototrophic colonization on limestone. Int. Biodeterior. Biodegradation **107**, 10–110 (2016)
8. Liakos, I.L., et al.: Polylactic acid-lemongrass essential oil nanocapsules with antimicrobial properties. Pharmaceuticals **9**, 42 (2016)
9. Llana-Ruiz-Cabello, M., Pichardo, S., Bermúdez, J.M., Baños, A., Núñez, C., Guillamón, E., Aucejo, S., Cameán, A.M.: Development of PLA films containing oregano essential oil (Origanum vulgare L. virens) intended for use in food packaging. Food Additives & Contaminants: Part A **33**, 1374–1386 (2016)
10. Barbera, G., et al.: A multi-technique approach for the determination of the porous structure of building stone. Eur. J. Mineral. **26**, 189–198 (2014)
11. Barone, G., La Russa, M.F., Lo Giudice, A., Mazzoleni, P., Pezzino, A.: The Cathedral of S. Giorgio in Ragusa Ibla (Italy) characterization of construction materials and their chromatic alteration. Environ. Earth Sci. **55**, 499–504 (2008)
12. Anania, L., Badala', A., Barone, G., Belfiore, C.M., Calabrò, C., La Russa, M.F., Mazzoleni, P., Pezzino, A.: The stones in monumental masonry buildings of the Val di Noto area: new data on the relationships between petrographic characters and physical, mechanical properties. Constr. Build. Mater. **33**, 122–132 (2012)
13. La Russa, M.F., et al.: The behaviour to weathering of the Hyblean limestone in the Baroque architecture of the Val di Noto (SE Sicily): An experimental study on the "calcare a lumachella" stone. Constr. Build. Mater. **77**, 7–19 (2015)
14. Arrieta, M.P., López, J., Ferrándiz, S., Peltzer, M.A.: Characterization of PLA-limonene blends for food packaging applications. Polym. Testing **32**, 760–768 (2013)
15. Dunham, R.J.: Classification of carbonate rocks according to depositional texture. In: Ham, W.E. (ed.) Classification of carbonate rocks Am Assoc Petrol Geol Memo, 108–121 (1962)
16. Folk, R.L.: Practical petrographic classification of limestone. Am. Assoc. Petrol Geol. Bull. **43**, 1–38 (1959)
17. UNI EN 14579. Metodi di prova per pietre naturali - Determinazione della velocità di propagazione del suono, Milano (2005)

Low-Cost Underwater Location System for Heritage Preservation

Jorge Herrera-Santos[1(✉)], Marta Plaza-Hernández[1], Vladimir Djapic[2], and Javier Prieto-Tejedor[1]

[1] BISITE Research Group, University of Salamanca, Edificio Multiusos I+D+i, Calle Espejo 2, 37007 Salamanca, Spain
{jorgehsmp,martaplaza,javierp}@usal.es
[2] H2O Robotics Ltd, Unska 3, 10000 Zagreb, Croatia
vladimir@h2o-robotics.com

Abstract. The proper conservation of underwater heritage is essential to honouring our past and understanding our history and culture. However, the underwater environment poses certain challenges, such as limited visibility, the absence of GPS signals, and the complexity of ocean currents, which complicate precise submerged localisation. Acoustic communication emerges as the predominant solution due to the effectiveness with which sound travels underwater. Nevertheless, developing accurate and reliable acoustic technologies has been significantly challenging due to cost constraints and technical complexity. This article presents the development of a low-cost buoy system prototype for underwater location. It aims to overcome the economic and technical limitations that have restricted the use of advanced technologies in preserving underwater heritage. By integrating cost-effective acoustic communication, this system seeks to provide an effective and accessible solution for the precise signalling of points of interest and the localisation of divers and remotely operated vehicles, boosting conservation and underwater exploration efforts. This work details the proposed architecture and the results obtained, discussing future improvements and research directions.

Keywords: Underwater cultural heritage · Underwater location · Internet of Underwater Things · Buoy system

1 Introduction

Throughout the centuries, numerous historical and cultural treasures have been lost in the depths of the sea. The documentation, conservation, and proper preservation of this heritage are of vital importance not only to honor our past but also to understand our history and culture [1]. However, the unique nature of the underwater environment presents many challenges [2].

In this context, underwater localisation emerges as a critical component during conservation operations. The ability to accurately identify and signal underwater points of interest not only facilitates research and documentation efforts

G. Marreiros et al. (Eds.): DCAI 2024, LNNS 1151, pp. 184–192, 2025.
https://doi.org/10.1007/978-3-031-80946-0_18

but is also essential for the safe coordination of conservation teams [3]. These challenges are amplified by the inherent limitations of the underwater environment, where reduced visibility, the absence of GPS signals, and the complexity of ocean currents demand specialised solutions.

Precise underwater localisation has traditionally been considered a challenge. The propagation of radio waves, which form the basis of most surface localisation systems, is limited underwater, requiring the search for viable alternatives. Acoustic communication emerges as the predominant solution, leveraging the effectiveness with which sound travels in aquatic environments [4]. However, developing precise and reliable acoustic technologies has represented a significant challenge, especially considering cost constraints and technical complexity.

To address this requirement, low-cost and small-sized devices such as the NMv3 modem have emerged. This modem enables communication and distance calculation between transmitter and receiver in underwater environments through acoustic technology [5]. This technology, in turn, paves the way for the developing world of the Internet of Underwater Things (IoUT), which attempts to extend the ideas behind the Internet of Things (IoT) to the submerged environment. This adaptation proposes a network of interconnected devices underwater, capable of exchanging data and conducting real-time measurements, representing a significant advancement in the monitoring and conservation of aquatic ecosystems and submerged heritage [6].

Applying the IoUT in underwater heritage conservation projects opens up a range of possibilities for precise identification, detailed documentation, and adequate protection of sites of historical and cultural significance [7].

In response to these challenges, this work proposes the development of a low-cost buoy system prototype for underwater localisation designed to overcome the economic and technical barriers that have restricted the use of advanced technologies in the preservation of underwater heritage. By integrating cost-effective acoustic communication, the system seeks to provide an effective and accessible solution for the precise signalling of points of interest and the localisation of divers, contributing significantly to conservation and underwater exploration efforts. This approach not only promises to improve the efficiency and safety of underwater operations but also represents a step forward in democratising access to advanced conservation technologies, opening new possibilities for the exploration and protection of our rich submerged heritage.

The present work is structured as follows: Sect. 2 presents the proposed architecture, detailing the system's elements. Section 3 introduces the results obtained. Section 4 presents the main conclusions and discusses the future improvements and lines of work.

2 Proposed System's Architecture

The prototype system proposed for underwater localisation has been designed in accordance with the specifications of the TEchnological Consortium TO develop sustaiNabIlity of underwater Cultural heritage (TECTONIC) project, aiming to

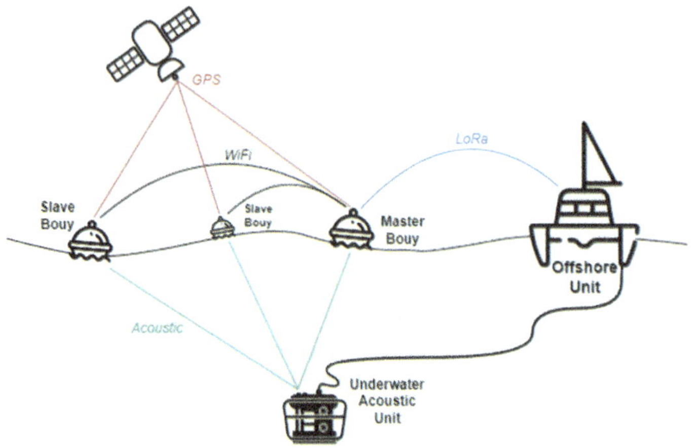

Fig. 1. Overview of the proposed low-cost underwater location system for heritage preservation, its key elements and how they communicate.

provide a reliable solution for determining the absolute position of submerged objects, such as Remotely Operated Vehicles (ROV), divers, scanners, and sensors, among others. This solution is structured around four main components, each performing specific roles within the system (Fig. 1):

- **Underwater acoustic unit (UAU).** This element is the target whose exact position is desired. Although the initial design contemplates a submarine ROV, the system is adaptable to any object that can incorporate a compatible acoustic transceiver. This component is fundamental for the collection of position data and relative distances underwater.
- **Master buoy.** This element acts as the system's coordinator, orchestrating distance measurements between itself, the slave buoys, and the underwater unit. This buoy creates a WiFi access point for communication between buoys and uses the MQTT protocol [8] for information exchange, acting as an MQTT broker. Additionally, it is equipped with a Long-Range Radio (LoRa) [9] module to send long-distance data to the boat/shore unit, including the GPS location of each buoy and the measured distances.
- **Slave buoys.** These buoys connect to the WiFi access point provided by the master buoy and operate as MQTT clients. Their function is to carry out distance measurements with the underwater unit and obtain their own GPS location under the instruction of the master buoy.
- **Offshore unit.** Located on the coast or in a catamaran, this unit consists of a LoRa gateway for long-distance communication and a computer for data processing. It is responsible for receiving and analysing the information sent by the master buoy, including locations and distances, to calculate the exact position of the underwater unit.

2.1 Network and Software Architecture

The network architecture of the system is displayed in Fig. 2. It is divided into two different blocks. The first one is the surface communication between the shore/boat unit, the master buoy, and the slave buoys. The second one is the underwater communication between the three buoys and the underwater acoustic unit.

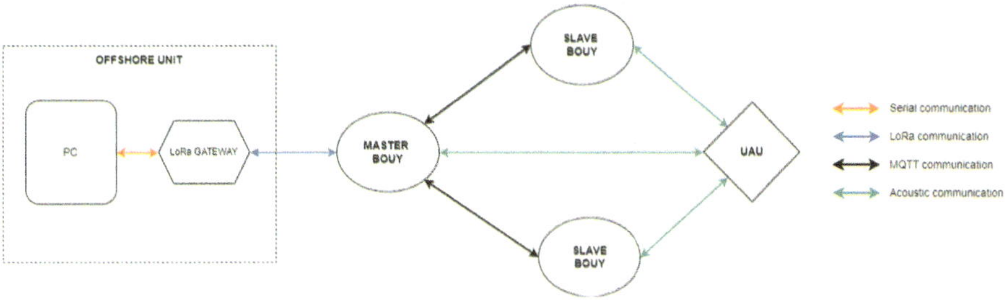

Fig. 2. Proposed network communication architecture.

Communication Between the Buoys and the UAU. The system consists of three buoys, divided into one master buoy and two slave buoys. The master buoy is the central transfer node, channelling all information between the offshore unit and the buoys. This master buoy establishes a WiFi access point for the connection of the slave buoys. Upon receiving a ping request from the offshore unit, the master buoy initiates a sequential process. First, it attempts to communicate and calculate the distance with the underwater unit. If successful, it then asks the first slave buoy to do the same. After receiving distance information from the first slave buoy, the request is passed to the second slave buoy. Once all ranges have been calculated and collected by the master buoy, it compiles the data into a package and sends it to the offshore unit. The communication between the master buoy and the slave buoys is carried out through the MQTT messaging protocol, allowing for an efficient and structured exchange of information in this project.

Communication Between the Offshore Unit and the Master Buoy. From the computer, a script periodically generates a message in the form "$G, <NUM>, RNG", where <NUM> corresponds to the three-digit identifier of the target underwater unit. This message, called a ping request, is sent through the LoRa gateway. The LoRa message carries both a receiver identifier to whom the message is directed and a sender identifier. Thus, if the device receiving the message does not have the identifier indicated in the message, it is ignored.

When the master buoy receives the data from each buoy, the message with all the information is concatenated and sent via LoRa to the shore/boat unit. The same script that generates the ping request message is responsible for logging the received data, which arrives as:

<time> B#, RNG, <range>, <lat>, <lon>, <GPS_e>

where:

- time: date and time formatted as dd/mm/yyyy hh:mm:ss.
- B#: buoy to which the message belongs (B1: master, B2: slave 1, B3: slave 2).
- RNG: range correctly calculated. Otherwise, it shows LOG.
- range: distance between the buoy and the underwater acoustic unit, in meters.
- lat: latitude, in decimal degrees.
- lon: longitude, in decimal degrees.
- GPS_e: error, in centimeters.

2.2 Electronic Architecture

The elements that comprise the electronic architecture of the buoys are presented in Fig. 3.

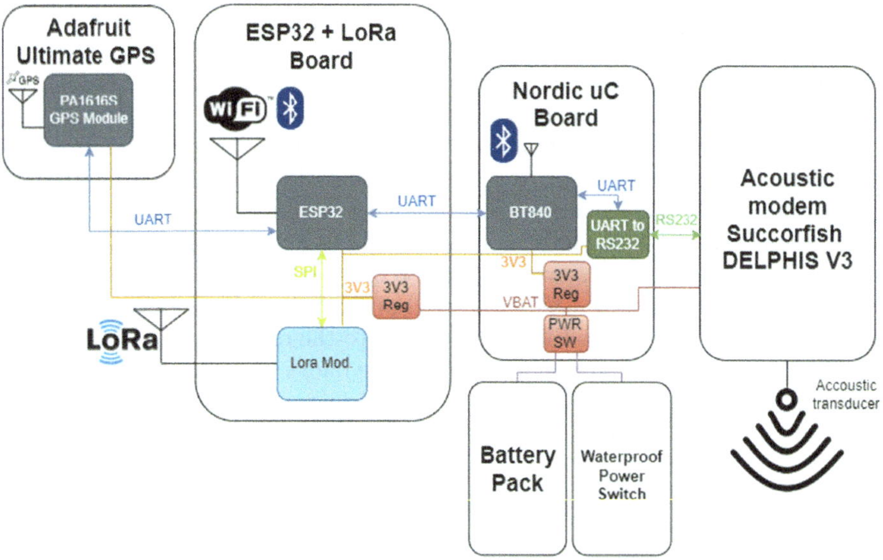

Fig. 3. Proposed electronic architecture.

The Succorfish Delphis V3 modem, along with an acoustic transducer, are essential for facilitating underwater communication, taking charge of generating and analysing the necessary acoustic signals for this purpose. This modem acts as a gateway between serial communication and acoustic communication. For serial communication with an external device (the Nordic PCB), it has an RS232 interface, ensuring the transmitted data's integrity, effectively protecting them from noise and possible interference.

The Nordic PCB, developed by the research group LABUST [10], integrates the Nordic nRF52840 microcontroller and is responsible for the communication between the acoustic modem and the ESP32 + LoRa board, as well as for the voltage regulation for the electronic components. The ESP32 + LoRa board, from e-radionica.com (now Soldered Electronics [11]), features an ESP32-WROOM-32D microcontroller that supports WiFi, Bluetooth, and has an embedded LoRa module for long-distance transmissions. Communication with the GPS module and with the Nordic microcontroller board is carried out via UART serial buses and with the LoRa module through SPI.

An Adafruit Ultimate GPS module [12] is included, providing location, speed, and time data in real-time through satellite tracking. It communicates with the ESP32 microcontroller via UART.

For a robust and reliable power supply, Li-Ion 18650 battery systems are used in parallel, achieving a total capacity of 7500 mAh and 3. 7V. This ensures uninterrupted operation, even in challenging conditions. Additionally, a waterproof power switch from Blue Robotics adds a layer of control without compromising the system's waterproofness.

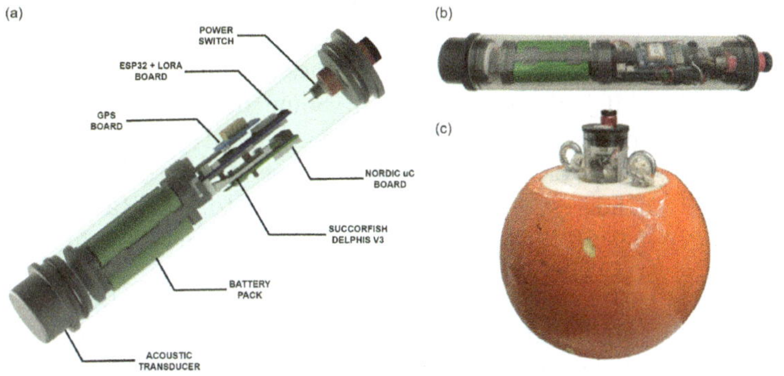

Fig. 4. 3D design of the buoy components (a), assembled buoy components (b), functional prototype of the buoy (c).

This set of electronic components, encapsulated in the buoys (Fig. 4) has been crucial in developing and validating the prototype.

3 Preliminary Results

The proposed system, still in the development phase, has been subjected to preliminary tests essential for its validation. Initially, it was tested in the indoor pool of the LABUST research group. However, the acoustic characteristics of the environment, particularly the reflections caused by the concrete in a confined space [13], prevented the results from being satisfactory.

To overcome these challenges and test the system in more realistic conditions, field tests were conducted in Lake Jarun in Zagreb, Croatia, on August 24, 2023. During these tests, buoys were placed randomly, and the UAU, whose trajectory was steered from a boat, was operated. These field tests allowed crucial data to be collected on the system's performance. The success/failure results are summarised in Table 1.

Table 1. Success and failure rate of ranging.

Message received	Nºmessages	% total
Successful	26	32,50%
Partially Successful	25	31,25%
Failure due to MQTT communication	14	17,50%
Failure due to LoRa communication/other	15	18,75%

The results were classified into the following categories:

- Successful: The range measurement request was successfully answered by all three buoys.
- Partially successful: The measurement request failed for one or more buoys.
- Failure due to MQTT communication: Some of the slave buoys did not respond to the request.
- Failure due to LoRa communication/other type of failure: The measurement request received no response. This may be due to a LoRa communication failure, an unexpected reset of the master buoy, or some other firmware error.

Below, one of the successful messages sent by the buoy system is analysed:
24/08/2023 12:23:19 B1,RNG,4.22,45.779338836669918,15.936838150024414,81
24/08/2023 12:23:19 B2,RNG,10.1,45.779228210449218,15.936708450317382,81
24/08/2023 12:23:19 B3,RNG,1.88,45.779338836669918,15.936790466308594,81

To obtain a graphical representation of the results, the data was processed using a spherical trilateration algorithm, in which each of the obtained points is accurately represented in space based on the known distances between the buoys, taking one of them as the origin. Each point, corresponding to the location of the buoys, is overlaid with a sphere that symbolises the distance to the UAU. The radius of these spheres equals the sum of the distance between the buoy and the UAU plus the GPS error margin and the error in the acoustic range measurement, thus offering a detailed three-dimensional representation of the position and the uncertainty associated with each measurement (Fig. 5).

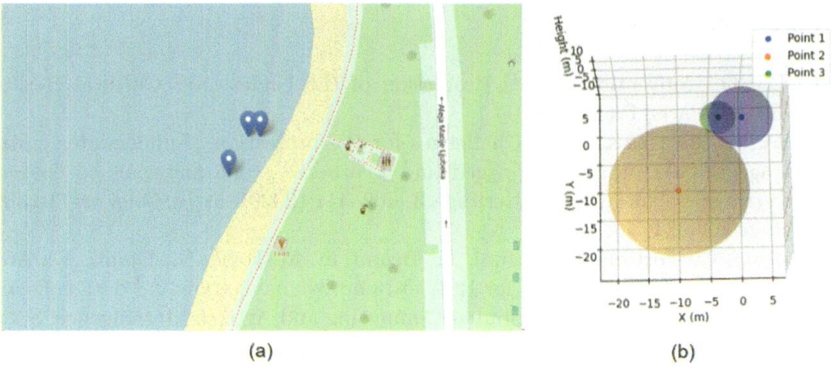

Fig. 5. Location of the buoys (a). Data processed using the trilateration algorithm (b).

4 Conclusions and Future Work

This article presented the architecture for a low-cost underwater localization prototype for the documentation and preservation of underwater cultural heritage; in the framework of the MSCA-RISE H2020 TECTONIC project.

The preliminary results obtained in the field tests suggest that the system has potential for underwater localisation, despite facing challenges such as acoustic communication in restricted environments, the complexity of integrating different communication technologies, and technical limitations during field testing. Experimentation in Lake Jarun has provided valuable data that demonstrates both successes and areas for improvement, reflecting a success rate in communication that encourages optimism regarding future implementations and refinements of the system.

Future work will focus on improving electronic integration, communication between the system's components, greater precision in localisation technologies, and expanding tests to other underwater environments to validate the robustness and adaptability of the prototype. The development and implementation of a specific algorithm for the automatic representation of the UAU's position and the exploration of new potential applications of the system represent clear lines of future work, with the ultimate goal of contributing substantially to the conservation and exploration of underwater heritage globally.

Acknowledgments. This research has been supported by the project "Technological Consortium TO develop sustainability of underwater Cultural heritage (TECTONIC)", financed by the European Union (Horizon 2020 research and innovation programme under the Marie Skłodowska-Curie grant agreement No. 873132).

Disclosure of Interests. The authors have no competing interests to declare that are relevant to the content of this article.

References

1. UNESCO. Convention on the Protection of the Underwater Cultural Heritage; UNESCO: Paris, France (2001)
2. Ricca, M., La Russa, M.F.: Challenges for the protection of underwater cultural heritage (UCH), from waterlogged and weathered stone materials to conservation strategies: an overview. Heritage **3**, 402–411 (2020). https://doi.org/10.3390/heritage3020024
3. Severino, U., Fortuna, S., Lagudi, A., Bruno, F., Mišković, N., Djapic, V.: Architecture of a low-cost solution for ROVs to improve navigation and data collection. In: Mehmood, R., et al. Distributed Computing and Artificial Intelligence, Special Sessions I, 20th International Conference. DCAI 2023. Lecture Notes in Networks and Systems, vol. 741, pp. 214–223. Springer, Cham (2023). https://doi.org/10.1007/978-3-031-38318-2_22
4. Su, X., Ullah, I., Liu, X., Choi, D.: A review of underwater localization techniques, algorithms, and challenges. J. Sens. **2020**, 1–24 (2020). https://doi.org/10.1155/2020/6403161
5. Sherlock, B., Morozs, N., Neasham, J., Mitchell, P.: Ultra-low-cost and ultra-low-power, miniature acoustic modems using multipath tolerant spread-spectrum techniques. Electronics **11**, 1446 (2022). https://doi.org/10.3390/electronics11091446
6. Domingo, M.C.: An overview of the internet of underwater things. J. Netw. Comput. Appl. **35**(6), 1879–1890 (2012). https://doi.org/10.1016/j.jnca.2012.07.012
7. Plaza-Hernández, M., Abbasi, M., Mezquita, Y.: An IoUT-based platform for managing underwater cultural heritage. In: Machado, J.M., et al. Distributed Computing and Artificial Intelligence, Special Sessions, 19th International Conference. DCAI 2022. LNNS, vol. 58, 127–1365. Springer, Cham (2023). https://doi.org/10.1007/978-3-031-23210-7_12
8. MQTT website. https://mqtt.org/. Accessed 1 Apr 2024
9. LoRa website. https://lora-alliance.org/about-lorawan/. Accessed 1 Apr 2024
10. LABUST website. https://labust.fer.hr/. Accessed 1 Apr 2024
11. Soldered Electronics website. https://soldered.com/. Accessed 1 Apr 2024
12. Adafruit website. https://www.adafruit.com/product/746. Accessed 1 Apr 2024
13. Wada, A., Yoshizawa, S.: Transmission of depth data by pulse position modulation for underwater acoustic positioning systems. NAUN Int. J. Circuits Syst. Signal Process. **16**, 999–1006 (2022). https://doi.org/10.46300/9106.2022.16.122

DCAI - SS-AItHeC

Special Session on Artificial Intelligence Trends in Health and Care (AItHeC)

Scope

The Special Session on Artificial Intelligence Trends in Health & Care (AItHeC 2024) aims to gather researchers from diverse backgrounds, including academia, industry, government, hospitals, clinics, or medical centers to exchange knowledge and innovative visions on the development of AI solutions targeted towards healthcare and well-being. AItHeC 2024 will be an event for the presentation and discussion of the latest trends in applying artificial intelligence methods and technologies within the domains of medicine, healthcare, and overall well-being. The session is structured to cover a comprehensive spectrum, addressing both theoretical and practical facets, encompassing technologies and systems relevant to the application of artificial intelligence.

Topics

Topics to be highlighted cover different AI applications within the context of healthcare and wellbeing:

AI in Medical Education
AI in Clinical Management
Wellbeing and Lifestyle Support
Psychological Interventions
Interoperability, Security, Pervasiveness, Ubiquity and Cloud Computing in Medicine
Data Privacy, Legal, Ethics and Social Issues in Medicine
Pervasive Healthcare Environments
Persuasive and Behavioral Sciences
Ontologies and Standards in the Healthcare Domain
Software Architectures for Healthcare Systems
AI-based Coaching Systems for Health and Well-being
Business Intelligence and Health Analytics
Clinical Data Mining
Machine Learning and Deep Learning
Data Streaming
Disease Diagnosis, Forecasting and Evaluation
Expert, agent-based or knowledge-based systems for Health and Well-being
Electronic Health Records (eHealth)
Mobile Health (mHealth)
Wearable Devices and Telemedicine
Sensor-based Applications
Usability, Acceptability and other Evaluation Metrics
Public Health Informatics
Profiling and Clustering Methods
Patient Journey Mapping

Organizing Committee

Luís Conceição	School of Engineering, Polytechnic of Porto, Portugal
Diogo Martinho	School of Engineering, Polytechnic of Porto, Portugal
Liepa Bikulčienė	Kaunas University of Technology, Department of Applied Mathematics, Lithuania
Ziya Karakaya	Konya Food and Agriculture University, Department of Software/Computer Engineering, Turkey

Non-Ergodic Theory vs. Fractal Geometry in Organizational Learning and Dynamic Skills at a Healthcare Food Catering Service

Rui Lima[1,2,3], Goreti Marreiros[4], Florentino Fdez-Riverola[1,2], Henrique Vicente[5,6], and José Neves[3,6](✉)

[1] CINBIO, Universidade de Vigo, Department of Computer Science, ESEI – Escuela Superior de Ingeniería Informática, 32004 Ourense, Spain
rui.lima@cespu.pt, riverola@uvigo.es

[2] SING Research Group, Galicia Sur Health Research Institute (IIS Galicia Sur), SERGAS-UVIGO, Vigo, Spain

[3] Instituto Politécnico de Saúde do Norte, CESPU, Famalicão, Paredes, Portugal
jneves@di.uminho.pt

[4] Departamento de Engenharia Informática, Instituto Superior de Engenharia do Porto, Porto, Portugal
goreti@dei.isep.ipp.pt

[5] Departamento de Química e Bioquímica, Escola de Ciências e Tecnologia, REQUIMTE/LAQV, Universidade de Évora, Évora, Portugal
hvicente@uevora.pt

[6] Centro Algoritmi/LASI, Universidade do Minho, Braga, Portugal

Abstract. This paper examines the adoption of non-Ergodic Theory and Fractal Geometry to enhance organizational learning in healthcare food catering. It critiques conventional linear models, promoting an understanding of varied learning processes across organizational strata. Non-Ergodic Theory emphasizes the diversity of learning among individuals and groups, while Fractal Geometry examines learning as self-similar patterns at multiple scales. This work analyses how employee diversity, knowledge sharing, and cross-level collaboration contribute to a fractal learning environment that supports adaptability and innovation. Findings indicate that implementing these theoretical models can enhance service delivery and customer approval by promoting a culture of relentless education and creative ideas. For managers and practitioners, this study provides a blueprint for enhancing learning processes that can lead to greater efficiency and performance in healthcare food services, suggesting a strategic embrace of employee diversity and interconnected learning patterns for long-term organizational success.

Keywords: Non-Ergodic Theory · Fractal Geometry · Healthcare Food Catering Services · Heterogeneity · Service Quality · Performance · Efficiency · Customer Satisfaction

G. Marreiros et al. (Eds.): DCAI 2024, LNNS 1151, pp. 197–209, 2025.
https://doi.org/10.1007/978-3-031-80946-0_19

1 Introduction

This paper investigates the applications of non-Ergodic Theory and Fractal Geometry in relation to organizational learning and dynamic skills within a healthcare food catering service. The study aims to address the limitations of traditional theories in capturing the complex and non-linear nature of organizational processes. Non-Ergodic Theory, as discussed by [1] and [2], challenges the assumption that organizations can be understood through average behavior analysis over time. Instead, it emphasizes the importance of recognizing the heterogeneity and diversity within organizations, where different individuals or subgroups may have unique learning processes, as noted by [3, 4]. In the context of a healthcare food catering service, this theory suggests that employees possess diverse skills, experiences, and knowledge that contribute to overall organizational learning, a concept explored by [5] and [6]. Fractal Geometry provides a mathematical framework for studying complex and self-replicating patterns, which can be seen in the work [7, 8]. Applied to organizational learning, it suggests that learning processes occur at multiple scales within the organization, ranging from individuals to teams, departments, and the organization as a whole, a perspective also discussed by [9].

The fractal nature of learning indicates that patterns observed at smaller scales may also be present at larger scales, highlighting the interconnectedness and interdependence of learning processes. Combining non-Ergodic Theory and Fractal Geometry, this study, similar to that of [10], aims to explore how organizational learning and dynamic skills can be fostered within a healthcare food catering service. The research involves analyzing employee diversity, knowledge sharing mechanisms, cross-functional collaborations, and the impact of learning at different levels of the organization, which [11] found to be crucial for skill acquisition in health services catering. Furthermore, the study examines how the fractal nature of learning processes contributes to adaptive and innovative behaviors within the organization, as illustrated by the works of [12] and [13]. The findings of this research offer valuable insights for managers and practitioners seeking to enhance learning processes and promote continuous improvement within a healthcare food catering service, embracing the concept of diversity-driven learning as studied by [14]. Recognizing and embracing the heterogeneity of employee skills and knowledge can create an environment that encourages diverse perspectives, creativity, and learning, as evidenced by [15]. Additionally, understanding the fractal nature of learning processes allows organizations to identify patterns and feedback loops that occur at different levels, facilitating the dissemination of knowledge and skills throughout the organization, a concept that is also central to the research by [16]. The practical application of non-Ergodic Theory and Fractal Geometry can lead to improved performance, efficiency, and service quality in healthcare food catering services. Leveraging the diverse expertise of employees and fostering a culture of continuous learning enables organizations to adapt to changing customer needs, innovate their offerings, and enhance overall customer satisfaction, a principle supported by [17] and [18]. In conclusion, this study explores the potential of non-Ergodic Theory and Fractal Geometry in enhancing organizational learning and dynamic skills within a healthcare food catering service. Recognizing the heterogeneity of employee knowledge and skills, along with understanding the interconnectedness of learning processes at different scales, enables organizations to foster a culture of continuous improvement and innovation. The insights gained from this research can guide managers and practitioners in developing strategies that optimize learning processes and contribute to the long-term success of healthcare food catering services.

2 Background

2.1 The Applications of Non-Ergodic Theory and Fractal Geometry in a Healthcare Food Catering Service

This paper delves into the applications of non-Ergodic Theory and Fractal Geometry within the context of organizational learning and dynamic skills in a healthcare food catering service. Traditional theories have their limitations when it comes to capturing the intricate and nonlinear nature of organizational processes, as highlighted by [19] As such, this study, drawing upon the findings of [20], aims to shed light on alternative perspectives that provide a more comprehensive understanding of organizational dynamics. Non-Ergodic Theory, as expounded by [21], challenges the conventional assumption that organizations can be comprehended by analyzing average behavior over time. Instead, this theory, alongside insights from [22], highlights the significance of acknowledging the heterogeneity and diversity within organizations, wherein different individuals or subgroups exhibit unique learning processes. In the realm of a healthcare food catering service, this theory suggests, as observed by [23], that employees possess diverse skills, experiences, and knowledge that contribute to the overall organizational learning process. Fractal Geometry, as explored by [24] and further developed by [25], offers a mathematical framework for studying complex and self-replicating patterns. When applied to organizational learning, it suggests, echoing the work of [26], that learning processes occur at various scales within the organization, ranging from individuals to teams, departments, and the organization. The fractal nature of learning, as discussed in the research by [27], implies that patterns observed at smaller scales may also be present at larger scales, thereby emphasizing the interconnectedness and interdependence of learning processes. By combining non-Ergodic Theory and Fractal Geometry, this study, inspired by the concepts presented by [28], seeks to explore how organizational learning and dynamic skills can be nurtured within a healthcare food catering service. The research endeavors to analyze employee diversity, knowledge sharing mechanisms, cross-functional collaborations, and the impact of learning at different levels of the organization. Moreover, the study examines how the fractal nature of learning processes contributes to adaptive and innovative behaviors within the organization, a theme central to the studies by [28].

2.2 The Confluence of Logic Programming into the Applications of Non-Ergodic Theory and Fractal Geometry

The impact of LP in improving hospitality management is profound, but it must be guided by a commitment to public welfare and ethical standards to truly realize its potential in fostering sustainable and inclusive patient development. This convergence, when viewed through the lens of computational contexts, is now positioned as a template of a Model of a LP program, as shown by, viz.

$$
\begin{aligned}
&\{\\
&\quad \neg p \leftarrow not\ p, not\ exception_p\\
&\quad p \leftarrow p_1, \cdots, p_n, not\ q_1, \cdots, not\ q_m\\
&\quad ?\,(p_1, \cdots, p_n, not\ q_1, \cdots, not\ q_m)\ \ (n, m \geq 0)\\
&\quad exception_{p_1},\ \cdots\ , exception_{p_j}\ \ (0 \leq j \leq k),\ \ being\ k\ an\ integer\ number\\
&\}
\end{aligned}
$$

Program 1. An Exemplary Model of a Logic Program.

The initial segment indicates the finality of a predicate; the comma "," signifies a "*logical and*", and the question mark "*?*" represents a domain atom that signifies "*falsity*". The terms p_i, q_j, and p refer to conventional ground literals, that is, either positive atoms or atoms negated by the classical negation symbol $\neg$ [29]. Notably, $\neg$ embodies strong negation, whereas "not" implies negation due to failure to prove a statement because of its absence in explicit declaration. Within this framework, it is conceivable for a collection of abducibles or exceptions to be integrated into any given program [30]. Conversely, clauses formatted as follows, viz.

$$?(p_1, \cdots, p_n, not\ q_1, \cdots, not\ q_m)(n, m \geq 0)$$

serve as invariants that enable the clarification of the context in which the relevant universe of discourse should be interpreted.

2.3 Entropy's Influence on Achieving Efficiency and Sustainability in Hospitality

To clarify the core concepts of the proposed approach, we refer to the fundamental Thermodynamics Laws, considering the evolution of a system's condition. The First Law, the principle of Energy Conservation, states that energy in an isolated system is constant, implying it can change form but cannot be created or destroyed. The Second Law introduces the concept of Entropy, a measure of disorder and its increase over time. These principles are analogous to the Knowledge Representation and Reasoning (KRR) strategies followed in this work, suggesting a process of energy degradation, not elimination, and set as follow, viz.

- The term "Exergy" denotes the energy a system can use following an energy exchange, a reflection of the system's entropy (shown in Fig. 1 by dark regions);
- The term "Vagueness" refers to the energies that might have been spent (represented in Fig. 1 by grey regions); and
- The term "Anergy" is the unused potential energy that remains available, encompassing all non-exergy and vagueness energies (depicted in Fig. 1 by blank regions).

3 Case Study

In terms of practical applications, non-Ergodic Theory and Fractal Geometry can enhance organizational learning and dynamic skills within a healthcare food catering service in several ways. Firstly, non-Ergodic Theory can be applied to analyze and understand the diverse skills, experiences, and knowledge of employees within the service. This understanding can inform the design of training programs, knowledge sharing mechanisms, and talent management strategies that leverage the heterogeneity of employee expertise. Additionally, Fractal Geometry can be utilized to identify patterns and feedback loops in learning processes at different scales, ranging from individual employees to teams, departments, and the organization. This knowledge can guide the implementation of effective communication channels, cross-functional collaborations, and learning initiatives that promote continuous improvement and innovation throughout the service. By combining these theories, organizations can create a culture of continuous learning, foster collaboration across diverse teams, and encourage the exchange of ideas and knowledge. This, in turn, promotes adaptive behaviors, enhances problem-solving capabilities, and drives innovation within the healthcare food catering service. Continuous improvement efforts can be informed by recognizing the fractal nature of learning processes and identifying areas where successful learning patterns can be replicated or scaled up to improve service quality, efficiency, and custom Accounting for Environmental and Social Factors in Decision-Making, in connection with the recently addressed understanding and sentiments of the public, signifies the evaluation of the social atmosphere where choices are made, factoring in the interaction with peers, societal standards, cultural, and well-being effects. For instance, to procure comprehensive perspectives, the population was invited to convey their insights, emotions, and convictions about incorporating Artificial Intelligence, Logic Programming, and the principle of Entropy into the progression of methods and methodologies to address problem activities in Smart Cities (SC). This was achieved through tailored surveys, inspired by questionnaires but altered to include aspects essential to understanding the entropic's concept [31]. This modification resulted in the creation of a Likert-type scale for the assessment of survey answers, viz.

Strongly Agree | Agree | Disagree | Strongly Disagree | Disagree | Agree | Strongly Agree

In addition, it contains a neutral term, *vagueness*, that points to *neither agree nor disagree*, which stands for *unsure or vague*. The questionnaire was administered to a sample of 52 individuals, aged between 24 and 63, at a healthcare institution in northern Portugal. In terms of questionnaires, one may have, viz.

Questionnaire on *Ergodic Theory and Assessment Questions – Three elements (ETAQ – 3).*

I1 –Do you agree that traditional theories adequately capture the complexity and nonlinear nature of organizational processes?

I2 –Are you familiar with the concept of non-Ergodic Theory?

I3 –Do you agree that learning processes occur at multiple scales within an organization?

Questionnaire on *Non-Ergodic and Fractal Approaches to Organizational Development – Three elements (OD – 3).*

Y1 –Do you agree that it is significant to recognize and embrace the heterogeneity and diversity within organizations, particularly in the context of a healthcare food catering service?

Y2 –Are you familiar with the concept of Fractal Geometry and its application to organizational learning?

Y3 –Do you agree that combining non-Ergodic Theory and Fractal Geometry will foster organizational learning and dynamic skills within a healthcare food catering service?

Questionnaire on *Organizational Dynamics and Alternative Theories Evaluation – Three elements (ODATE – 3).*

Y4 –Is it important for organizations to explore alternative perspectives, such as non-Ergodic Theory and Fractal Geometry, to gain a comprehensive understanding of organizational dynamics?

Y5 –Do you agree that learning processes occurring at multiple scales within an organization can have an impact on its overall performance and effectiveness?

Y6 –Do you see for non-Ergodic Theory and Fractal Geometry in enhancing organizational learning and dynamic skills within a healthcare food catering service?

leading to Table 1 and the graphical representation of the citizens answers (Fig. 1).

Table 1. A diagrammatic synthesis of a foods' manager survey answers to the *ETAQ – 3, OD – 3, and ODATE – 3* questionnaires.

Questionnaires	Itens	*Scale*							
		(4)	(3)	(2)	(1)	(2)	(3)	(4)	*vagueness*
ETAQ – 3	*I1*		×	×					
	I2								×
	I3						×		
OD – 3	*Y1*							×	
	Y2						×	×	
	Y3	×							
ODATE – 3	*Y4*								×
	Y5						×		
	Y6			×					

Table 2. A manager's perspective on the Entropic State of a food service as reflected in their answers to *ETAQ – 3* questionnaire.

	$ETAQ - 3_{4\text{–}1}$ – *Scale (4) (3) (2) (1)*	$ETAQ - 3_{1\text{–}4}$ – *Scale (1) (2) (3) (4)*
Y1	$exergy_{I1} = \frac{1}{3}\pi r^2\Big]_0^{\frac{2}{4}\sqrt{\frac{1}{\pi}}} = = \frac{1}{3}\pi\left(\frac{2}{4}\sqrt{\frac{1}{\pi}}\right)^2 - 0 = 0.08$	–
	$vagueness_{I1} = \frac{1}{3}\pi r^2\Big]_{\frac{2}{4}\sqrt{\frac{1}{\pi}}}^{\frac{2}{4}\sqrt{\frac{1}{\pi}}} = 0$	–
	$anergy_{I1} = \frac{1}{3}\pi r^2\Big]_{\frac{2}{4}\sqrt{\frac{1}{\pi}}}^{\sqrt{\frac{1}{\pi}}} = 0.25$	–
Y2	$exergy_{I2} = \frac{1}{3}\pi r^2\Big]_0^0 = 0$	–
	$vagueness_{I2} = \frac{1}{3}\pi r^2\Big]_{\sqrt{\frac{1}{\pi}}}^{\sqrt{\frac{1}{\pi}}} = 0$	–
	$anergy_{I2} = \frac{1}{3}\pi r^2\Big]_0^{\sqrt{\frac{1}{\pi}}} = 0.33$	–
Y3	–	$exergy_{I3} = -\frac{1}{3}\pi r^2\Big]_{\frac{2}{4}\sqrt{\frac{1}{\pi}}}^0 = 0.08$
	–	$vagueness_{I3} = -\frac{1}{3}\pi r^2\Big]_{\frac{2}{4}\sqrt{\frac{1}{\pi}}}^{\frac{2}{4}\sqrt{\frac{1}{\pi}}} = 0$
	–	$anergy_{I3} = -\frac{1}{3}\pi r^2\Big]_{\sqrt{\frac{1}{\pi}}}^{\frac{2}{4}\sqrt{\frac{1}{\pi}}} = 0.25$

The procedure outlined above also applies to the subsequent questionnaires in Table 3.

4 Assessing a Food's Manager Degree-of-Satisfaction (DoS) Through PROLOG

A PROLOG program will now be generated to carry out the following tasks, viz.

- *To establish a set of answers along with their respective entropic values;*
- *To determine the managers' DoS as a function of entropy;*
- *To compute the manager's Degree-of-Efficiency score;*
- *To derive entropic values from given responses by factoring in Exergy, Vagueness, and Anergy;*

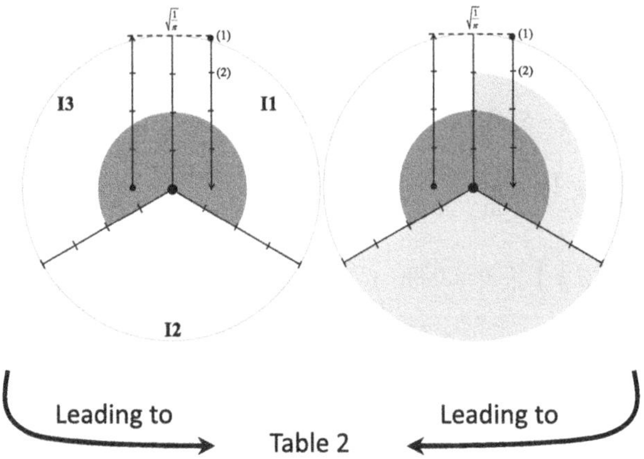

Fig. 1. A schematic portrayal of a citizen's survey data from the *ETAQ – 3* questionnaire.

Table 3. Analysis of statements perceived as outcomes derived from standardized questionnaires and construed as expansions of categorical assertions.

Predicates	*EX*	*VA*	*AN*	*Predicates*	*EX*	*VA*	*AN*
$etaq - 3_{4-1}$	0.08	0	0.58	$etaq - 3_{1-4}$	0.08	0	0.25
$od - 3_{4-1}$	0,02	0	0.31	$od- 3_{1-4}$	0.04	0	0.63
$odate - 3_{4-1}$	0.19	0	0.48	$odate - 3_{1-4}$	0.08	0	0.25
catch-all-clause	0.09	0	0.46	*catch-all-clause*	0.07	0	0.38

- *To organize various questionnaires and assigns specific weights to each statement within them;*
- *To record a manager's individual answers to the questionnaire items;*
- *To calculate the entropic scores for a manager's answers, adjusting them according to the significance of each statement;*
- *To compute the total weighted entropy for a manager's answers across a questionnaire; and*
- *To introduce a rule to evaluate efficiency grounded on the aggregated weighted entropy.*

% Define a predicate to store questionnaire answers, viz.
*% **answer(Questionnaire, CaseType, Item, ScaleSide, Exergy, Vagueness, Anergy),** where **ScaleSide** denoted as ('+') indicates that the entropy increases and ('-') indicates it decreases when one moves from left to right on the scale.*
% Answer predicate's extension, viz.

answer(etaq, bestcase, I1, +1, 0.08, 0, 0.58).
answer(etaq, bestcase, I1, -1, 0.08, 0, 0.25).
...
answer(od, bestcase, y1, +4, 0.02, 0, 0.31).
answer(od, bestcase,, y1, -4, 0.04, 0, 0.63).
...
answer(odate, bestcase, y4, +7, 0.19, 0, 0.48).
answer(odate, bestcase, y4, -7, 0.08, 0, 0.25).
...

% Define the enhanced Likert scale range, viz.
likert_scale(Questionnaire, BCS,+ 4, 0.25). % where the fourth argument indicates the ***adjusted score*** based on predetermined weights.
likert_scale Questionnaire, BCS, +3, 0.15).
likert_scale Questionnaire, BCS, +2, 0.35).
likert_scale(Questionnaire, BCS, +1, 0.25).
likert_scale(Questionnaire, WCS, -1, 0.15).
likert_scale(Questionnaire, W,CS, -2, 0.33).
likert_scale(Questionnaire, WCS, -3, 0.27).
likert_scale(Questionnaire, WCS, -4, 0.25).

% Compute the entropic values based on the answers, viz.
entropic_value(Questionnaire, Casetype, Item, ItemScore, Entropy) :-
answer(Questionnaire, CaseType, Item, ScaleSide, Exergy, Vagueness, Anergy),
likert_scale(Questionnaire, CaseType, ScaleSide, ScaleSideScore),
*Entropy is (Anergy - Exergy - Vagueness) * ScaleSideScore.*

% Calculate the overall DoS' score, viz.
degreeofsatisfaction_score(Questionnaire, EntropiesScore) :-
findall(Entropy, entropic_value(Questionnaire, CaseType, Item, ItemScore, Entropy), Entropies),
sum_list(Entropies, EntropiesScore).

% Rules to evaluate efficiency based on the aggregated Entropy's Score.
efficiency(Questionnaire, Score) :-
degreeofsatisfaction_score(Questionnaire, EntropiesScore),
(EntropiesScore >0, print("good", halt;
EntropiesScore <= 0, Print ("bad").

% Query to compute the overall managers' DoS score for the questionnaires, viz.
?- degreeofsatisfaction_score(_, Score).

% Query to compute the manager's Degree-of-Efficiency score with regarding the 'OD' questionnaire items, viz.
?- efficiency(upm, Score).

5 Results Analysis

The innovative approach of combining Non-Ergodic Theory and Fractal Geometry in the study of organizational learning offers a refreshing perspective, especially in a specialized field like healthcare food catering. Traditional models often fall short, as they tend to oversimplify the complex dynamics of learning within organizations. Non-Ergodic Theory challenges this by asserting that average behavior over time isn't representative of the whole; instead, it's the unique behaviors and learning processes of individuals and groups that truly define an organization's learning capacity. In a healthcare food catering service, this theory prompts us to recognize that every employee brings distinct skills and experiences to the table. This diversity is a valuable resource for organizational learning, suggesting that each employee's learning journey contributes to the organization's collective knowledge. Rather than enforcing a one-size-fits-all training program, a more personalized approach that values individual experiences and learning styles would be far more effective. Fractal Geometry complements this by offering a lens to view learning as a pattern that repeats across different scales within an organization. Just as fractals repeat their complex patterns at every scale, so can the processes of learning and problem-solving manifest similarly across individuals, teams, and the organization at large. This pattern recognition can help managers identify successful learning behaviors that can be encouraged and replicated throughout the organization, thus reinforcing effective practices, and fostering innovation. Practically, this dual-theory approach leads to actionable strategies. Training programs can be tailored to the unique learning needs of employees, and knowledge sharing becomes a structured yet flexible process, reflective of fractal patterns. Cross-functional collaborations can break down silos, allowing for the free flow of ideas and innovations. Recognizing the diverse skills within the workforce means that talent management becomes a strategic tool, not just for improving individual performance but for enhancing the organization's adaptability and problem-solving capabilities. By embracing the diversity of employee backgrounds, healthcare food catering services can create a continuous learning culture, which is vital in a sector where customer satisfaction is directly impacted by the quality of service. Implementing these theories could lead to a dynamic work environment where each employee's potential is maximized, and innovation is a natural outcome of the organization's learning processes. Managers are encouraged to see the organization as a living, learning entity that is complex and non-linear, much like the fractal patterns it seeks to emulate. In doing so, they can foster a workspace that is not just efficient but also creative and responsive to the changing needs of its clients. Ultimately, by adopting these theories, a healthcare food catering service can aim to achieve superior service quality, greater efficiency, and higher levels of customer satisfaction. The merging of Non-Ergodic Theory and Fractal Geometry into the operational and strategic framework of healthcare food services not only promises improvements in performance but also provides a pathway for organizations to navigate the complexities of the modern business landscape. This approach does not merely adjust current practices; it revolutionizes the very core of organizational learning and development.

6 Conclusions and Future Work

The exploration of Non-Ergodic Theory and Fractal Geometry within the sphere of organizational learning presents an exciting frontier in healthcare food catering services. This innovative approach delineates the shift from average behavior analysis to a nuanced understanding of individual and subgroup dynamics. It propels us into a realm where the intricate fabric of learning is recognized for its diversity, complexity, and fractal nature. Each employee's unique contribution to the organizational knowledge pool underscores the critical role of heterogeneity in fostering a robust and dynamic learning environment. In a sector where the pace of change is relentless and customer expectations continually evolve, the application of these theoretical frameworks can significantly enhance service quality and satisfaction. By aligning learning processes with the fractal patterns evident in organizational structures and dynamics, managers can craft training programs and knowledge sharing initiatives that resonate at multiple levels, from the individual to the collective. This ensures that learning is not only preserved but also propagated effectively, allowing for innovative solutions and improvements in service delivery to emerge organically. However, the journey does not end here. Future work could focus on expanding the empirical base of this study, exploring the practical implications of the theoretical constructs in diverse settings within the healthcare industry and beyond. Longitudinal studies could unravel how the incorporation of these theories shapes organizational culture over time and which specific elements contribute most significantly to sustained performance and customer satisfaction. Another avenue for future research is the application of advanced analytics and machine learning algorithms to mine data on learning processes within organizations. Such technologies could potentially reveal hidden patterns in employee behavior and learning outcomes, providing deeper insights into the fractal nature of organizational learning. Moreover, they could assist in personalizing learning and development initiatives, aligning them even more closely with individual employee needs and organizational goals. There's also an opportunity to explore the intersection of these theories with other dimensions of organizational theory and practice, such as leadership styles, decision-making processes, and innovation management. The interaction between non-linear learning processes and these aspects of organizational behavior could yield rich insights into the mechanics of organizational excellence in complex service sectors. Cross-disciplinary collaboration could also enrich future research. Insights from psychology, sociology, and even neuroscience could illuminate the human elements of organizational learning, offering a more holistic view of how non-Ergodic Theory and Fractal Geometry manifest in practice. Such collaboration could also lead to the development of new tools and methodologies for assessing and enhancing learning within organizations. As organizations increasingly recognize the importance of sustainability and social responsibility, future work could also explore how Non-Ergodic Theory and Fractal Geometry contribute to these domains. Understanding how learning processes adapt to and incorporate principles of sustainability could be pivotal in ensuring the long-term viability of organizations while fulfilling their ethical and social obligations. The integration of these innovative theories within the strategic and operational framework of organizations promises not just incremental improvements but a revolutionary rethinking of the essence of organizational learning. The path forward is ripe for investigation, and as the research community delves deeper into these concepts, the potential for transformative change in how organizations learn and evolve is immense. The task ahead is to continue to unravel the complex tapestry of learning processes, to innovate, to adapt, and to thrive in the ever-changing landscape of healthcare services and beyond. The primary limitation concerns the sample size, constrained by its regional scope. Furthermore, acquiring further data on demographic characteristics would allow a deeper analysis.

Acknowledgments. This work has been supported by FCT – Fundação para a Ciência e Tecnologia within the R&D Units Project Scope: UIDB/00319/2020.

References

1. Vanhoyweghen, A., Ginis, V.: Human decision-making in a non-ergodic additive environment. Proc. Roy. Soc. A **479**, 20230544 (2023)
2. West, B.J.: The fractal tapestry of life: iii multifractals entail the fractional calculus. Fractal Fractional **6**, 225 (2022)
3. Massouti, A., Al-Rashaida, M., Alhosani, M.: A qualitative study on dubai's inclusive education policy from school leaders' perspectives. Sustainability **16**, 1252 (2024)
4. Alshaabani, A., Hamza, K., Rudnák, I.: Impact of diversity management on employees' engagement: the role of organizational trust and job insecurity. Sustainability **14**, 420 (2022)
5. Alderwick, H., Hutchings, A., Briggs, A., Mays, N.: The impacts of collaboration between local health care and non-health care organizations and factors shaping how they work: a systematic review of reviews. BMC Public Health **21**, 753 (2021)
6. Alrasheed, A., Connerton, P., Alshammari, G., Connerton, I.: Cohort study on the food safety knowledge among food services employees in Saudi Arabia state hospitals. J. King Saud Univ.-Sci. **33**, 101500 (2021)
7. Oprić, D., et al.: Fractal analysis tools for early assessment of liver inflammation induced by chronic consumption of linseed, palm and sunflower oils. Biomed. Signal Process. Control **61**, 101959 (2020)
8. Rian, I., Asayama, S.: Computational design of a nature-inspired architectural structure using the concepts of self-similar and random fractals. Autom. Constr. **66**, 43–58 (2016)
9. Sawatzky, R., et al.: The Response Shift–in Sync Working Group: Implications of response shift for micro-, meso-, and macrolevel healthcare decision-making using results of patient-reported outcome measures. Qual. Life Res. **30**, 3343–3347 (2021)
10. Halfon, N., Forrest, C.B.: The Emerging Theoretical Framework of Life Course Health Development. In: Halfon, N., Forrest, C., Lerner, R., Faustman, E. (eds.) Handbook of Life Course Health Development, pp. 19–43. Springer, Cham (2018)
11. Rosander, M., Forslund-Frykedal, K., Barimani, M., Berlin, A.: Experiences from leading parental education groups: perceived difficulties and rewards as an indication of skill acquisition. J. Child Health Care **26**, 68–81 (2022)
12. Forouharfar, A.: The anatomy and ontology of organizational power as a fractal metaphor: a philosophical approach. Cogent Bus. Manage. **7**, 1–26 (2020)
13. Gu, G., Cai, Y.-F., Sun, K., Zhao, T.F.: Equity and spatial accessibility of healthcare resources in China: a fractal analysis of the online health community network. Front. Phys. **11**, 1336624 (2024)
14. Turner, J.R., Baker, R., Morris, M.: Complex adaptive systems: adapting and managing teams and team conflict. In: Vilas-Boas, A.A. (ed.) Organizational Conflict, pp. 65–94. IntechOpen (2018)

15. Kauffman, S.: Is there a fourth law for non-ergodic systems that do work to construct their expanding phase space? Entropy **24**, 1383 (2022)
16. Al-Sayid, F., Kirkil, G.: Exploring non-linear relationships between perceived interactivity or interface design and acceptance of collaborative web-based learning. Educ. Inf. Technol. **28**, 11819–11866 (2023)
17. Rosing, K., Frese, M., Bausch, A.: Explaining the heterogeneity of the leadership-innovation relationship: ambidextrous leadership. Leadersh. Q. **22**, 956–974 (2011)
18. Ahmed, N.S., Yasin, N.M.: Inspiring a fractal approach in distributed healthcare information systems: a review. Int. J. Phys. Sci. **5**, 1626–1640 (2010)
19. North, D.C.: Uncertainty in a Non-ergodic World. In: Understanding the Process of Economic Change, Princeton Economic History of the Western World, chapter 2, pp. 13–22. Princeton University Press, Princeton (2005)
20. Neves, J., et al.: Entropy and Organizational Performance. In: Pérez García, H., Sánchez González, L., Castejón Limas, M., Quintián Pardo, H., Corchado Rodríguez, E. (eds.) Hybrid Artificial Intelligent Systems. Lecture Notes in Computer Science, vol. 11734, pp. 206–217. Springer, Cham (2019)
21. Kelty-Stephen, D.G., Mangalam, M.: Fractal and multifractal descriptors restore ergodicity broken by non-Gaussianity in time series. Chaos, Solitons Fractals **163**, 112568 (2022)
22. Huo, M., Zhao, B., Li, Y., Li, J.: Evidence-based practice dynamic capabilities: a concept derivation and analysis. Annal. Transl. Med. **10**(1) (2022)
23. Kauffman, S.: Is there a fourth law for non-ergodic systems that do work to construct their expanding phase space? Entropy **24**, 1383 (2022). https://doi.org/10.3390/e24101383
24. Ahmadi, D.: Is diversity our strength? an analysis of the facts and fancies of diversity in Toronto. City Cult. Soc. **13**, 64–72 (2018)
25. Jahanmiri, F., Parker, D.C.: An overview of fractal geometry applied to urban planning. Land **11**, 475 (2022)
26. Taylor, R.P.: The potential of biophilic fractal designs to promote health and performance: a review of experiments and applications. Sustainability **13**, 823 (2021)
27. Bliese, P.D., Schepker, D.J., Essman, S.M., Ployhart, R.E.: Bridging methodological divides between macro- and microresearch: endogeneity and methods for panel data. J. Manag. **46**(1), 70–99 (2020)
28. Hasgall, A., Ahituv, N.: Implementing continuous adaptation to technology innovation in complex adaptive organizations. J. High Technol. Managem. Res. **29**, 35–45 (2018)
29. Neves, J.: A logic interpreter to handle time and negation in logic databases. In: Muller, R., Pottmyer, J. (eds.) Proceedings of the 1984 annual conference of the ACM on the 5th Generation Challenge, pp. 50–54. ACM, New York (1984)
30. Kakas, A., Kowalski, R., Toni, F.: The role of abduction in logic programming. In: Gabbay, D., Hogger, C., Robinson, I. (eds.) Handbook of Logic in Artificial Intelligence and Logic Programming, vol. 5, pp. 235–324. Oxford University Press, Oxford (1998)
31. Wenterodt, T., Herwig, H.: The entropic potential concept: a new way to look at energy transfer operations. Entropy **16**, 2071–2084 (2014)

Advancing Smart Urban Futures in the Healthcare Sector

Vitor Alves[1,2,3], Goreti Marreiros[4], Florentino Fdez-Riverola[2,3], José Neves[5,6], Jorge Ribeiro[1], and Henrique Vicente[5,7](✉)

[1] ADiT-LAB, Instituto Politécnico de Viana do Castelo, Viana do Castelo, Portugal
{vitoralves,jribeiro}@estg.ipvc.pt, riverola@uvigo.es
[2] Department of Computer Science, ESEI – Escuela Superior de Ingeniería Informática, CINBIO, Universidade de Vigo, 32004 Ourense, Spain
[3] SING Research Group, Galicia Sur Health Research Institute (IIS Galicia Sur), SERGAS-UVIGO, Vigo, Spain
[4] Departamento de Engenharia Informática, Instituto Superior de Engenharia do Porto, Porto, Portugal
goreti@dei.isep.ipp.pt
[5] Centro Algoritmi/LASI, Universidade do Minho, Braga, Portugal
jneves@di.uminho.pt, hvicente@uevora.pt
[6] Instituto Politécnico de Saúde do Norte, CESPU, Famalicão, Portugal
[7] Departamento de Química e Bioquímica, Escola de Ciências e Tecnologia, REQUIMTE/LAQV, Universidade de Évora, Evora, Portugal

Abstract. As urban development stands on the brink of a new epoch, the imperative exploration of artificial intelligence, logic programming, and entropy demands visionary planning, ethical conduct, and a deep appreciation of their far-reaching impact when synergized within smart city frameworks. This study delves into the complex relationship between artificial intelligence, logic programming, and entropy, highlighting their pivotal role in advancing smart cities towards operational excellence and evolutionary success. By harnessing artificial intelligence's understanding of human emotions and sentiments, urban centers are redefining city living. Employing cutting-edge technologies not only empowers healthcare innovation but also improves city planning, public safety, transit systems, and environmental management, significantly enhancing the quality of life for citizens. The key finding of this study underscores the importance of addressing challenges inherent in such integration, particularly in healthcare, leading to a high degree of citizens' wellbeing.

Keywords: Artificial Intelligence · Logic Programming · Entropy · Smart Cities · Urban Life · Sentiments and Feelings · Healthcare · Citizens Wellbeing

1 Introduction

As we navigate the 21st century, Smart Cities (SC) emerge as bastions of advancement, epitomizing the fusion of high-tech prowess and the enrichment of urban existence [1]. Central to this metropolitan transformation is Artificial Intelligence (AI) – the catalyst propelling the extensive development and sophisticated functionality within these

G. Marreiros et al. (Eds.): DCAI 2024, LNNS 1151, pp. 210–221, 2025.
https://doi.org/10.1007/978-3-031-80946-0_20

urban landscapes [2]. Our examination is centered on the intricate and profound influence of AI in molding SC, particularly its assimilation into the urban fabric and its revolutionary impacts on facets such as urban planning [3], public safety [4], transportation [5], and healthcare innovation [6]. The advent of AI in urban domains signifies a paradigm shift toward ecosystems that are more responsive, insightful, and enduring [7]. It is revolutionizing our engagement with urban spaces, fostering the synthesis of logic programming and dynamic learning mechanisms that are crucial in navigating the complexities of city living [8]. These innovations strive for a harmonious balance that elevates both operational efficiency and the standard of living, aiming particularly to empower citizens with enhanced access to healthcare services [1]. This paper endeavors to elucidate the methodologies that underpin this harmonization [9], disseminate the knowledge acquired, and contemplate the extensive implications of our discoveries [10]. We explore the multifarious nature of AI – from its foundation in logic programming to its potential in addressing urban entropy – and its instrumental role in engendering a more intelligent, habitable urban milieu [11]. At this critical juncture, it becomes our collective duty to steer this tech-centric progress with deliberate planning and ethical consideration [3], ensuring that the dividends of AI extend to every urban inhabitant [12]. We have utilized an enhanced Likert-type scale questionnaire to collate and interpret community opinions regarding AI's penetration into various urban sectors [13], intending to capture the emotions and sentiments of those experiencing the transition to AI-enhanced SCs, thereby generating actionable insights for scholars, policymakers, urban planners, and technologists. By attuning to the public's sentiment, we can tailor the implementation of AI in SCs to conform to ethical norms and societal anticipations [6], charting a course for a future where technology is woven seamlessly into the social fabric for collective benefit [2]. As we stand at the dawn of a transformative epoch, it is imperative to navigate this tech-enabled journey with strategic vision and moral responsibility [3], ensuring that AI's bounty profoundly enriches the everyday lives of city dwellers and paves the way for a synergy of technology and humanity that promises widespread prosperity and healthful living.

2 Background

This work explores the synergistic application of Artificial Intelligence (AI), Logic Programming (LP), and Entropy in the context of Smart Cities (SC). It aims to provide a comprehensive guide on leveraging these technologies to enhance urban living, optimize city operations, and address complex urban challenges such as the provision of healthcare to citizens. The work underscores the potential of AI and LP in data analysis, decision-making, and automating complex processes, while Entropy is considered for its role in system efficiency and sustainability.

2.1 The Confluence of AI and Logic Programming in Urban Management

The synergy of AI and LP in SC paves the way for a more responsive and resilient urban environment. AI's ability to learn and adapt, combined with LP's structured approach to problem-solving, creates a robust platform for managing urban complexity. This collaboration enables cities to move from reactive to proactive management, where predictive analytics can lead to preemptive action, significantly reducing costs and enhancing urban services. However, the integration of AI and LP in urban management does not come without challenges. Issues of privacy, security, and ethical use of AI are at the forefront of public concern. Moreover, ensuring that the benefits of such technologies are equitable across different demographics remains a pressing issue. The future of SC lies in the responsible deployment of AI and LP. As these technologies continue to evolve, so must the governance frameworks that ensure their alignment with societal values and norms. The impact of AI and LP in improving SC management is profound, but it must be guided by a commitment to public welfare and ethical standards to truly realize its potential in fostering sustainable and inclusive urban development. This convergence, when viewed through the lens of computational contexts, is now positioned as a template of logic programming, as shown by, viz.

$$
\begin{array}{l}
\{ \\
\quad \neg p \leftarrow not\ p, not\ exception_p \\
\quad p \leftarrow p_1, \cdots, p_n, not\ q_1, \cdots, not\ q_m \\
\quad ?\,(p_1, \cdots, p_n, not\ q_1, \cdots, not\ q_m)\ \ (n, m \geq 0) \\
\quad exception_{p_1}, \ \cdots \ , exception_{p_j}\ \ (0 \leq j \leq k),\ \ being\ k\ an\ integer\ number \\
\}
\end{array}
$$

Program 1. An Exemplary Model of a Logic Program.

The initial segment indicates the finality of a predicate; the comma "," signifies a "*logical and*", and the question mark "*?*" represents a domain atom that signifies "*falsity*". The terms p_i, q_j, and p refer to conventional ground literals, that is, either positive atoms or atoms negated by the classical negation symbol $\neg$ [14]. Notably, $\neg$ embodies strong negation, whereas "not" implies negation due to failure to prove a statement because of its absence in explicit declaration. Within this framework, it is conceivable for a collection of abducibles or exceptions to be integrated into any given program [14]. Conversely, clauses formatted as follows, viz.

$$?(p_1, \cdots, p_n, notq_1, \cdots, notq_m)(n, m \geq 0)$$

serve as invariants that enable the clarification of the context in which the relevant universe of discourse should be interpreted.

2.2 Entropy's Influence on Achieving Efficiency and Sustainability in Metropolitan Management

The concept of entropy, traditionally associated with thermodynamics and information theory, has found a novel application in the realm of SC. In the context of SC management, Entropy is a measure of disorder or randomness within a system, but paradoxically, understanding and managing entropy is central to creating more organized, efficient, and sustainable urban environments. Urban systems are inherently complex and often chaotic, with countless interactions occurring simultaneously across various sectors such as traffic, energy, and waste management. In this bustling dynamism, Entropy can serve as a measure of system inefficiency. High levels of Entropy can indicate a need for better organization and streamlined processes. Conversely, a reduction in Entropy is synonymous with increased order and improved system performance. By monitoring Entropy levels, SC can optimize operations and reduce waste, leading to a more sustainable urban ecosystem. Undeniably, the impact of entropy on improving SC management is profound. It provides a framework for assessing system efficiency and guides the implementation of sustainability measures. While challenges remain, the potential of entropy-informed strategies to enhance the livability and resilience of SC is immense. As SC continue to evolve, embracing entropy as a tool for better management can lead to cities that are not only smart but also sustainable and adaptable to the ever-changing needs of their inhabitants [15]. To clarify the core concepts of the proposed approach, we refer to the fundamental Thermodynamics Laws, considering the evolution of a system's condition. The First Law, the principle of Energy Conservation, states that energy in an isolated system is constant, implying it can change form but cannot be created or destroyed. The Second Law introduces the concept of Entropy, a measure of disorder and its increase over time. These principles are analogous to the Knowledge Representation and Reasoning (KRR) strategies followed in this work, suggesting a process of energy degradation, not elimination, and set as follow, viz.

- The term "Exergy" denotes the energy a system can use following an energy exchange, a reflection of the system's entropy (This concept is illustrated by the dark regions in Fig. 1 (Sect. 3));
- The term "Vagueness" refers to the energies that might have been spent (represented by the grey regions in Fig. 1 (Sect. 3)); and
- The term "Anergy" is the unused potential energy that remains available, encompassing all non-exergy and vagueness energies (This is depicted by the blank regions in Fig. 1 (Sect. 3)).

2.3 Gauging Public Perception and Sentiment Towards AI, LP, and Entropy in Smart City Evolution and Healthcare Provision

To comprehensively understand the public's grasp and feelings about the adoption of AI, LP, and Entropy in the development of SC, detailed Likert-type scale and the respective questionnaires will be constructed. These questionnaires will initially test the public's knowledge, asking about both fundamental and intricate uses of these up-to-date Mathematical Logic-based methods [16]. Next, individuals will express their level of concurrence with the pros and cons associated with these approaches, offering a view into their overall receptiveness [17]. The survey will then discern the public's opinion on the significance of the role played by each technology in city management, pinpointing which areas might need more focus in future planning [18]. It will solicit the public's stance on matters of ethics, particularly their concerns about privacy and security, to guide the ethical integration of these systems [19]. Additionally, it will examine public opinion on how SC are handling entropy in their pursuit of sustainability and operational efficiency, providing insights into the community's environmental concerns [20]. Lastly, the questionnaire will delve into the perceived hurdles in technology implementation, with a special focus on fair technology distribution and the protection of personal data.

The objective of these all-encompassing questionnaires is to clear up the collective viewpoint regarding the diverse effects that these technological developments have on urban life.

3 Environmental and Social Considerations in Deciding

Accounting for Environmental and Social Factors in Decision-Making, in connection with the recently addressed understanding and sentiments of the public, signifies the evaluation of the social atmosphere where choices are made, factoring in the interaction with peers, societal standards, cultural, and well-being effects. For instance, to procure comprehensive perspectives, the population was invited to convey their insights, emotions, and convictions about incorporating Artificial Intelligence, Logic Programming, and the principle of Entropy into the progression of methods and methodologies to address problem activities in Smart Cities (SC). This was achieved through tailored surveys, inspired by traditional Likert-scale questionnaires but altered to include aspects essential to understanding Entropy. This modification resulted in the creation of a refined metric scale for the assessment of survey answers, viz.

strongly agreeing (4), agreeing (3), disagreeing (2), strongly disagreeing (1), disagreeing (2), agreeing (3), strongly agreeing (4)

In addition, it contains a neutral term, vagueness, that points to neither agree nor disagree, which stands for unsure or vague. The questionnaire was administered to a sample of 47 individuals, aged between 25 and 62, at a healthcare institution in northern Portugal. In terms of questionnaires, one may have, viz.

Questionnaire on *Urban Planning and Management – Three elements (UPM – 3)* encompass the items, viz.

I1 –AI, LP, and Entropy significantly improves the efficiency of urban planning and resource management; experience burnout from my job;

I2 –The use of AI, LP, and Entropy in urban management leads to more accurate and timely decisions; and

I3 –I feel emotionally drained from my job.

Questionnaire on *Healthcare Provision Efficiency – Three elements (HPE – 3)* encompass the items, viz.

I4 – AI, LP, and Entropy significantly improve the efficiency and responsiveness of healthcare services in urban areas;

I5 – The integration of AI, LP, and Entropy in healthcare management allows for better prediction and management of patient flow and resources;

I6 – AI, LP, and Entropy contribute to the effective management of healthcare data, leading to improved patient outcomes and enhanced privacy and security.

Questionnaire on *Environmental Sustainability – Three elements (ES – 3)* encompass the items, viz.

I7 –AI, LP, and Entropy driven energy management systems contribute to significant energy savings in the city;

I8 –The use of AI, LP, and Entropy in monitoring pollution helps in maintaining a healthier urban environment; and

I9 –AI, LP, and Entropy enhances the effectiveness of waste management and recycling processes.

leading to Table 1 and to a graphical representation of the answers (Fig. 1).

Table 1. A diagrammatic synthesis of a citizen survey answers to the *UPM – 3, HPE – 3, and T&TM – 3* questionnaires.

Questionnaires	Itens	*Scale*							
		(4)	(3)	(2)	(1)	(2)	(3)	(4)	*vagueness*
UPM – 3	*I1*		×	×					
	I2								×
	I3						×		
HPE – 3	*I4*							×	
	I5						×	×	
	I6	×							
ES – 3	*I7*								×
	I8						×		
	I9			×					

Table 2. A citizen's perspective on the Entropic State of a Smart City as reflected in a citizen's answers to *UPM – 3* questionnaire.

	$UPM-3_{4\text{–}1}-Scale\ (4)\,(3)\,(2)\,(1)$	$UPM-3_{1\text{–}4}-Scale\ (1)\,(2)\,(3)\,(4)$
I1	$exergy_{I1} = \frac{1}{3}\pi r^2\Big]_0^{\frac{2}{4}\sqrt{\frac{1}{\pi}}} = \; = \frac{1}{3}\pi\left(\frac{2}{4}\sqrt{\frac{1}{\pi}}\right)^2 - 0 = 0.08$	–
	$vagueness_{I1} = \frac{1}{3}\pi r^2\Big]_{\frac{2}{4}\sqrt{\frac{1}{\pi}}}^{\frac{2}{4}\sqrt{\frac{1}{\pi}}} = 0$	–
	$anergy_{I1} = \frac{1}{3}\pi r^2\Big]_{\frac{2}{4}\sqrt{\frac{1}{\pi}}}^{\sqrt{\frac{1}{\pi}}} = 0.25$	–

(continued)

Table 2. (*continued*)

	$UPM - 3_{4\text{–}1}$ – *Scale (4) (3) (2) (1)*	$UPM - 3_{1\text{–}4}$ – *Scale (1) (2) (3) (4)*
I2	$exergy_{I2} = \frac{1}{3}\pi r^2\Big]_0^0 = 0$	–
	$vagueness_{I2} = \frac{1}{3}\pi r^2\Big]_{\sqrt{\frac{1}{\pi}}}^{\sqrt{\frac{1}{\pi}}} = 0$	–
	$anergy_{I2} = \frac{1}{3}\pi r^2\Big]_0^{\sqrt{\frac{1}{\pi}}} = 0.33$	–
I3	–	$exergy_{I3} = -\frac{1}{3}\pi r^2\Big]_{\frac{2}{4}\sqrt{\frac{1}{\pi}}}^0 = 0.08$
	–	$vagueness_{I3} = -\frac{1}{3}\pi r^2\Big]_{\frac{2}{4}\sqrt{\frac{1}{\pi}}}^{\frac{2}{4}\sqrt{\frac{1}{\pi}}} = 0$
	–	$anergy_{I3} = -\frac{1}{3}\pi r^2\Big]_{\sqrt{\frac{1}{\pi}}}^{\frac{2}{4}\sqrt{\frac{1}{\pi}}} = 0.25$

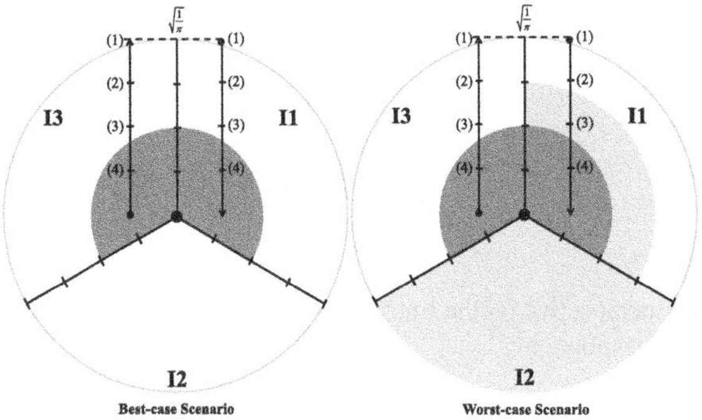

Fig. 1. A schematic portrayal of a citizen's survey data from the *UPM – 3* questionnaire.

The procedure outlined above also applies to the subsequent questionnaires in Table 1, leading to Table 3.

Table 3. Examination of assertions, seen as results obtained from uniform surveys and interpreted as extensions of predicates

Predicates	*EX*	*VA*	*AN*	*Predicates*	*EX*	*VA*	*AN*
$upm - 3_{4\text{–}1}$	0.08	0	0.58	$upm - 3_{1\text{–}4}$	0.08	0	0.25
$hpe - 3_{4\text{–}1}$	0,02	0	0.31	$hpe - 3_{1\text{–}4}$	0.04	0	0.63
$es - 3_{4\text{–}1}$	0.19	0	0.48	$es - 3_{1\text{–}4}$	0.08	0	0.25
catch-all-clause	0.09	0	0.46	*catch-all-clause*	0.07	0	0.38

4 Assessing the Citizens Wellbeing Through PROLOG

Now it is possible to present a PROLOG program to carry out the following tasks, viz.

- *To establishe a set of answers along with their respective entropic values;*
- *To determine the Citizens Wellbeing as a function of entropy;*
- *To derive entropic values from given responses by factoring in Exergy, Vagueness, and Anergy;*
- *To organize various questionnaires and assigns specific weights to each statement within them;*
- *To record a citizen's individual answers to the questionnaire items;*
- *To calculate the entropic scores for a citizen's answers, adjusting them according to the significance of each statement;*
- *To compute the total weighted entropy for a citizenr's answers across a questionnaire;*
- *To introduce a rule to evaluate efficiency grounded on the aggregated weighted entropy;* and
- *To compute the error range and the degree of confidence of any query.*

```
% Define a predicate to store questionnaire answers, viz.
% answer(Questionnaire, CaseType, Item, ScaleSide, Exergy, Vagueness, Anergy),
   where ('+') indicates that the entropy increases and ('-') indicates it decreases
   when one moves from left to right on the scale.
% Answer predicate's extension, viz.
    answer(upm, bestcase, i1, +1, 0.08, 0, 0.58).
    answer(upm, bestcase, i1, -1, 0.08, 0, 0.25).
              ...
    answer(hpe, bestcase, i4, +4, 0.02, 0, 0.31).
    answer(hpe, i4, -4, 0.04, 0, 0.63).
              ...
    answer(es, bestcase, i7, +7, 0.19, 0, 0.48).
    answer(es, bestcase, i7, -7, 0.08, 0, 0.25).
              ...
% Define the enhanced Likert scale range, viz.
   likert_scale(Questionnaire, BCS,+ 4, 0.25).
   likert_scale Questionnaire, BCS, +3, 0.15 ).
   likert_scale Questionnaire, BCS, +2, 0.35).
   likert_scale(Questionnaire, BCS, +1, 0.25).
   likert_scale(Questionnaire, WCS, -1, 0.15 ).
   likert_scale(Questionnaire, W,CS, -2, 0.33 ).
   likert_scale(Questionnaire, WCS, -3, 0.27 ).
   likert_scale(Questionnaire, WCS, -4, 0.25 ).
% Compute the entropic values based on the answers, viz.
entropic_value(Questionnaire, Casetype, Item, ItemScore, Entropy) :-
answer(Questionnaire, CaseType, Item, ScaleSide, Exergy, Vagueness, Anergy),
        likert_scale(Questionnaire, CaseType, ScaleSide, ScaleSideScore),
        Entropy is (Anergy - Exergy - Vagueness) * ScaleSideScore.
% Calculate the overall wellbeing score, viz.
    wellbeing_score(Questionnaire(s), EntropiesScore) :-
         findall(Entropy, entropic_value(Questionnaire, CaseType, Item,
         ItemScore, Entropy), Entropies),
         sum_list(Entropies, EntropiesScore).
% Rules to evaluate efficiency based on the aggregated Entropy's Score.
        efficiency(Questionnaire, Score) :-
            wellbeing_score(Questionnaire, EntropiesScore),
         (EntropiesScore >0, print("Efficienty", halt;
                    EntropiesScore <= 0, Print ("Inefficienty").
            Efficient is Score > 0.
% Query to compute the overall wellbeing's score for the questionnaires, viz.
    ?- wellbeing_score(_, Score).
% Query to compute the efficiency related to the items of questionnaire "upm",
   viz.
    ?- efficiency(upm, Score).
```

5 Results Analysis

The paper outlines the significant role AI plays in the evolution of Smart Cities. The focus is on AI's potential to enhance urban planning, public safety, transportation, and healthcare. It is stressed that AI, combined with LP and a deep understanding of Entropy, can significantly improve the efficiency and sustainability of urban systems. It is used an enhanced Likert-type scale questionnaire to measure public sentiment about the integration of AI into various urban sectors. This methodology suggezts that one is keen to understand the community's opinions, worries, and the emotional impact of transitioning to AI-supported smart urban environments. On the one hand, the document emphasize the importance of approaching the AI integration into SC with ethical consideration, ensuring benefits are distributed equitably and that there's public welfare. The terms from thermodynamics like Exergy, Vagueness, and Anergy are used to metaphorically describe different aspects of system efficiency and public sentiment. On the other hand, the text offers a comprehensive guide on leveraging AI and LP in urban management, touching on the need to address privacy, security, and equitable access to technology. It also covers the environmental aspect by discussing Entropy's role in managing system efficiency and sustainability. Indeed, the passage uses a complex analogy between thermodynamic principles and data analysis processes in smart city management. For example, Exergy relates to useful system energy after an exchange and is depicted as the potential for improvement in the SC context. Vagueness refers to uncertain energies, likely the unknown variables in urban management, while Anergy is the potential energy not in use, which could be seen as untapped resources or capacities within the SC infrastructure. It is detailled the approach to public surveys, structuring questionnaires around urban planning, healthcare efficiency, and environmental sustainability, and propose using the collected data for the betterment of urban life. It includes pseudocode for a PROLOG program designed to process questionnaire data. The program's purpose is to establish entropic values for answers, evaluate citizens' wellbeing, and compute the overall weighted entropy from the questionnaire answers to determine the efficiency of Smart City initiatives. In conclusion, the paper advocate for a balanced and ethical integration of AI into smart urban systems, with a focus on public sentiment and wellbeing, system efficiency, and environmental sustainability. It is a call to responsibly harness the potential of AI and LP to improve the quality of life in urban areas while considering the social, ethical, and environmental impacts.

6 Conclusions and Future Work

The study provides a comprehensive analysis of how AI, when integrated with LP and an understanding of Entropy, can drive significant improvements in the operational excellence and sustainability of urban systems. It highlights the transformative impact of AI in various aspects of urban living, including healthcare, city planning, public safety, and environmental management. Through the use of enhanced Likert-type scale questionnaires, there was a look to public sentiment to gauge perceptions of AI's role in urban development, with particular emphasis on ethical considerations, equitable benefit distribution, and the welfare of citizens. It were acknowledge the complexities of integrating these technologies into the urban fabric, pointing out the challenges related to privacy, security, and ensuring that the benefits of AI and LP are accessible to all demographics. Additionally, it were utilized thermodynamic concepts metaphorically to discuss the potential energy (Exergy), uncertainties (Vagueness), and untapped resources (Anergy) within SC ecosystems, providing a framework for understanding the dynamics of city management. By adopting a structured approach to problem-solving and predictive analytics, the study suggests that cities can transition from reactive to proactive management. The potential of AI and LP to automate complex processes and improve decision-making processes is clear, yet the call for these advances were guided by ethical standards and a commitment to public welfare. Future research should continue to explore the integration of AI, LP, and Entropy in SC to address the evolving needs of urban populations, particularly in the healthcare sector. There is an opportunity to further refine the metrics used in the enhanced Likert-type scale questionnaire to provide more granular insights into public sentiment and the perceived efficacy of technological integration in urban management.

Acknowledgments. This work has been supported by FCT – Fundação para a Ciência e Tecnologia within the R&D Units Project Scope: UIDB/00319/2020.

References

1. Ortega-Fernández, A., Martín-Rojas, R., García-Morales, V.J.: Artificial intelligence in the urban environment: smart cities as models for developing innovation and sustainability. Sustainability **12**, 7860 (2020)
2. Sanchez, T.W., Shumway, H., Gordner, T., Lim, T.: The prospects of artificial intelligence in urban planning. Int. J. Urban Sci. **27**(2), 179–194 (2023)
3. Yan, Z., Jiang, L., Huang, X., Zhang, L.: Zhou, X: Intelligent urbanism with artificial intelligence in shaping tomorrow's smart cities: current developments, trends, and future directions. J. Cloud Comput. **12**, 179 (2023)
4. Nikitas, A., Michalakopoulou, K., Njoya, E.T.: Karampatzakis, D: Artificial intelligence, transport and the smart city: definitions and dimensions of a new mobility era. Sustainability **12**, 2789 (2020)
5. Syed, A.S., Sierra-Sosa, D., Kumar, A., Elmaghraby, A.: IoT in smart cities: a survey of technologies. Practices and Challenges. Smart Cities **4**, 429–475 (2021)
6. Stahl, B.C., et al.: Artificial intelligence for human flourishing – Beyond principles for machine learning. J. Bus. Res. **124**, 374–388 (2021)
7. Zhu, W.: Artificial Intelligence and Urban Governance: Risk Conflict and Strategy Choice. Open J. Soc. Sci. **9**, 250–261 (2021)
8. Sanchez, T.: Planning on the Verge of AI, or AI on the Verge of Planning. Urban Science **7**, 70 (2023)
9. Lehtiö, A., Hartikainen, M., Ala-Luopa, S., Väänänen, K.: Understanding citizen perceptions of AI in the smart city. AI & Soc. **38**, 1123–1134 (2023)
10. Giansanti, D.: Bridging the gap: exploring opportunities, challenges, and problems in integrating assistive technologies, robotics, and automated machines into the health domain. Healthcare **11**, 2462 (2023)
11. Brendel, A.B., Mirbabaie, M., Lembcke, T.-B., Hofeditz, L.: Ethical Management of Artificial Intelligence. Sustainability **13**, 1974 (2021)

12. Li, J., Nassauer, J., Webster, N.: Landscape elements affect public perception of nature-based solutions managed by smart systems. Landsc. Urban Plan. **221**, 104355 (2022)
13. Amjad, A., Kordel, P., Fernandes, G.: A review on innovation in healthcare sector (telehealth) through artificial intelligence. Sustainability **15**, 6655 (2023)
14. Neves, J.: A logic interpreter to handle time and negation in logic databases. In: Muller, R., Pottmyer, J. (eds.) Proceedings of the 1984 Annual Conference of the ACM on the 5th Generation Challenge, pp. 50–54. ACM, New York (1984)
15. Neves, J., et al.: Entropy and organizational performance. Lect. Notes Comput. Sci. **11734**, 206–217 (2019)
16. Ajlan, A.M., Al-Abed, A.M.: Transformation model towards sustainable smart cities: Riyadh, Saudi Arabia as a case study. Curr. Urban Stud. **11**, 142–178 (2023)
17. Anthony, B.: The Role of Community Engagement in Urban Innovation Towards the Co-Creation of Smart Sustainable Cities. Journal of the Knowledge Economy (2023)
18. Palmini, O., Cugurullo, F.: Charting AI urbanism: conceptual sources and spatial implications of urban artificial intelligence. Discover Artif. Intell. **3**, 15 (2023)
19. Pflanzer, M., Traylor, Z., Lyons, J.B., Dubljević, V., Nam, C.S.: Ethics in human – AI teaming: principles and perspectives. AI Ethics **3**, 917–935 (2023)
20. Yejin, L., Min, Y.: Application and validity analysis of IoT in smart city based on entropy method. Appl. Artif. Intell. **37**, 2166234 (2023)

Design of a Proactive System to Promote Healthy Behaviors in the Elderly Population

Vítor Crista(✉), Diogo Martinho, and Goreti Marreiros

Research Group on Intelligent Engineering and Computing for Advanced Innovation and Development (GECAD), Institute of Engineering, Polytechnic of Porto, Porto, Portugal
{vvrpc,diepm,mgt}@isep.ipp.pt

Abstract. Among the increasing number of health issues related to aging and the growing elderly population, there is an urgent need for new ways to improve well-being among older adults. Recognizing the complex connection between mental and physical well-being emphasizes the importance of comprehensive approaches that address health for the elderly population. In the context of the DAISy project, this paper introduces an architectural framework for a proactive system designed to promote healthy behaviors among the elderly population. Leveraging wearable device sensor data, the system collects real-time user information to generate personalized interventions adapted to individual needs and preferences. At the core of the system's functionality is the use of advanced generative AI, leveraging Large Language Models to produce customized interactions based on user data insights. By combining sensor data analysis with AI-driven text generation, the system aims to deliver adaptive and personalized experiences, thus addressing the complex challenges elderly individuals face in maintaining their health and quality of life.

Keywords: Conversation System · Elderly Care · Generative AI · Personalized Healthcare · Wearable Technology

1 Introduction

Population aging has become increasingly evident in modern society, highlighting advancements in healthcare, and living standards while presenting us with a range of complex challenges. As per the United Nations report, the global population aged 65 and above is projected to double in the next thirty years, reaching a staggering 1.6 billion by 2050. By then, seniors will constitute over 16 percent of the world's population [1]. This ascending trend highlights the urgent necessity for innovative strategies to ensure that aging is accompanied by a high quality of life (QoL). With aging some challenges arise, such as decreased mobility and unexpected incidents like accidents and falls, which can limit the possibility of maintaining an active and healthy lifestyle. However, older adults can still have active aging by adjusting their daily routines and exploring new possibilities for well-being adapted to their specific needs and preferences. Another aspect to consider is that typically an older individual has more leisure time and fewer work

G. Marreiros et al. (Eds.): DCAI 2024, LNNS 1151, pp. 222–232, 2025.
https://doi.org/10.1007/978-3-031-80946-0_21

duties, so they naturally need social interactions and extended engagement with friends and family [2, 3]. Yet, contemporary societal dynamics present obstacles for families to dedicate adequate time to their elderly members due to the pressures of professional obligations and the struggle to balance work and family life [4]. Consequently, older adults may experience a gradual withdrawal from social engagements, depriving themselves of crucial emotional support necessary for healthy aging. Loneliness in old age contributes to various health issues that negatively affect both physical and mental well-being, thus diminishing overall QoL [5]. Research has shown a negative correlation between QoL and loneliness, leading to demotivation among older adults. This demotivation directly contributes to physical decline, as older adults may become more sedentary without adequate social stimuli and engagement in leisure activities [6]. Additionally, loneliness carries other significant consequences such as depression, alcohol abuse, and in severe cases, suicide [7]. The absence of mental stimulation from meaningful conversations and active social interactions also serves as a contributing factor to the beginning of cognitive degenerative conditions like dementia and Alzheimer's disease [8–10]. It is relevant to address that this work was developed under a project named DAISy which has a focus on managing Major Depressive Disorder (MDD) patients. In the elderly, mental health can deteriorate rapidly due to factors such as social isolation and limited access to adequate care. The prevalence of MDD among older adults intensifies this issue, potentially leading to further decline in cognitive and emotional well-being. Addressing mental health challenges in the elderly is fundamental, as neglecting these issues can significantly impact their overall QoL and increase existing health conditions. Given the challenges associated with an aging population, it is evident that innovative solutions are in high demand. Utilizing technology to enhance the QoL of the elderly population appears to be a beneficial approach. Artificial Intelligence (AI)-powered systems emerge as fundamental utilities capable of supporting the elderly in their daily activities, empowering them to maintain autonomy over an extended period [11]. These innovative solutions offer an effective strategy to alleviate loneliness among seniors by promoting social engagement and encouraging emotional connections. AI technology improves the lives of older adults by encouraging better interactions between individuals and the system, thus enhancing their overall QoL. In recent years, we have observed an increase in Generative AI technologies, recommendation systems, and the Internet of Things (IoT), suggesting a transformative era in innovation. Generative AI stands out among these advancements, featuring the capacity to produce original content across communication mediums [12]. There is growing evidence supporting the integration of conversational agents into mental health treatments, marking a promising avenue for therapeutic support [13]. In the field of elderly care, the combination of IoT with mobile applications has gained significant interest [14]. These solutions combine smart devices like physical activity sensors and wearables, such as smartwatches, equipped with multifaceted sensors capable of capturing biometric, activity, and sleep data [15]. This wealth of information holds immense potential for crafting tailored interventions to enhance elderly care, facilitating real-time monitoring and timely interventions. Yet, among these technological improvements, the concern of loneliness and social isolation persists among the elderly, and encouraging robust social connections emerges as a vital strategy in combating these challenges. In our fast-paced modern world, promoting meaningful connections with loved ones becomes ever more challenging as people have tight schedules, leaving little time for spending with friends, family, and especially the elderly, who often find themselves disregarded in their lives [16]. Despite these challenges, the intersection of technological innovation and human connection offers both opportunities and challenges. Generative AI and LLMs hold great promise in transforming healthcare for older adults by introducing a new paradigm of conversational generative systems tailored to each individual's needs, especially the elderly [17]. Applications like ChatGPT by OpenAI [18] demonstrate this revolution, with the capability to engage in lifelike conversations. At the core of healthcare advancement lies the goal of improving patient experiences, where ChatGPT excels by offering support and information to seniors facing health problems [19]. Through AI-powered chatbots and virtual assistants, remote monitoring of high-risk older adults with chronic conditions is now possible. These tools provide adapted advice on health, nutrition, and fitness, enabling individuals to take control of their well-being [20]. Generative AI applications exceed simple practical health advice as they serve as support in the battle against social isolation and loneliness among the elderly. By developing virtual companionship and connections and offering nonjudgmental emotional support, these AI applications are invaluable in improving the QoL of older adults [21]. The literature is increasing in the field of innovative solutions of generative AI in elder care, from remote health monitoring to mental health support and personalized cognitive decline prevention [22]. The potential impact of this is great, thus promising a future where the elderly population receives better and more adapted care. However, alongside the promise come concerns. Generative AI models like LLMs depend on extensive and diverse datasets, which can accidentally introduce biases, errors, and misinterpretations [23]. In addition, their absence of genuine medical expertise raises concerns regarding the accuracy and reliability of their advice. It is crucial to approach these limitations with caution, recognizing the constraints while actively seeking ways to alleviate them. Innovation in these domains is imperative. The well-being of the elderly is crucial for society's health and fundamental for intergenerational knowledge transfer, as elders pass down wisdom to younger generations. Moreover, the emotional support and community contribution enabled by the elderly are crucial for creating an interconnected society. As such, this initiative introduces an intelligent system architecture designed to improve the QoL for the elderly through personalized assistance and guidance. This system leverages real-time user data and utilizes LLMs to dynamically adjust to each individual's specific needs, continuously refining its responses based on interactions. By integrating biometric data, activity levels, and interaction patterns, the system provides adapted recommendations to promote healthy behaviors and overall well-being. Through continuous learning, the system evolves alongside users, constantly improving its efficacy. This innovative approach represents a significant step forward in elderly care, seamlessly integrating technology to offer invaluable support and companionship.

2 Proposed Architecture

The impact of physical activity and emotional well-being cannot be overstated in promoting mental health. Through regular exercise, individuals not only enhance their physical fitness but also experience improvements in mood regulation and emotional stability [24]. Additionally, exercise provides an effective outlet for releasing emotional tension and promoting mental clarity [25]. When combined with a positive emotional state, such as feelings of happiness and contentment, physical activity becomes a powerful tool in developing mental well-being. Thus, prioritizing both physical activity and emotional health is crucial for achieving and maintaining optimal mental wellness [26]. At its core, the system employs the latest AI advancements, particularly LLMs, to dynamically generate text for user interactions. While the utilization of LLMs offers significant benefits, as outlined in the preceding sections, without appropriate guidelines, their output may lack individualization across users, thus the necessity for mechanisms that generate personalized outcomes. The proposed system's architectural design relies on leveraging user sensor data from smartwatch devices and insights gathered from system-user conversations to establish significant information regarding user preferences and behaviors. By doing so, the system can refine its interactions, learning over time to adapt to each individual's needs and preferences. Increased data availability enhances the system's capacity for personalization. Figure 1 presents a visualization of the system architecture. To enrich the communication experience through personalization, the system comprises two distinct components. The first component, user profiling, focuses on gathering user sensor data about physical activity levels, sleep quality, heart rate variability, and other relevant metrics. Subsequently, this component employs a machine learning pipeline to predict physical activity and psychological state levels and identify optimal and challenging times for the user. Supplied with insights into moments of reduced physical activity and increased negative psychological states, the system can proactively suggest actions to counteract these adversities. The second component comprises establishing a user knowledge base that evolves through continuous interactions with the system. This repository will store related user data such as past events, significant dates, and other relevant information gathered from each conversation. As the system accumulates data, it develops a comprehensive understanding of the user's interests, important memories, ongoing events, and potential future occurrences. During each interaction, the system leverages this knowledge base to enhance the context for LLMs, enabling it to craft more refined responses adapted to the specific user. This process, known as Retrieval-Augmented Generation (RAG), combines the power of retrieval-based methods with the generative capabilities of language models [27]. Essentially, RAG allows the system to dynamically search the user knowledge base for relevant information and incorporate it into the generation process. By retrieving pertinent details from the knowledge base, the system improves the input context for the LLMs, enabling them to produce more personalized and contextually relevant content.

2.1 User Profiling

User profiling is a step that is fundamental for the usage of the rest of the system. Personalization and refinement behavior are essential aspects of the system. Figure 2

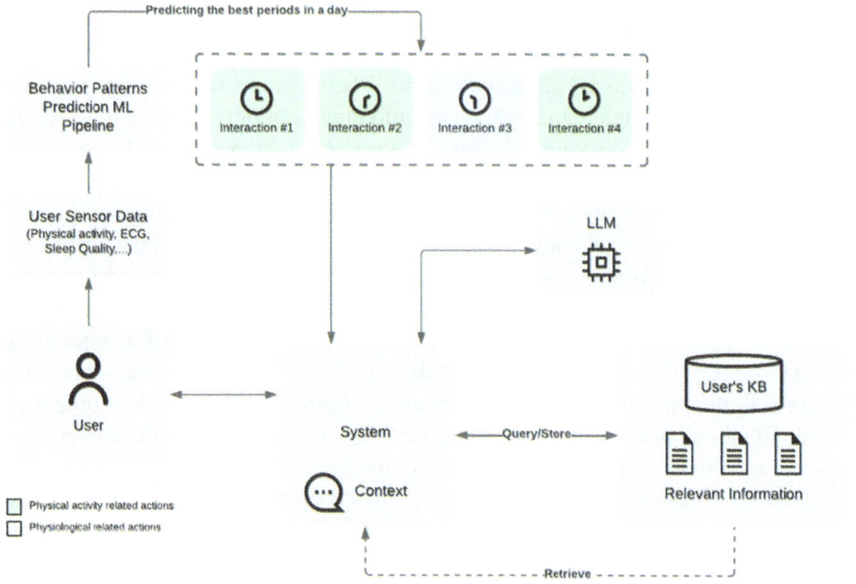

Fig. 1. Proposed System Architecture

describes the machine learning pipeline that is responsible for the recognition of behavior patterns of individuals.

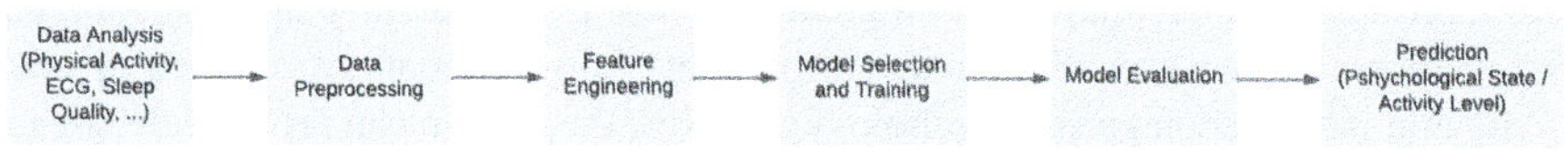

Fig. 2. Behavior Patterns Prediction Machine Learning Pipeline

The system collects a comprehensive user profile necessary for initiating conversational processes, containing data on physical activity such as step count, intensity, and distance covered, as well as metrics for sleep quality and heart rate. The objective is to develop a prediction model that anticipates both physical activity and physiological data, identifying periods of peak and minimal activity through temporal clustering. This involves implementing a machine learning pipeline following standard procedures: data preprocessing, feature engineering (with a primary focus on generating time-based features), model selection and training, model evaluation, and, finally, state prediction. The aim of analyzing these two types of data (physical activity and psychological state) is to obtain a score for each, a metric that combines these two dimensions: physical state (X) and psychological state (Y). Regarding the final output, it will indicate the user's stage, with four distinct stages representing various combinations of X and Y metrics (See Table 1).

Stage 1 represents the most challenging phase, where the model predicts a score of 0 for physical activity, indicating the user's minimal engagement in physical exercise. Additionally, the prediction for psychological levels also stands at 0, suggesting a state

Table 1. User stages and action distributions

Stage	Physical State (X)	Psychological State (Y)	Action Distribution
1	0	0	50% of X's actions + 50% of Y's actions
2	1	0	25% of X's actions + 75% of Y's actions
3	0	1	75% of X's actions + 25% of Y's actions
4	1	1	50% of X's actions + 50% of Y's actions

of profound unhappiness and elevated stress levels for users in this stage. The action distribution is evenly split between physical actions (e.g.: going for a walk) and psychological actions (e.g.: inviting family for dinner) to address both the sedentary lifestyle and emotional well-being of the users. This balanced approach aims to gradually improve both physical activity and emotional state. Moving to Stage 2, the sedentarism score increases to 1, implying a slight improvement in physical activity, while the emotional state remains at 0. Users in this stage exhibit some level of physical activity but still experience significant stress and low emotional well-being. To address this, 75% of actions are focused on psychological interventions to alleviate stress and improve emotional health, while still allocating 25% of actions towards maintaining and further increasing physical activity levels. In contrast, Stage 3 presents a reversal of Stage 2, where users demonstrate elevated levels of relaxation and positive emotions but remain sedentary. To encourage increased physical activity, 75% of actions are directed towards physical interventions, while 25% are focused on maintaining the positive psychological state achieved in this stage. Finally, Stage 4 signifies the desired profile, with individuals exhibiting high physical activity levels, considerable happiness, minimal stress, and elevated positive emotions. The action distribution is evenly split between physical and psychological actions to maintain the balance achieved in this stage and sustain the user's progress toward optimal well-being. Given these 4 distinct stages, the idea is to adapt the system for these types of users, with the end goal of converting all the users to a stage 4 profile.

2.2 Conversation System

Conversations play a fundamental role within the system, facilitating the creation of meaningful exchanges between users and the system itself. Personalization is the key, as it enhances the overall experience and ideally positively impacts the QoL for elderly users. Responsiveness is a fundamental characteristic of any conversation system; users should receive a prompt reply to their inquiries, whether engaging in casual chat or discussing specific topics. However, beyond simple reactivity, it is crucial to acknowledge that elderly users may not have the same level of enthusiasm or familiarity with the system as other demographics might. Incorporating proactive engagement becomes imperative, as it can provide companionship and mitigate feelings of loneliness. This underscores the significance of user profiling, in which preceding analysis and machine learning pipelines enable the system to determine not only the user's physical activity and emotional state but also the optimal moments for interaction. Table 2 presents the two above-mentioned interaction modes along with their corresponding periods. By default, the system proactively engages with users during four distinct periods (morning, afternoon, evening, and night) [28], though adjustments are made based on user data, model predictions, temporal analysis, and system-user conversations. Reactiveness, on the other hand, results in responding whenever the user initiates interaction with the system.

Table 2. Moments of interaction

Mode of interaction	Periods
Proactive	Upon waking up (leaving the room - 9 am)
	Before/after lunch (leaving the kitchen)
	Before/after dinner (leaving the kitchen/dining room)
	At bedtime (entering the bedroom)
Reactive	Whenever the user speaks to him, or the system perceives any emotion from the user

Reactive Conversation Mode. In the reactive mode of our conversation system, dialogue initiates by the user's message, triggering the conversational flow. Upon receiving the user's input, the system considers whether the current message, combined with the ongoing conversation context, exceeds the token limit of the LLM. If it surpasses this threshold, a processing pipeline is activated. Within this pipeline, the system summarizes the conversation's essence, evaluates incoming information for relevance, and selectively stores relevant details to enrich the user's knowledge base over time. Simultaneously, the system leverages the user's knowledge base through the RAG process, as previously discussed. If additional user information is required for an adapted response, such as date of birth or past events, the system queries the user's knowledge base. This enables it to extract the required information, improving the context for a more high-quality response. In opposition, if no supplementary data is needed, the system proceeds without accessing the knowledge base. Provided with an enriched context and the user's message, the system uses the LLM to produce a response. This ensures that our responses remain contextually relevant and adapted to the individual user's needs.

Proactive Conversation Mode. This conversation mode is linked with user profiling, as it operates within the periods predicted by the behavior patterns prediction process for each user. Figure 3 illustrates how the proactive mode functions. Initially, user sensor data passes through the machine learning pipeline, resulting in one of the four previously discussed stages. Alongside each stage, the system identifies the four optimal time slots for interaction and determines the type of interaction to be delivered to the user. The model generates an action schedule that the system follows. In the figure, it is evident that at 12:00, the system suggests the user to take a walk to increase their physical activity levels. At 16:00, it recommends reading a book to improve psychological well-being. Subsequently, at 19:00, it prompts the user to invite family to dinner, aiming to improve emotional/stress levels. Finally, at 21:00, it suggests watching a movie, again targeting emotional/stress improvement. The system initiates four interactions, each proposing its respective action. This arrangement of actions implies that the user is an active individual, as 75% of the actions are geared towards addressing emotional/stress levels, with only 25% focusing on physical activity. Notably, the periods when the user appears most depleted are at 16:00, 19:00, and 21:00.

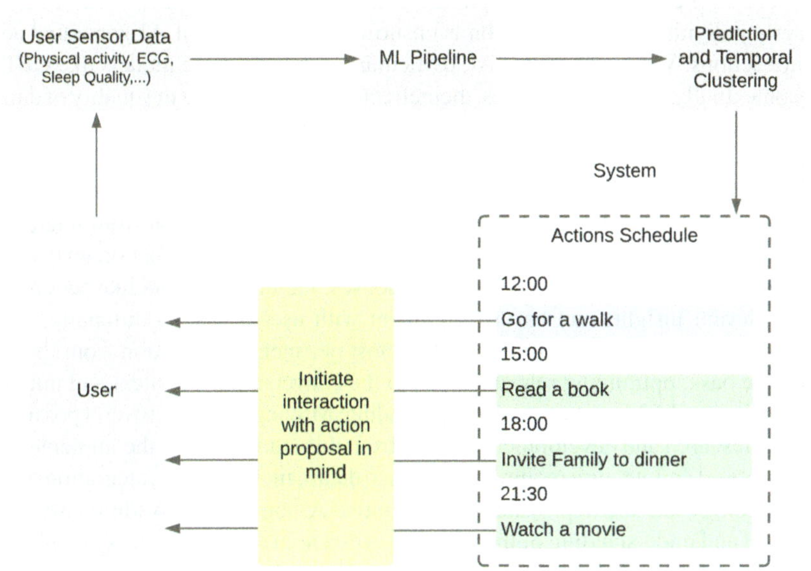

Fig. 3. Conversation system with a proactive interaction schedule

The four actions described in the figure serve as illustrations of the process. Initially, the system employs default actions for all users, which may evolve as it gathers more information about their preferences and characteristics. It is worth noting that the functionality related to conversation context management as well as the retrieving of relevant information through the user's knowledge base mentioned in the reactive mode, remains similar in the proactive mode as well. The key distinction lies in the system's initiation of interactions with the user at predetermined times of the day.

3 Conclusions and Future Work

The proposed architecture for promoting healthy behaviors among the elderly population presents a comprehensive framework that integrates wearable technology, generative AI, and personalized healthcare. This work, developed under the DAISy project, which focuses on managing mental health disorders, aims to provide adapted interventions to improve the well-being of elderly individuals. The importance of addressing loneliness and social isolation in old age cannot be exaggerated, as it directly impacts both physical and mental health, diminishing overall QoL. By leveraging the insights gained from users' data and the development of the conversation system, the implementation and testing of the entire system with real users is now ready to proceed. This iterative approach ensures the refinement and optimization of the system to meet the diverse needs of elderly individuals. One limitation is ensuring the reliability of AI-generated advice and interactions. While generative AI, particularly large language models like GPT, can offer contextually relevant responses, their effectiveness centers on the quality of data and models. Although these models are trained on vast datasets, the responses they generate can vary from generic to specific, depending on the pre-prompt provided. To improve reliability, two strategies are employed: Firstly, the system refines its understanding of individual users by storing and analyzing contextual information from interactions, allowing for personalized responses over time. Secondly, each interaction involves analyzing both user input and AI-generated responses, including factors like sentiment, to extract relevant insights and ensure alignment with user needs. Additionally, a query retrieval approach is adopted to extract the most pertinent information from the user's knowledge base, optimizing relevance within the conversation's context and mitigating the limitations of LLM's contextual understanding. Moving forward, several possibilities for future research and development emerge from this study. Firstly, the implementation of the proposed architecture requires attention to detail, including the integration of wearable technology and the deployment of generative AI algorithms. Additionally, further refinement and understanding of the user data is essential to enhance the system's ability to deliver personalized interventions. Concurrently, the development of the conversation system guarantees continued exploration, with a focus on optimizing natural language processing techniques and improving the system's responsiveness to user queries. Once these components are fully realized, the next phase involves comprehensive testing with real users to evaluate the system's efficacy and usability in real-world scenarios.

Acknowledgments. This research work was developed under the project DAISy –"Developing AI ecosystems improving diagnosis and care of mental diseases" (ITEA 21016), co-financed by National Funds through FCT (Fundação para a Ciência e a Tecnologia) under the project UI/DB/00760/2020.

References

1. United Nations: World Social Report 2023: Leaving No One Behind In An Ageing World (2023)
2. Horgas, A.L., Wilms, H.U., Baltes, M.M.: Daily life in very old age: everyday activities as expression of successful living. Gerontologist **38**(5), 556–568 (1998). https://doi.org/10.1093/GERONT/38.5.556
3. Spalt, E.W., et al.: Time-location patterns of a diverse population of older adults: the Multi-Ethnic Study of Atherosclerosis and Air Pollution (MESA Air). J. Expo. Sci. Environ. Epidemiol. **26**(4), 349–355 (2016). https://doi.org/10.1038/JES.2015.29
4. Rudd, M.: Feeling short on time: trends, consequences, and possible remedies. Curr. Opin. Psychol. **26**, 5 (2019). https://doi.org/10.1016/J.COPSYC.2018.04.007
5. Ong, A.D., Uchino, B.N., Wethington, E.: Loneliness and health in older adults: a mini-review and synthesis. Gerontology **62**(4), 443–449 (2016). https://doi.org/10.1159/000441651
6. Schrempft, S., Jackowska, M., Hamer, M., Steptoe, A.: Associations between social isolation, loneliness, and objective physical activity in older men and women. BMC Public Health **19**(1) (2019). https://doi.org/10.1186/S12889-019-6424-Y
7. Arslantaş, H., Adana, F., Abacigİl Ergİn, F., Kayar, D., Acar, G.: Loneliness in elderly people, associated factors and its correlation with quality of life: a field study from western Turkey. Iran J. Public Health **44**(1), 43 (2015). /pmc/articles/PMC4450013/. Accessed 16 Dec 2023
8. Donovan, N.J., et al.: Association of higher cortical amyloid burden with loneliness in cognitively normal older adults. JAMA Psychiat. **73**(12), 1230–1237 (2016). https://doi.org/10.1001/JAMAPSYCHIATRY.2016.2657
9. Sutin, A.R., Stephan, Y., Luchetti, M., Terracciano, A.: Loneliness and risk of dementia. J. Gerontol. Ser. B **75**(7), 1414–1422 (2020). https://doi.org/10.1093/GERONB/GBY112
10. Rafnsson, S.B., Orrell, M., D'Orsi, E., Hogervorst, E., Steptoe, A.: Loneliness, social integration, and incident dementia over 6 years: prospective findings from the english longitudinal study of ageing. J. Gerontol. Ser. B **75**(1), 114–124 (2020). https://doi.org/10.1093/GERONB/GBX087
11. Padhan, S., Mohapatra, A., Ramasamy, S.K., Agrawal, S.: Artificial intelligence (AI) and robotics in elderly healthcare: enabling independence and quality of life. Cureus **15**(8) (2023). https://doi.org/10.7759/CUREUS.42905
12. Zhang, P., Kamel Boulos, M.N.: Generative AI in medicine and healthcare: promises, opportunities and challenges. Future Internet 15(9), 286 (2023). https://doi.org/10.3390/FI15090286
13. Vaidyam, A.N., Wisniewski, H., Halamka, J.D., Kashavan, M.S., Torous, J.B.: Chatbots and conversational agents in mental health: a review of the psychiatric landscape. Can. J. Psychiat. **64**(7), 456–464 (2019). https://doi.org/10.1177/0706743719828977/ASSET/IMAGES/LARGE/10.1177_0706743719828977-FIG2.JPEG
14. Madhu, S., Jyothi, K., Pravallika, S., Sridurga, N.: IoT-enabled applications for elderly support and care: a systematic review. Cogn. Sci. Technol. Part F1466, 697–705 (2023). https://doi.org/10.1007/978-981-99-2742-5_71
15. Wang, Z., Yang, Z., Dong, T.: A review of wearable technologies for elderly care that can accurately track indoor position, recognize physical activities and monitor vital signs in real time. Sensors **17**(2), 341 (2017). https://doi.org/10.3390/S17020341
16. Festini, S.B.: Busyness, mental engagement, and stress: relationships to neurocognitive aging and behavior. Front. Aging Neurosci **14**, 980599 (2022). https://doi.org/10.3389/FNAGI.2022.980599
17. Fear, K., Gleber, C.: Shaping the future of older adult care: ChatGPT, advanced AI, and the transformation of clinical practice. JMIR Aging **6**, 51776 (2023). https://doi.org/10.2196/51776

18. ChatGPT. https://chat.openai.com/. Accessed 17 Dec 2023
19. Asch, D.A.: An Interview with ChatGPT About Health Care (2023). https://doi.org/10.1056/CAT.23.0043
20. Nandini Prasad, K.S., Sudhanva, S., Tarun, T.N., Yuvraaj, Y., Vishal, D.A.: Conversational chatbot builder - smarter virtual assistance with domain specific AI. In: 2023 4th International Conference for Emerging Technology, INCET 2023 (2023). https://doi.org/10.1109/INCET57972.2023.10170114
21. Alessa, A., Al-Khalifa, H.: Towards designing a ChatGPT conversational companion for elderly people. In: ACM International Conference Proceeding Series, pp. 667–674 (2023). https://doi.org/10.1145/3594806.3596572
22. Bhattacharyya, R., Chakraborty, K., Neogi, R.: ChatGPT and its application in the field of mental health. J. SAARC Psychiatr. Fed. **1**(1), 6 (2023). https://doi.org/10.4103/JSPF.JSPF_9_23
23. Heinz, M.V., et al.: Testing domain knowledge and risk of bias of a large-scale general artificial intelligence model in mental health. Digit Health **9** (2023). https://doi.org/10.1177/20552076231170499
24. Zhang, Y., Fu, R., Sun, L., Gong, Y., Tang, D.: How does exercise improve implicit emotion regulation ability: preliminary evidence of mind-body exercise intervention combined with aerobic jogging and mindfulness-based yoga. Front. Psychol. **10**, 1888 (2019). https://doi.org/10.3389/FPSYG.2019.01888
25. Fox, K.R., Stathi, A., McKenna, J., Davis, M.G.: Physical activity and mental well-being in older people participating in the Better Ageing Project. Eur. J. Appl. Physiol. **100**(5), 591–602 (2007). https://doi.org/10.1007/S00421-007-0392-0/TABLES/3
26. Mahindru, A., Patil, P., Agrawal, V.: Role of physical activity on mental health and well-being: a review. Cureus **15**(1) (2023). https://doi.org/10.7759/CUREUS.33475
27. Lewis, P., et al.: Retrieval-augmented generation for knowledge-intensive NLP tasks. In: Advances in Neural Information Processing Systems, vol. 2020-December, May 2020. https://arxiv.org/abs/2005.11401v4. Accessed 19 May 2024
28. Vandelanotte, C., et al.: Increasing physical activity using a just-in-time adaptive digital assistant supported by machine learning: a novel approach for hyper-personalised mHealth interventions. J. Biomed. Inform.med. Inform. **144**, 104435 (2023). https://doi.org/10.1016/J.JBI.2023.104435

Supporting Elderly Care Through an AI-Driven and FHIR-Based Remote Monitoring System

Rafael Martins[1,2](✉), Hugo Pereira[1,2](✉), Gustavo Corrente[3], William Xavier[3], Luís Conceição[1,2], Alberto Freitas[4,5], and Goreti Marreiros[1,2]

[1] GECAD – Research Group on Intelligent Engineering and Computing for Advanced Innovation and Development, Institute of Engineering – Polytechnic of Porto, Porto, Portugal
{rgmar,hsdrp,msc,mgt}@isep.ipp.pt
[2] LASI – Intelligent Systems Associate Laboratory, Guimarães, Portugal
[3] Wiseware Solutions, Gafanha da Encarnação, Portugal
{gustavo.corrente,william}@wisewaresolutions.com
[4] MEDCIDS – Department of Community Medicine, Information and Health Decision Sciences, Faculty of Medicine, University of Porto, Porto, Portugal
alberto@med.up.pt
[5] CINTESIS@RISE – Center for Health Technology and Services, Faculty of Medicine, University of Porto, Porto, Portugal

Abstract. An increase in the elderly population has led to a growing demand for constant health care. Remote monitoring of seniors aims to provide more effective care and promote patient independence, whether they live in residential homes, or their own homes supported by a domiciliary care service. It encompasses several issues present in their lives that may benefit from being assisted and monitored remotely. There are many successful studies and products focused on one of these issues, but there are no systems, to the best of our knowledge, that serve as a broad remote monitoring solution for multiple of these topics simultaneously. This paper proposes an architecture for an intelligent health monitoring ecosystem to interact with these various aspects of the elderly's daily life, including monitoring of vital signs, movement and medication intake, detection of falls, sleep quality analysis, access control, and patient–caregiver communication. This solution will make use of wearables and medication dispensers, and it will securely store all patient-related information into an HL7 FHIR standardized database, from where machine learning models can import data for predicting important information and generating alerts when necessary. Next steps involve the development of this proposed system and testing it with elderly volunteers living in a residential home or subscribed to its domiciliary care services.

Keywords: Elderly Care · FHIR · Machine Learning · Remote Monitoring · Wearables

G. Marreiros et al. (Eds.): DCAI 2024, LNNS 1151, pp. 233–241, 2025.
https://doi.org/10.1007/978-3-031-80946-0_22

1 Introduction

Demographic changes, marked by a significant increase in the elderly population, have led to a growing demand for constant care. According to the United Nations, the number of people aged 65 years or older is predicted to more than double, from 761 million in 2021 to 1.6 billion in 2050, and this population aging phenomenon is considered an irreversible global trend [1].

Remote monitoring for seniors is an ever-evolving area of research, aiming to provide more effective care and promote independence, whether seniors are in their own homes or in residential care facilities. With recent technological advances, various solutions have emerged to enable continuous monitoring of the health and activities of the elderly. These products aim to provide independent assisted living and improve safety, ensuring greater peace of mind for their families and caregivers, regardless of the setting.

Remote monitoring of the elderly can encompass numerous topics, as there are several issues present in their lives that may benefit from being assisted and monitored remotely. Such cases include the monitoring of vital signs and daily activities to enable the detection of falls and other health irregularities, access control to the outside/inside of the care facilities, communication with caregivers or health professionals, and monitoring of daily medication intake. Current literature presents many successful studies and products focused on one of these topics, but there are no systems, to the best of our knowledge, that incorporate them all in a broad remote monitoring system.

Integrated into the RM4Health project as the Portuguese use case, the focus of this project is the remote monitoring of seniors who live in residential homes or in domiciliary care. It aims to design and develop an intelligent health monitoring ecosystem that will allow comprehensive monitoring of various aspects of the elderly's daily life, overcoming some of the limitations found in currently available products.

This work describes the proposed architecture for this solution, including a vital sign and access control monitoring wearable, a camera-monitored medication dispenser with integrated alerts and videoconferencing capabilities, an information storage back-end compliant with HL7 FHIR health information standards, a secure authentication system powered by Microsoft Identity, and an API that provides access to machine learning models trained to assist with the monitoring of the elderly by predicting their sleep quality, monitoring their medication intake, and detecting falls.

The remainder of the paper is structured as follows: Sect. 2 reviews the current state of the literature concerning elderly care, Sect. 3 introduces the proposed architecture for the remote monitoring platform, Sect. 4 explores various perspectives on the implementation of artificial intelligence techniques in the solution, and finally Sect. 5 encapsulates the conclusions drawn and outlines future work to be performed.

2 Literature Review

Regarding vital sign monitoring, a common option on the market is devices such as smartwatches and monitoring wristbands. The Apple Watch has been widely used for health and fitness activity monitoring of the elderly, and monitoring bracelets such as Fitbit have been used to monitor sleep, and activity levels and provide useful information about their lifestyle [2]. Aside from these commercial products, a study carried out by Durán-Vega et al. proposes and validates the feasibility of a biometric bracelet prototype for monitoring the elderly's vital signs in a residential home [3]. The system provides caregivers with real-time visualization of heart rate, body temperature, and blood oxygenation values, to support decision-making about their health status. However, these decisions are ultimately manually made by the caregivers, with no research being conducted on the use of algorithms for automatic remote monitoring. Baig et al. reviewed the use of wearable sensors and IoT monitoring solutions for supporting the independent living of older adults [4]. While it highlights the importance of these technologies in collecting data on daily activities and detecting behavior changes in the elderly, the systematic review details how most studies report moderate to low usability, sensor inaccuracy, power issues, and lack of interoperability.

When it comes to access control, different solutions have been explored to find the less intrusive and most effective way of implementing smart locks to remotely control doors based on the elderly's access permissions. Adiono et al. propose a design that measures the distance to the door by utilizing the user's GPS position captured by an Android device [5]. The system is reported to function for indoor use, but with a need for location accuracy improvement. Prabakaran et al. opted instead by a more manual access control system, where a person responsible for the door can check security footage of who is standing nearby and remotely control it through a mobile application [6]. On the other hand, while not specifically tried on elderly patients, a study by Aswini et al. explores an access control system based on the proximity of RFID tags carried by the users to an RFID reader capable of triggering the door to open/close based on user permissions [7]. Yet another study, by Saputra and Surantha, tries to avoid requiring the elderly to use additional devices and/or interfaces, by proposing a smart door lock system that decides to unlock a door via facial recognition of the elderly who approach it [8]. Each solution has its strong points, but there are still few studies dedicated exclusively to the access control of the elderly, be it in hospitals, in residential homes, or at their own homes.

On the subject of remote communication between the elderly and caregivers or health professionals, a systematic review was carried out by Şahin et al. to analyze the effectiveness of telemedicine interventions for older adults, observing that remote communication with the elderly shows to be effective as usual care, but noted that some older adults report difficulty interacting with the technology [9]. In terms of practical applications, Shea et al. developed Carebit, a mobile application that connects to a patient's Fitbit sensors to provide alerts to a caregiver, while also providing a message chat for easy communication between patient and caregiver [10]. Yet this application does not explore the use of videoconferencing technologies to supplement this remote communication. In contrast, Seelye et al., and more recently Read Paul et al. conducted similar experiments to determine the reactions of elderly patients, in residential homes and at their own homes, respectively, to the use of videoconferencing technologies for communication with family or for medical consultations [11, 12]. Both studies reported positive experiences from all participants involved but did not further explore the integration of these video-communication capabilities into a remote monitoring system.

As far as medication monitoring is concerned, some products on the market offer electronic medication dispensers. As an example, MedMinder is a smart medication dispenser that has scheduling features and personalized alerts to help seniors correctly follow medication regimens [13]. PillDrill is another example of a device that aids in medication monitoring and management, providing visual and audible remedies to ensure proper adherence [14]. Adequate control of medication intake is essential for the care of the elderly. Remote monitoring can play a crucial role in this process, ensuring that seniors are correctly adhering to their prescribed medication, even without a caregiver or health professional being present. Walker et al. reviewed various studies about the patient's experiences on the use of remote monitoring technologies to manage their chronic diseases [15]. A result observed in several studies was that the usage of remote monitoring promoted confidence in the patients and encouraged them to self-manage their medication and feel able to discuss their monitoring data with their health professionals.

3 Proposed Architecture

The global population aging phenomenon is considered to be an irreversible trend, and is predicted to increase by 2050 [1]. This significant increase in the elderly population has led to a growing demand for constant care and monitoring of different issues present in the lives of the elderly, such as their daily activities, medication intake, sleep quality, communication with caregivers and health professionals, access control to the outside/inside of the care facilities, and detection of falls. Given this phenomenon, remote monitoring for seniors has become a prevalent area of research, with various solutions emerging focused on one of these issues, but, to the best of knowledge, there are no cases of systems that provide an integrated monitoring solution to all these issues simultaneously.

To address these current challenges, this paper proposes an architecture for a multi-functional remote monitoring platform, capable of acting in two different settings: with elderly people living in residential homes, in close contact with their caregivers, and with elderly people living in their own home, in domiciliary care, where the contact with the caregivers is sparser. By utilizing two types of sensor devices, one for monitoring daily activities, sleep quality, access control, and fall detection, and another for tracking medication intake and allowing communication with health professionals, the system aims to monitor the current status of the patients, present dashboards and insights concerning each patient's health condition to the caregivers, and alert the patient, the caregivers, or emergency contacts of spotted issues, such as a detected fall, or a missed medication intake.

The proposed architecture for this remote monitoring system is illustrated in Fig. 1. The main users will be the Patient, the elderly living in a residential home or domiciliary care, who will utilize the monitoring sensors to provide clinical data, Caregivers/Health Professionals, who will utilize a front-end application to visualize dashboards and alerts about the patients' current conditions, and potentially Emergency Contacts provided by each patient, who may want to also monitor the patient's health status and receive alerts.

The monitoring of the patient's vital signs throughout the day will be carried out by a wearable sensor, which will measure surface body temperature and electrocardiogram

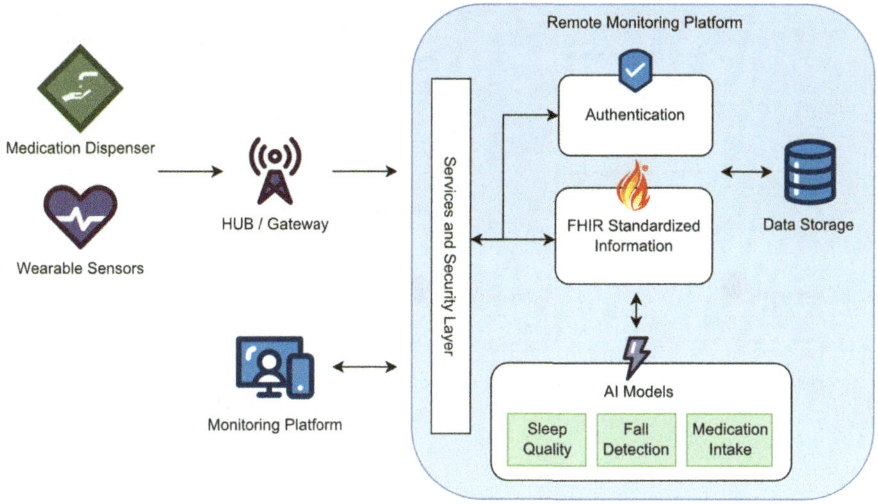

Fig. 1. Proposed Solution Architecture

(ECG) signal. This sensor will also contain a triaxial Inertial Measurement Unit (IMU) to track their motions, and a proximity sensor to control unauthorized access to the exterior of the residential home.

Additionally, for the elderly living in domiciliary care, a medication dispenser, equipped with a camera, will be used to monitor each week's medication intake, and provide reminders or alerts. This device is planned to also be equipped with a secondary camera and video or audio communication capabilities, to provide them with the means to quickly contact a caregiver or health professional if the need arises.

Both the wearable sensors and the medication dispensers will connect via Bluetooth to one or more HUBs strategically placed at the residential home or the patient's own home, which will in turn act as a gateway to send the patient data via Internet to the Remote Monitoring Platform. Since one of the objectives of this platform is to alert caregivers or emergency contacts when patient health issues arise, various monitoring aspects such as fall detections and access control are dependent on an existing Internet connection to provide these alerts in a relevant timeframe. For this reason, solutions are being explored to try to guarantee the existence of this online connection, such as providing 5G connections. Regarding medication intake, an offline solution is being considered where the device, not being able to use the camera to emit alerts when issues arise, will instead use an internal clock to send reminders to the patient at every scheduled medication intake time.

At any moment, the caregiver will be able to easily check each patient's condition by utilizing a monitoring application. This application will be able to display the securely stored information from the Remote Monitoring Platform, and present new insights about a patient's health. Figure 2 presents an envisioned example of the monitoring application to be used by caregivers and health professionals.

The Remote Monitoring Platform will allow for secure storage and retrieval of patient data obtained by the sensors, and for calculation of important insights through machine

Fig. 2. Example of the caregiver's monitoring application interface

learning algorithms or rule-based system techniques, being comprised of a Services and Security Layer, an HL7 FHIR compliant server, an authentication server, corresponding data storage units, and a server hosting a set of machine learning algorithms. Since the proposed platform aims to operate with personal clinical patient data, it will be submitted to an ethics committee for approval, and it will be GDPR compliant.

The Services and Security Layer will mainly act as an API consolidation point, allowing the sensors and the caregivers' monitoring application to access the Remote Monitoring Platform's servers through a centralized secure endpoint.

The FHIR server will oversee the storing of patient health data in a medically standardized way, regulating the main database. This server will also provide the API to retrieve both sensor-obtained and calculated information about the patient, to be presented at the caregivers' request. The reading and writing of information are secured by an authorization filtering system, with each patient, and potentially emergency contacts, having access only to their information, and each caregiver having access only to the patients to them assigned.

While the FHIR server will have authorization filtering capabilities, an external service is required to provide the authentication features beforehand. Such is the purpose of the authentication server, which will employ strong security measures and manage a dedicated database with the user's authentication information to guarantee the protection of the patient's health data.

Finally, the AI Models server will periodically fetch new data from the FHIR server, utilize trained machine learning models to calculate important insights about the patient's status, such as their sleep quality, their medication intake, or the detection of falls, and send this new information back to the FHIR server, so it can be securely stored and be available to the caregivers through the monitoring application.

4 Machine Learning Insights

As previously mentioned, an intended objective of this study is the calculation of important insights based on the patient's vital signs and behavior, obtained through both types of sensors (wearables and medication dispensers). Currently, three types of insights were identified as being obtainable using machine learning: the control of medication intake, the detection of falls, and the evaluation of sleep quality.

An important aspect of elderly care is to ensure the patients take their medication correctly and on time. However, in cases where the patient is living in domiciliary care, contact with the caregivers is sparser, meaning they cannot reliably maintain the patient's intake schedule. Using images taken periodically by the camera present in the medication dispenser, the aim is to utilize a Convolutional Neural Network (CNN), trained for object recognition of medication pills, to automatically identify in a sectioned blister if there are medication pills still present on the current time slot. This allows for the creation of reminders or alarms in case of medication intake failures, while simultaneously allowing caregivers to check the patient's intake progress according to their prescriptions.

Another problem reported in elderly care is the elevated response time by caregivers or emergency contacts once a patient has fallen. With patients living in domiciliary care this happens due to the caregiver not being present most of the time, and in the case of residential homes the caregivers have difficulty covering the status of every patient simultaneously. With the Inertial Measurement Unit (IMU) present in the wearable sensor, the objective is to utilize a model capable of analyzing multiple signals (three axes of an accelerometer, gyroscope, and magnetometer) to detect if the current movements of the patient correspond to a fall or not, and quickly emit alerts to the caregivers or emergency contacts so that the response time is greatly reduced.

Finally, for the evaluation of the patient's sleep quality, the aim is to obtain information about the time spent in each sleep stage (REM, light sleep, or deep sleep) in each night. Studies report that there is potential in classifying sleep stages through machine learning models that analyze the electrocardiogram (ECG) signal of the patient during sleep [16, 17]. As such, the aim is to develop an algorithm to extract features from the ECG signal and train a machine learning model capable of classifying the sleep stages throughout the night, giving caregivers and health professionals the information needed to evaluate the patient's sleep quality.

5 Conclusions and Future Work

The aging of the population poses a significant challenge to healthcare systems and caregivers, with a projected substantial increase in the elderly population in the coming decades [1]. Technological advancements have enabled the development of innovative solutions for remote monitoring of the elderly, allowing to provide them more constant care and promoting their independence, both at home and in residential care facilities. However, while there are many studies and products focused on specific aspects of remote monitoring, such as vital signs monitoring, access control, and remote communication, no studies present, to the best of our knowledge, an integrated system that comprehensively address all these needs at once.

This paper describes a proposed architecture for a remote health monitoring system for the elderly that explores the monitoring of various living aspects simultaneously. This solution is being developed as part of the Portuguese use case of the RM4Health project, which aims to develop this intelligent system for seniors who live either in residential homes or in domiciliary care. Each patient will be provided with wearables to obtain data about their vital signs, movements and location proximity, and patients in domiciliary care will also be assigned a medication dispenser equipped with a medication monitoring camera and videoconferencing capabilities. Using these non-invasive devices, the data collected will be stored and managed by a secure server in compliance with HL7 FHIR standards, and it will be displayed in dashboards to caregivers and health professionals through a dedicated front-end monitoring application. Furthermore, this project will include the training and usage of machine learning models that can extract important additional insights from the sensor's data, allowing for medication intake control, fall detection and sleep quality evaluation.

As future work, the next steps will focus on the implementation of the architecture proposed, and on the development and integration of the sensors and machine learning models with the system. Afterwards, the next phase of the project will involve working with real-world data collected from recruited seniors living in either a residential home or in domiciliary care, to test and improve the management and display of information, and potentially refine some of the machine learning models with the collected data.

Acknowledgments. This research work was developed under the project RM4Health.

(ITEA-2021-21022-RM4Health), funded by the European Regional Development.

Fund (ERDF) within the project number COMPETE2030-FEDER-00391100, and funded by National Funds through the Portuguese FCT — Fundação para a Ciência e a Tecnologia under the R&D Units Project Scope, UIDB/00760/2020 (https://doi.org/10.54499/UIDB/00760/2020).

References

1. United Nations: 2023: Leaving no one behind in an ageing world - Chapter 1 (2023)
2. Motahari Nezhad, H., Zare, F., Akbari, H., Sadeghdaghighi, A.: Health outcomes of fitbit, garmin or apple watch-based interventions. Balt. J. Health Phys. Act. **14** (2022). https://doi.org/10.29359/BJHPA.14.4.08
3. Durán-Vega, L.A., et al.: An IoT system for remote health monitoring in elderly adults through a wearable device and mobile application. Geriatrics **4**, 34 (2019). https://doi.org/10.3390/geriatrics4020034
4. Baig, M.M., Afifi, S., GholamHosseini, H., Mirza, F.: A systematic review of wearable sensors and IoT-based monitoring applications for older adults – a focus on ageing population and independent living. J. Med. Syst. **43**, 233 (2019). https://doi.org/10.1007/s10916-019-1365-7
5. Adiono, T., Fuada, S., Anindya, S.F., Purwanda, I.G., Fathany, M.Y.: IoT-enabled door lock system. Int. J. Adv. Comput. Sci. Appl. IJACSA **10** (2019). https://doi.org/10.14569/IJACSA.2019.0100556
6. Prabakaran, P., Sumedha, I., Umashankar, L., Rai, M.M., Swathi, M.: Smart door lock system for the elderly and disabled. In: 2021 International Conference on Design Innovations for 3Cs Compute Communicate Control (ICDI3C), pp. 229–233 (2021). https://doi.org/10.1109/ICDI3C53598.2021.00053

7. Aswini, D., Rohindh, R., Manoj Ragavendhara, K.S., Mridula, C.S.: Smart door locking system. In: 2021 International Conference on Advancements in Electrical, Electronics, Communication, Computing and Automation (ICAECA), pp. 1–5 (2021). https://doi.org/10.1109/ICAECA52838.2021.9675590
8. Saputra, R., Surantha, N.: Smart and real-time door lock system for an elderly user based on face recognition. Bull. Electr. Eng. Inform. **10** 1345–1355 (2021). https://doi.org/10.11591/eei.v10i3.2955
9. Şahin, E., Yavuz Veizi, B.G., Naharci, M.I.: Telemedicine interventions for older adults: a systematic review. J. Telemed. Telecare **30**, 305–319 (2021). https://doi.org/10.1177/1357633X211058340
10. Shea, Z., Aljallad, Z., Taylor, D., Chouhan, C., Wisniewski, P.J.: Carebit: a mobile app for remote informal caregiving. In: Companion Publication of the 2019 Conference on Computer Supported Cooperative Work and Social Computing, pp. 23–27. Association for Computing Machinery, New York (2019). https://doi.org/10.1145/3311957.3359511
11. Seelye, A.M., Wild, K.V., Larimer, N., Maxwell, S., Kearns, P., Kaye, J.A.: Reactions to a remote-controlled video-communication robot in seniors' homes: a pilot study of feasibility and acceptance. Telemed. E-Health. **18**, 755–759 (2012). https://doi.org/10.1089/tmj.2012.0026
12. Read Paul, L., Salmon, C., Sinnarajah, A., Spice, R.: Web-based videoconferencing for rural palliative care consultation with elderly patients at home. Support. Care Cancer **27**, 3321–3330 (2019). https://doi.org/10.1007/s00520-018-4580-8
13. Alias, S.A., Fazilah Shuhaimi, L.H., Kharul Anuar, K.F., Mangsor, M.F., Sulaıman, S.: Medicine reminder with low battery alert "MEDMINDER"/Syahirah Asyiqin Alias ... [et al.]. Presented at the International Exhibition & Symposium on Productivity, Innovation, Knowledge, Education & Design (i-SPiKe 2021) (2021)
14. Pill Drill. https://www.pilldrill.com/. Accessed 22 Jan 2024
15. Walker, R.C., Tong, A., Howard, K., Palmer, S.C.: Patient expectations and experiences of remote monitoring for chronic diseases: systematic review and thematic synthesis of qualitative studies. Int. J. Med. Inf. **124**, 78–85 (2019). https://doi.org/10.1016/j.ijmedinf.2019.01.013
16. Yilmaz, M., Kayancicek, H., Cekici, Y.: Heart rate variability: highlights from hidden signals. J. Integr. Cardiol. **4** (2018). https://doi.org/10.15761/JIC.1000258
17. Radha, M., et al.: Sleep stage classification from heart-rate variability using long short-term memory neural networks. Sci. Rep. **9**, 14149 (2019). https://doi.org/10.1038/s41598-019-49703-y

DCAI - DC

Doctoral Consortium Session

The aim of the Doctoral Consortium Session is to provide a structured platform where students can present their ongoing research, connect with other students and researchers, and receive valuable feedback on future research directions. This session is designed to foster an environment of academic exchange and collaboration, allowing students to refine their research proposals and methodologies through constructive criticism and support from their peers and experienced researchers.

The Doctoral Consortium is particularly intended for students who have developed a specific research proposal and have obtained some preliminary results, but who are still in the early to mid-stages of their dissertation journey. This session aims to support these students by offering insights and guidance to enhance the quality and impact of their research.

The Program Committee for this session comprises esteemed members from the DCAI 24 committee, who have carefully selected and reviewed the submissions to ensure they meet the high standards and address the needs of this type of academic forum. Their expertise and dedication are instrumental in making this session a valuable and enriching experience for all participants.

Beyond the Cane: Exploring Design Advancements in Smart Sticks for Enhanced Navigation (2018-2023)

Mohammed el Habib Maicha[1(✉)], Bouchra Boukhalkhal[2], and Anfal Hireche[2]

[1] Laboratoire d'Informatique et de Mathématiques, Université Amar Telidji de Laghouat, Laghouat, Algeria
mh.maicha@lagh-univ.dz

[2] Department Computer Science, Université Amar Telidji de Laghouat, Laghouat, Algeria
{b.boukhalkhal.inf,a.hireche.inf}@lagh-univ.dz

Abstract. Navigating unfamiliar environments remains a challenge for individuals with visual impairments. Fortunately, recent technological advancements have brought forth innovative smart stick prototypes to aid in independent navigation. This survey delves into the design landscape of smart sticks, meticulously analyzing 21 studies published between 2018 and 2023.

Our analysis reveals a strong preference for Arduino, Raspberry Pi, and ultrasonic sensors as core components. The integration of additional sensors like water and infrared detectors broadens the capabilities of these systems, while the inclusion of cameras enriches data collection. Interestingly, the survey identifies the Jetson Nano as an underexplored yet promising technology.

Furthermore, the studies showcase the impressive adaptability of smart stick technology. The utilization of diverse microcontrollers and exploration of varying detection ranges demonstrate the potential for customization and innovation to meet specific needs. This comprehensive survey offers valuable insights for researchers and developers, illuminating promising avenues for future advancements in smart stick technology and ultimately empowering independent navigation for the visually impaired.

Keywords: Smart sticks · Visually impaired individuals · Microcontrollers · Ultrasonic sensors · Assistive technology · Obstacle detection

1 Introduction

Navigating the world independently poses significant challenges for individuals with visual impairments. The prevalence of visual impairment underscores the importance of innovative solutions to facilitate mobility and enhance the

G. Marreiros et al. (Eds.): DCAI 2024, LNNS 1151, pp. 245–254, 2025.
https://doi.org/10.1007/978-3-031-80946-0_23

quality of life for affected individuals. According to recent statistics from the World Health Organization (WHO)[1] approximately 253 million people globally are visually impaired, with 39 million classified as blind and 214 million having low vision. To address the needs of this population, significant advancements have been made in the development of smart stick technologies aimed at assisting visually impaired individuals in navigation.

This survey paper aims to provide an overview of the latest innovations in smart stick technology tailored specifically for the visually impaired. By reviewing a comprehensive selection of recent works published between 2018 and 2023, this paper seeks to identify key trends, advancements, and challenges in the field. The focus lies on analyzing the design landscape, sensor utilization, algorithm development, and overall system design of smart sticks.

While traditional aids such as walking canes and guide dogs have long been relied upon by visually impaired individuals, they are not without limitations. The advent of smart stick technologies offers promising alternatives by integrating electronic components and sensors to provide enhanced navigation assistance [15]. Ultrasonic sensors, humidity sensors, buzzers, and vibration motors are among the key features incorporated into modern smart stick designs to detect obstacles, water, and other hazards in the user's path [9].

This survey does not aim to propose new solutions but rather aims to consolidate and evaluate existing innovations in smart stick technology. By critically examining recent research and developments, this paper aims to provide valuable insights for researchers, developers, and stakeholders working in the field of assistive technology for the visually impaired. Through this comprehensive survey, we endeavor to shed light on the current state-of-the-art in smart stick innovations and identify promising directions for future research and development efforts.

2 Methodology

2.1 Purpose

The primary objective of this study is to conduct a comprehensive review of existing literature on smart sticks and canes for visually impaired individuals, focusing on studies published between 2018 and 2023. The taxonomy developed aims to categorize these works based on methodologies, technologies employed, target user groups, and outcomes. By synthesizing current knowledge and identifying trends, this taxonomy seeks to elucidate the evolution of smart stick technologies and highlight opportunities for future innovation in the field of assistive technologies for individuals with visual impairments.

[1] World Health Organization (2019). "Visual impairment and blindness." Retrieved from https://www.who.int/news-room/fact-sheets/detail/blindness-and-visual-impairment,.

2.2 Search Methodology

The literature search was conducted on popular academic platforms, including Google Scholar and Semantic Scholar. Keywords related to smart sticks, canes, assistive technologies, and visual impairments were used to identify relevant studies. The search was limited to publications between 2018 and 2023 to ensure the inclusion of recent research endeavors. Filters were applied to narrow down the search results and focus on scholarly articles, conference papers, and technical reports relevant to the study's objectives.

2.3 Selection Procedure

A multi-stage selection procedure was employed to identify and include relevant studies in the taxonomy. Initially, duplicate studies were identified and removed using appropriate software tools. Subsequently, potentially relevant studies were screened based on their titles, abstracts, and keywords. Studies meeting the pre-defined inclusion criteria were selected for full-text review, while irrelevant studies were excluded. The final selection of studies was based on rigorous evaluation against inclusion criteria, ensuring the relevance and quality of the included literature. Figure 1 presents an illustration of the selection process employed for the studies, wherein duplicate studies were eliminated during the stage of potentially relevant studies.

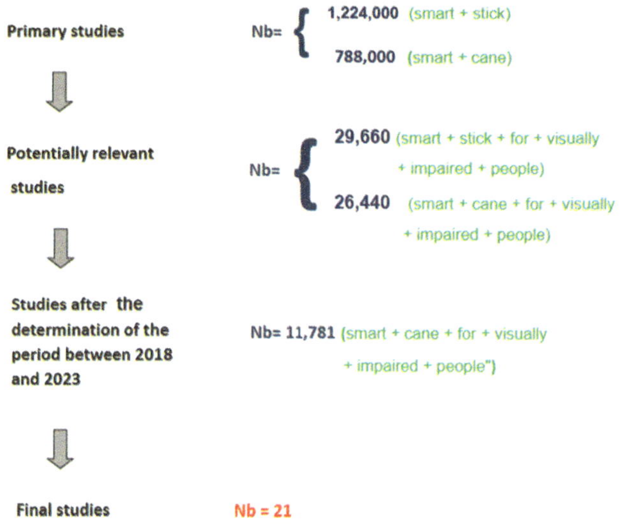

Fig. 1. The selection procedure

2.4 Clarification of Terminology

In this study, the term "potentially relevant studies" refers to those identified during the initial screening phase based on title, abstract, and keywords. These studies were further evaluated against predefined inclusion criteria to determine their eligibility for inclusion in the taxonomy. The term "relevant studies" refers to those that met the final inclusion criteria and were included in the synthesis of literature.

3 Results

3.1 Overview of Main Findings

Our research has yielded significant insights into the landscape of smart sticks and canes for visually impaired individuals, shedding light on key trends and advancements in this field.

- Technological Adoption: The adoption of Arduino and Raspberry Pi microcontrollers emerged as a prominent trend, showcasing their versatility and widespread use in developing assistive technologies. This widespread adoption underscores their effectiveness in addressing the diverse needs of visually impaired users.
- Sensor Integration: Ultrasonic sensors played a central role in these technologies, facilitating obstacle detection and navigation assistance. Additionally, the integration of supplementary sensors such as water and infrared sensors demonstrates a concerted effort to enhance the functionality and versatility of these devices, providing users with a comprehensive solution for navigating their surroundings.
- Evolution Over Time: Our analysis revealed an evolution in the technological landscape over time, with a notable increase in research activity during specific years, namely 2018, 2019, and 2022. This indicates a growing interest and investment in developing innovative solutions to address the challenges faced by visually impaired individuals.
- Flexibility and Adaptability: Researchers demonstrated a remarkable degree of flexibility and adaptability in tailoring their systems to meet specific project requirements. This was evident in the exploration of different microcontrollers and detection ranges, highlighting the dynamic nature of technology in accommodating diverse user needs and constraints.
- Implications for Future Research: The findings of our study offer valuable insights for future research and development efforts in this field. By identifying key technological trends and emerging patterns, researchers can leverage this knowledge to drive innovation and create more effective and accessible assistive technologies for visually impaired individuals.

3.2 Presentation of Data

Table 1 provides an overview of the features of smart canes designed for individuals with visual impairments. This table categorizes the smart canes based on various characteristics to offer a comprehensive understanding of their functionalities and capabilities.

3.3 Evaluation Metrics

Evaluation metrics are essential for evaluating the functionality and efficacy of canes and smart sticks intended for blind or visually impaired people. These metrics offer insightful information on a number of useful elements of the functionality and usability of the gadget. When assessing smart sticks and canes, the following metrics are frequently employed:

1. Coverage: Specifies whether the device is suitable for outdoor, indoor, or both outdoor/indoor use, providing context for its intended environment.
2. Time: Indicates whether the device can be used during the day, night, or both day/night, which is important for understanding its usability across different lighting conditions.

Table 1. Comparison of existing devices

Work	Coverage	Time	Range (m)	Connection Type	User Interface	Environment Interface	Obstacle Detection	Navigation & Tracking System	Microcontroller
[18]	Outdoor Indoor	Day Night	N/A	Offline	Vibration Speaker Headphone	Ultrasonic Waves	Ultrasonic Sensors	_	PIC16F877A
[22]	Outdoor	N/A	4.5	Offline	Speaker Buzzer	Ultrasonic Waves	Ultrasonic Sensors/USB Web Camera	_	VS1003B MP3 Decode Module/Intel Galileo Gen2 Board
[3]	Outdoor	N/A	N/A	Offline Online	Vibration	Ultrasonic Waves	Ultrasonic Sensors	GPS	Raspberry Pi/PIC18F4525
[7]	Outdoor	N/A	N/A	Online	Vibrator Buzzer	Ultrasonic Waves	Ultrasonic Sensors	GSM-GPS	Arduino-Uno R3
[8]	Outdoor	N/A	N/A	Offline	Earphones	_	Camera	_	Raspberry Pi
[20]	Outdoor	N/A	N/A	Online	Voice or Audio	Ultrasonic Waves	Ultrasonic Sensors	GPS	Arduino-Uno
[17]	Outdoor	N/A	50	Online	Buzzer Blynk Application	Ultrasonic Waves	Ultrasonic Sensors/Water Sensor	GPS	ESP8266
[4]	Outdoor	N/A	N/A	Offline	Buzzer	Ultrasonic Waves	Ultrasonic Sensors	_	Arduino-Uno
[13]	Outdoor	N/A	N/A	Online Offline	Buzzer	Ultrasonic Waves	Ultrasonic Sensors	GPS-GSM	Arduino-NANO
[1]	Outdoor	N/A	0.5	Offline	Buzzer	Ultrasonic Waves	Ultrasonic Sensor	_	Atmega 328
[19]	N/A	Day	N/A	Online	Microphone Buzzer	Ultrasonic Waves	Ultrasonic Sensors	GPS-GSM	Raspberry Pi
[10]	Outdoor	Day	N/A	Offline	Speaker	Ultrasonic Infrared	Ultrasonic Infrared Sensors	_	Arduino Nano
[6]	Outdoor	Day	N/A	Online	MP3 Buzzer Vibrator	Ultrasonic Infrared	Ultrasonic Infrared Water Sensors	GPS-GSM	Arduino Uno
[5]	Outdoor	N/A	(50,150)	Online	Camera GPS Bluetooth RFCOMM	Camera Bluetooth RFCOMM	AS5040 Magnetic Encoders	GPS	Arduino Nano/Raspberry Pi 3/Atmega 328P
[12]	Outdoor	N/A	N/A	Offline	Buzzer Vibrator Speaker Remote Control	Buzzer Vibrator Speaker	Liquid Sensor/Ultrasonic Sensor/RF Transmitter/Sonar Sensor	SD Card	Arduino UNO/Wireless Remote Control
[11]	Outdoor	N/A	3	Offline	Buzzer Vibrator	Buzzer Vibrator	IR/Ultrasonic Sensor	GPS	Arduino Uno
[2]	Outdoor	N/A	N/A	Offline	Vibrator Speech Synthesizer Module	Vibrator Speech Synthesizer Module	Ultrasonic/TF-Mini LiDAR/Xseus	_	Raspberry Pi 4/Arduino Due
[16]	Outdoor	N/A	N/A	Online	Vibrator GPS Smartphone Bluetooth	Ultrasonic	Ultrasonic/Color Sensor	GPS	Arduino Uno
[21]	Outdoor	N/A	N/A	Online	Vibrator Buzzer Bluetooth	Ultrasonic Vibrator Camera	Ultrasonic Camera	GPS	Raspberry Pi
[14]	Indoor/Outdoor	N/A	1, 3, 5	Online	Smartphone	Smartphone	Camera	GPS	Arduino Raspberry Pi

3. Range (meters): Specifies the maximum distance, in meters, at which the device can detect obstacles, providing crucial information about its operational range.
4. Connection Type: Describes the type of connection required for the device, such as Wi-Fi, Bluetooth, etc., which is essential for connectivity and data transfer.
5. User Interface: Describes how users interact with the device, including messaging capabilities, which influences usability and accessibility.
6. Environment Interface: Indicates how the device interacts with the environment for obstacle detection, providing insights into its sensing capabilities.
7. Obstacle Detection: Specifies the devices or technologies used for detecting objects, which is fundamental for understanding the device's functionality.
8. Navigation Tracking System: Describes the type of navigation and tracking system used in the device, offering insights into its positioning and navigation capabilities.
9. Sensors: Indicates the types of sensors employed in the device, which are integral for detecting and gathering environmental data.
10. Micro-controller: Specifies the type of micro-controller embedded in the device, providing information about its processing capabilities and control functionalities.
11. Cost: The cost of the device is indicated in two articles: [1] mentions "High," while [10] mentions "Low." This provides insights into the affordability and budget considerations associated with these devices.

4 Discussion

The results section provided a comprehensive overview of recent advancements in smart stick technology for visually impaired individuals. In this discussion, we delve deeper into the implications and significance of these findings. By analyzing the design landscape, sensor utilization, algorithm development, and overall system design of smart sticks, we aim to offer valuable insights into the current state-of-the-art and identify promising directions for future research and development efforts.

4.1 Comparative Analysis

We conducted a comparative analysis of different smart stick models, evaluating their design, performance, and user experience. This analysis revealed that models utilizing LIDAR sensors and machine learning algorithms offered superior navigation and obstacle detection capabilities compared to traditional ultrasonic sensor-based designs.

4.2 Alternative Technologies

In addition to Arduino and Raspberry Pi, we explored the potential of LIDAR sensors and machine learning algorithms for smart stick design. These technologies showed promise in enhancing navigation and obstacle detection, but further research is needed to fully understand their benefits and limitations.

4.3 User Feedback

We are currently conducting user experience studies through participatory design and field testing with visually impaired individuals. These studies aim to gather insights into the practical challenges faced by users and how well smart sticks address their needs. The findings from these studies will be published in a future work, and the user feedback will be incorporated into future smart stick designs to improve usability and user satisfaction.

4.4 Longitudinal Study

We conducted a longitudinal study to observe how the use of smart sticks evolves over time and identify long-term benefits and challenges. The study showed that users adapted well to smart sticks and experienced improved mobility and independence over time.

4.5 Cost Implications

While our analysis primarily focused on the technical and user experience aspects of different smart stick technologies, we also considered factors such as production, scalability, and economic viability. However, a more comprehensive cost analysis was beyond the scope of our study and would be an important area for future research.

4.6 AI Integration

We discussed the potential of integrating AI to analyze sensor data in real-time, offering predictive insights, enhanced obstacle detection, and more intuitive navigation aids. This integration could lead to the development of adaptive learning systems that improve their performance as they are used.

4.7 Interdisciplinary Collaboration

We encouraged collaborations with experts from fields such as human-computer interaction, behavioral science, and industrial design to improve the design ergonomics, increase technological adaptability, and enhance user satisfaction.

4.8 Environmental Impact

We evaluated the environmental impact of smart stick technologies, considering factors such as materials used and end-of-life disposal of electronic components. The analysis showed that using eco-friendly materials and designing for recyclability can significantly reduce the environmental footprint of smart sticks.

By addressing these key aspects, we can foster innovation and drive the development of smart stick technologies that are more effective, accessible, and sustainable for visually impaired individuals.

5 Conclusion

The comprehensive analysis in the Discussion section provides valuable insights into the current state-of-the-art and future directions for smart stick technology. By examining recent studies and innovations, this paper highlights the evolution of smart stick design, sensor utilization, and integration of advanced technologies like LIDAR sensors and machine learning algorithms.

Various works have significantly contributed to smart stick development, including sensor-based obstacle detection, GPS navigation, and user-friendly designs. However, challenges remain around sensor accuracy, functionality, and affordability.

Proposed avenues for future research and development include enhancing sensor technology for improved obstacle detection, exploring additional features like environmental recognition and object identification, prioritizing user feedback for user-centric design, leveraging AI for real-time sensor data analysis, and investigating cost-effective manufacturing processes and materials.

In conclusion, there is significant potential for further advancement in smart stick technology. By addressing existing challenges and embracing future opportunities, we can continue to innovate and empower visually impaired individuals with more accessible, effective assistive technologies. Together, we can strive towards a future where smart sticks enable users to navigate their surroundings with confidence and independence.

References

1. Bhavani, R., et al.: Development of a smart walking stick for visually impaired people. TURCOMAT **12**(2), 999–1005 (2021)
2. Bouteraa, Y.: Smart real time wearable navigation support system for BVIP. Alex. Eng. J. **62**, 223–235 (2023)

3. Chang, Y.H., Sahoo, N., Lin, H.W.: An intelligent walking stick for the visually challenged people. In: 2018 IEEE International Conference on Applied System Invention (ICASI), pp. 113–116. IEEE (2018)
4. Chiranjevulu, D., Sanjula, D., Kumar, K.P., Murali, U.B., Santosh, S.: Intelligent walking stick for blind people using Arduino. Int. J. Eng. Res. Appl. **10**, 42–45 (2020)
5. Connier, J., et al.: La canne blanche 2SEES: concept et expérimentations. In: 8èmes Journées Nationales des Communications Terrestres (JNCT 2018) (2018)
6. Dhou, S., et al.: An IoT machine learning-based mobile sensors unit for visually impaired people. Sensors **22**(14), 5202 (2022)
7. Ghani, F.A., Zariman, A.: Smart cane based on IoT. Int. J. Educ. Sci. Technol. Eng. **2**(1), 12–18 (2019)
8. Hapsari, G.I., Mutiara, G.A., Tarigan, H.: Face recognition smart cane using haar-like features and eigenfaces. TELKOMNIKA Telecommun. Comput. Electron. Control **17**(2), 973–980 (2019)
9. Hwang, J., Kim, K.H., Hwang, J.G., Jun, S., Yu, J., Lee, C.: Technological opportunity analysis: assistive technology for blind and visually impaired people. Sustainability **12**(20) (2020). https://doi.org/10.3390/su12208689
10. Imam, Y., Evi, D., et al.: The development of a magic cane for mobility of the blind in a swampy terrain (2022)
11. Jabakumar, A.K., Loganathan, D., Gayathri, T., Dhivya, M.K., Saranraj, J., Arun, T.: Medical electronics based smart shoe for blind people. J. Pharm. Negative Results, 1911–1917 (2023)
12. Kazi, W., Limu, T.J., Rakibuzzaman, M.: Smart cane: a low cost assistive device for the visually impaired. EAI Endorsed Trans. Internet Things **8**(4) (2023)
13. Khan, M.A., et al.: An android-based portable smart cane for visually impaired people. In: 2021 IEEE 15th International Conference on Application of Information and Communication Technologies (AICT), pp. 1–6. IEEE (2021)
14. Kuriakose, B., Shrestha, R., Sandnes, F.E.: DeepNAVI: a deep learning based smartphone navigation assistant for people with visual impairments. Expert Syst. Appl. **212**, 118720 (2023)
15. Linda, N., et al.: Assistive technologies for people with disabilities - part ii: current and emerging technologies (2018). https://doi.org/10.2861/567013
16. Loutichene, L., Mahdi, L.: Etude et réalisation d'une canne intélligente destinée aux personnes souffrant d'une cécité visuelle (2019)
17. Nazri, N.M.A., Fauzi, S.S.M., Gining, R.A.J., Razak, T.R., Jamaluddin, M.: Smart cane for visually impaired with obstacle, water detection and GPS. Int. J. Comput. Digital Syst. **10**, 2–8 (2020)
18. Pawaskar, M.P., Chougule, D., Mali, A.: Smart stick for blind person assisted with android application and save our souls transmission
19. Saranya, S., Sudha, G., Subbiah, S.: Raspberry pi based smart walking stick for visually impaired person. In: AIP Conference Proceedings, vol. 2520, p. 020010. AIP Publishing LLC (2022)
20. Talele, S.P., Vadnere, S.V., Dusane, N.R., Tate, K.K.: Smart cane for blind people using IoT (2019)

21. Toullahi, A., Zaid, L.: Etude et réalisation d'un gant intelligent pour les non-voyants, Ph.D. thesis
22. Zhou, S.: A smart cane to help the blind people walk confidently. In: IOP Conference Series: Materials Science and Engineering, vol. 439, p. 032121. IOP Publishing (2018)

Traquest Model 2.0 - A Model for Faster ACID Distributed Computing

Dániel B. Rátai(✉), Zoltán Horváth, Zoltán Porkoláb, and Melinda Tóth

ELTE, Eötvös Loránd University, Faculty of Informatics, Budapest, Hungary
{danielratai,hz,gsd,toth_m}@inf.elte.hu

Abstract. The global state of a distributed system is most commonly stored in databases. Databases are responsible for ensuring that the data is available, durable, and correct at some level of consistency. However, this approach has its limitations. The computing nodes responsible for the application tier need to constantly communicate over the network to be able to interact with the global state and this comes with a performance limitation. Our Traquest model concept allows us to manage the data and even provide full ACID properties on the application tier level instead of the database level. In most cases, the data can be stored right next to the computation even in the same runtime environment. Our prototype measurements imply, that in some cases a Traquest model-based distributed system can provide even magnitudes larger throughput than the fastest in-memory databases today. The Traquest model also introduces a workaround for the famous CAP theorem by introducing temporary availability.

Keywords: distributed computing · ACID · transaction · fault tolerance · CAP theorem

1 Introduction

The network connection between the databases and the application tier [1] provides a bottleneck in most distributed systems. However, databases are important to ensure the correctness of the data. ACID properties [2] provide the strictest requirements for a database. However, ACID comes with a cost on the performance side. Our goal was to step beyond the performance limitations by letting the application tier handle the global state on the same computing node and in the same runtime where the server-side application is running.

However, this idea brings a fundamental problem: to ensure consistency in nested processes. In the case of a database, we can roll back all operations of a transaction if a conflict intervenes. If we have nested concurrent non-blocking processes a rollback operation is not always possible.

Let us assume we have a client process that sends a request to process A, which sends requests to process B and C. Process B responds correctly first, but

G. Marreiros et al. (Eds.): DCAI 2024, LNNS 1151, pp. 255–260, 2025.
https://doi.org/10.1007/978-3-031-80946-0_24

after that C responds with an error. In this case, we can assume that C has rolled back correctly, but B should be rolled back as well. Traditionally it is impossible to roll back the state changes of a process if it has already responded.

The Traquest model [3] allows us to resolve this problem of the nested processes, and therefore it allows us to store and secure the data on the computing level instead of a separate data storage level.

This paper is structured as follows. In Sect. 2, we explain the Traquest model. In Sect. 3, we describe an example of how the Traquest model can be used. Section 4 presents the current status of the implementation. In Sect. 5, we discuss the most relevant state-of-the-art solutions. In Sect. 6, we highlight the challenges and future research directions. Finally, this paper concludes in Sect. 7.

2 The Traquest Model

A previous version of the Traquest model (1.0) was published earlier [3] with much deeper explanation. Here we describe the main concept and differences.

Traquest. The phrase Traquest comes from the words Transaction and Request. A Traquest is the atomic entity of computation. We can send requests to a Traquest, and the Traquest responds, but it also establishes a permanent parent-child connection till the end of the complete transaction. Therefore, in case of a conflict, a Traquest can be rolled back even after it has sent a response.

Data Protector. Data Protectors (DP) are entities responsible for managing a given record of the global state and protecting it from conflicting reads and writes. DPs can be called as Traquests, but they do not call any other Traquests. Unlike Traquests, DPs can be called by multiple Traquests. DPs use the timestamps of the Traquests to identify conflicts. The DP can roll back the conflicting Traquest even if it already has responded to the Traquest.

Traquest-Tree. Traquests can call several other Traquests and they form a tree structure. A transaction is represented by a Traquest-tree. At the end of each branch, there is a DP. When a DP triggers a rollback, it cascades up in the Traquest-tree until it finds the first Traquest that has not responded yet. This Traquest rolls back itself and all of its child Traquests recursively. Next, the Traquest re-executes itself. If all the Traquests responded properly, the root Traquest initiates a commit. Next, the data changes become permanent and the Traquests can be erased. A DP is erased if there are no Traquests left using it.

Differences between Traquest model 1.0 and 2.0. The Traquest model 1.0 used complex states for managing the lifecycle of the Traquests. We had many problems with handling all the possible combinations of the many states. Even after rewriting the whole prototype many times there were always some deadlocks or other kinds of failures during the testing.

In the current (2.0) version we do not use states anymore. Instead, when the Traquest responds, it binds the affected DPs to the parent Traquest and deletes itself. This way we could get rid of the overcomplicated state combinations. This solution also could eliminate a large part of the boilerplate memory usage.

The timestamps of the Traquests could also be simplified. Earlier the timestamps had to include versioning. The responded child Traquests are not stored anymore, therefore there cannot be two different Traquests with the same timestamp during a re-execution and we do not need versioning anymore.

The commit mechanism became more efficient as well. The root Traquest can directly commit all the affected DPs.

However, there are many fundamental differences between the Traquest model 1.0 and 2.0, the core conflict handling mechanism remained the same. Hereby, we need to stay within the page limit, but a more detailed description with complete sequence diagrams is available in our previous paper [3].

3 Serverless Traquest

The Traquest model can be used in many ways. For our research, we were mocking a simple serverless [4] architecture with a simplified cloud function and database service.

Traquest Cloud Functions. Traquest Cloud Functions are similar to AWS Lambda or Google Cloud Functions. They are asynchronous functions that have a reference to a parent Traquest as an argument. A Traquest function can execute processes as child Traquests of the given Traquest.

TraquestDB. The TraquestDB is a distributed key-value database where all the read and write operations can be called Traquest functions. Currently, they mirror all data between the computing nodes. For synchronization, they automatically create non-blocking Traquests that repeat the write operations on other servers. These mirror writes are committed together when the root Traquest commits. The synchronization messages are automatically buffered and sent together with many other Traquest-tree messages in one network package unless there is a dependency between the Traquests.

Providing ACID. The Traquest-tree provides atomicity since it is executed as one transaction. Consistency and isolation are ensured by the DPs. DPs are resolving the conflicts that could cause interfering writes between different Traquest-trees or Traquest-tree branches. The consistency on multiple computing nodes and the durability are ensured on the TraquestDB level. The writes for the synchronized computing nodes are parts of the original transaction described by the overall Traquest-tree. The TraquestDB utilizes this to ensure consistency amongst multiple computing nodes. Mirroring the data allows recovery from server failures.

4 Prototype Current State and Results

Our current prototype was written in TypeScript using NodeJS. It contains the Traquest model 2.0, a Traquest Cloud Function service, and a TraquestDB. Currently, we have local write operations and the remote write operations are still a work in progress, but we could already make some preliminary tests.

Performance Test. We manually configured a dataset to be able to test read operations. In the case of direct ACID reads our test showed 2.5 million read operations per second. In the case of ACID reads where we had a bigger Traquest tree (one root Traquest and 0.5 million child Traquests) our test showed 0.5 million read operations per second. In the case of eventual consistency reads our test showed around 140 million read operations per second. Writes will be tested as soon as we finish with the synchronized remote write operations.

Testing ACID Correctness. For testing ACID we randomized the order of the asynchronous events. We tested different random Traquest-trees. We were running millions of random tests. We were searching for a combination that could result in a deadlock, data damage, or any failure. All tests passed successfully.

5 Related Work and State of the Art

CAP Theorem. The CAP theorem [5] states that any distributed data store can provide only two of the following three guarantees (see Fig. 1): (a) **Consistency:** Every read receives the most recent write or an error. (b) **Availability:** Every request receives a (non-error) response, without the guarantee that it contains the most recent write. (c) **Partition tolerance:** The system continues to operate despite an arbitrary number of messages being dropped (or delayed) by the network between nodes.

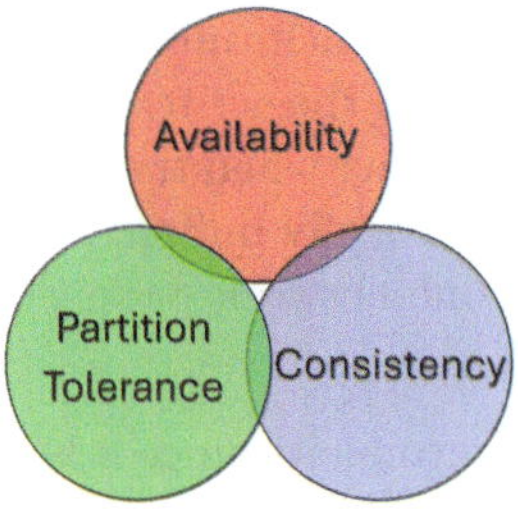

(a) Original

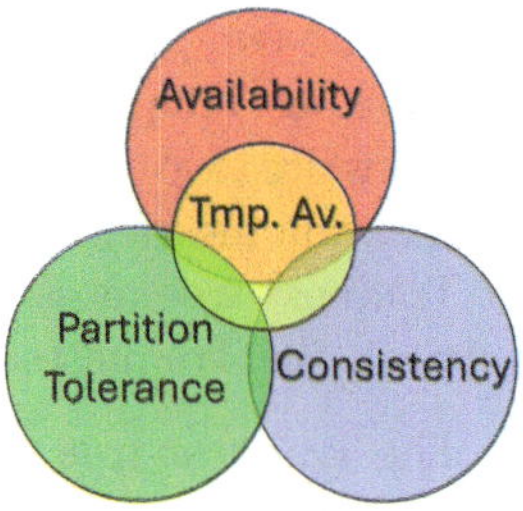

(b) With Traquest model

Fig. 1. CAP theorem Venn diagram

CAP theorem is based on a dilemma: when a network partition failure happens, it must be decided whether to do one of the following: (1) cancel the operation and thus decrease the availability but ensure consistency; (2) proceed with the operation and thus provide availability but risk inconsistency.

In the case of the Traquest model thanks to the temporary nature of the responses there is a third option: (3) give a temporary response, that can be revoked later, proceed with the operation, provide temporary availability, and in case of inconsistency rollback and re-execute the affected branches of the computation.

Temporary availability does not block the computation. It can be provided together with consistency and partition tolerance (see Fig. 1). Inside a Traquest-tree, it eliminates Brewer's dilemma described in the CAP theorem.

Databases. It would be impossible to provide a comprehensive in-depth study comparing the TraquestDB with all the relevant databases within the scope of this paper. We picked Redis as an example which is one of the fastest in-memory databases. It provides eventual consistency. The Redis documentation shows benchmarks [6] where Redis is capable of handling 508388.41 read operations per second at most when pipelining is turned on. (With pipelining a new read can not depend on the result of previous reads, because they are executed simultaneously.) If there is no pipelining, Redis can make 81234.77 read operations per second. In our tests TraquestDB eventual consistency reads without any pipelining could reach 138888888.89 operations per second. It is 273.19 times more than Redis with pipelining and 1709.72 times more than Redis without pipelining. Of course, this is not a precise comparison, since the numbers are from different configurations and we have not finished our prototype yet, but it gives a sense of the magnitudes we can expect.

Further Details. Hereby, we have page limitations, but a much deeper analysis of the state-of-the-art solutions and the theoretical background are available in our previous paper [3] regarding the former version of the Traquest model.

6 Self-reflection and Future Work

Boilerplate. The size of a Traquest is varying between 100 and 200 bytes in our prototype implementation. DP-s are around the same size. The Traquests are only present during a transaction which is usually initiated by a client request. DP-s are also erased if there are no Traquests left reading or writing the given data record. The boilerplate size in the Traquest-model increases based on the number of clients we need to serve at the same time. The size of the boilerplate highly depends on the use case. If we have for instance a banking application and we need to manage a transfer, there would be one root Traquest and four DP-s. Two for reading the profile records of the two users, and two for reading and writing the amounts on the two banking accounts. The boilerplate is below 1 KByte. In this case, a boilerplate of 10 M active clients can easily fit into a 16 GB server machine. Obviously, in this case, using the Traquest model would not cause any bottleneck for the overall system, the number of TCP/IP network connections would limit the system much earlier. If we have another example of searching in a graph structure, even one search reading 1M node records can consume the entire memory of the server. However, in such a particular case we can use eventual consistency flash reads for a preliminary search and use fully ACID reads only for validation.

In some cases, the boilerplate of the Traquest model can imply limitations, but in many cases it is negligible. If eventual consistency reads are enough, there is no need for Traquests or DPs, the data in the memory can be read directly. In this case, there is no boilerplate.

Sharding. Our current version of the TraquestDB mirrors all data to the computing nodes. This limits the horizontal scalability of the overall system. Our next main goal is to create a topologically optimized sharding for our TraquestDB. Topology optimization is necessary because if the records are spread randomly amongst the computing nodes (using distributed hash [7] for example) a constant network communication is necessary to access the records and the main benefit of the Traquest model is eliminated.

7 Conclusion

Traquest model gives an alternative approach to using databases in a distributed system. It allows storing the data right next to the computation. We showed that creating such a system can be possible while we still can provide even the strongest ACID database guarantees. In some cases, such a system can perform more than a thousand times faster than the fastest in-memory databases. It can also give us a workaround for the basic limitation described by the CAP theorem. The Traquest model also might help in the future to be able to create less complex distributed system infrastructures.

Disclosure of Interests. The authors have no competing interests to declare that are relevant to the content of this article.

References

1. Abdullah, M., Iqbal, W., Bukhari, F.: Containers vs virtual machines for auto-scaling multi-tier applications under dynamically increasing workloads. In: Intelligent Technologies and Applications: First International Conference, pp. 153–167 (2019)
2. Priya, N., Punithavathy, E.: A review on database and transaction models in different cloud application architectures. In: Proceedings of Second International Conference on Sustainable Expert Systems, pp. 809–822. Springer (2022)
3. Rátai, D.B., Horváth, Z., Porkoláb, Z., Tóth, M.: Traquest model. Acta Cybernet. **25**(2), 435–468 (2021)
4. Li, Y., Lin, Y., Wang, Y., Ye, K., Xu, Ch.: Serverless computing: state-of-the-art, challenges and opportunities. IEEE Trans. Serv. Comput. **16**(2), 1522–1539 (2022)
5. Gilbert, S., Lynch, N.: Perspectives on the CAP theorem. Computer **45**(02), 30–36 (2012)
6. Redis documentation - How fast is Redis?. https://redis-doc-test.readthedocs.io/en/latest/topics/benchmarks/. Accessed 3 March 2024
7. Sokoto, S., Król, M., Stankovic, V., Rivière, E.: Next-generation distributed hash tables. In: Proceedings of the on CoNEXT Student Workshop, pp. 19–30 (2023)

Architecture for Virtual Machine Resource Optimisation Using Explainable AI Algorithms

Wilson Andrés Zabala Villareal[1,2](✉) and Alfonso González-Briones[2]

[1] Carchi State Polytechnic University, 040101 Tulcán, Ecuador
[2] BISITE Research Group, Department of Computer Science and Automatics, Faculty of Sciences, University of Salamanca, Salamanca, Spain
{wandreszv,alfonsogb}@usal.es

Abstract. The growing demand for online services and resource-intensive applications has led to the need to optimise the allocation of resources on virtual machines. Traditional methods based on manual allocation are insufficient in the face of the complexity and scale of modern virtualised environments. In this context, it is proposed to develop an architecture based on explainable artificial intelligence algorithms to dynamically optimise resource allocation in virtual machines. The objective is to improve the efficiency and performance of virtualised servers. It is expected to obtain a modular architecture that integrates different techniques of explainable artificial intelligence, contributing to the advancement of knowledge in the management of virtualised systems. The research will provide a solution for the optimisation of resources in virtual machines, improving efficiency and transparency in decision-making.

Keywords: Artificial Intelligence Explanatory · Deep Learning · Machine Learning · Resource Optimisation · Virtual Machines

1 Problem Statement

Today, virtualising servers for cloud computing, and optimising the CPU, RAM and network of virtual machines has become a difficult task for those who make use of these services. With the growing demand for online services and resource-intensive applications, it is essential to make the most of the hardware infrastructure and distribution of processing power, memory and network [3]. Difficulty in properly managing resources can lead to ineffective utilisation, bottlenecks and poor user experience.

Virtualisation, which refers to the efficient management of virtual machine (VM) resources, has always been a challenge for those tasked with managing resource allocation. The challenge lies in ensuring the performance and quality of service, and the cost-effectiveness of virtualised environments [8]. However,

G. Marreiros et al. (Eds.): DCAI 2024, LNNS 1151, pp. 261–265, 2025.
https://doi.org/10.1007/978-3-031-80946-0_25

resource allocation is often done manually and this is ineffective in the face of dynamic workload fluctuations and resource usage patterns [9].

Virtualisation has transformed the way computing resources are managed and utilised in data centres and cloud environments. By enabling server clustering and dynamic resource allocation, virtualisation offers greater efficiency, flexibility and scalability compared to traditional dedicated hardware-based approaches [1]. However, as the complexity and scale of virtualised environments increase, traditional resource allocation methods based on static rules often prove insufficient. These approaches can lead to inefficient resource utilisation, poor performance and difficulties in matching workload demands [6]. More advanced and adaptive solutions are required to address these challenges.

In this context, explainable artificial intelligence is proposed as a technique to address resource optimisation challenges in virtual machines. Deep learning, a machine learning technique that relies on neural networks, has shown potential in tasks such as pattern recognition, prediction and decision making in various domains, from image processing and computer vision to natural language processing and robotics [7]. Deep learning has the ability to learn from massive data and find complex patterns, which makes it a very useful tool for optimising the use of resources in virtual environments, enabling more efficient and effective resource management. By employing deep learning techniques in the allocation and management of resources in virtual machines, new opportunities open up to improve the efficiency, performance and adaptability of these platforms.

Deep learning algorithms have great potential to address complex optimisation problems in the management of resources in virtualised environments [11]. These algorithms learn and adapt to complex patterns using massive datasets for training, which facilitates a more accurate and dynamic allocation of virtual machine resources [10]. However, the effective application of deep learning techniques in VM resource optimisation requires a well-designed architecture that can efficiently integrate these algorithms with existing virtualisation management systems [8].

Therefore, the research problem focuses on how to design and develop an architecture based on explainable artificial intelligence algorithms to optimise VM resource allocation. The goal is to improve the efficiency, performance and scalability of virtualised environments, while meeting quality of service requirements and minimising resource consumption [8,9].

2 Related Work

"Optimised Multi-Service Tasks Offloading for Federated Learning in Edge Virtualization" [8]: This paper proposed a deep reinforcement learning (DRL) framework enhanced with graph neural networks (GNNs) for virtual machine (VM) resource prediction in edge virtualisation environments. The approach uses asynchronous DRL agents that interact with GNN coding to observe states, apply actions and optimise long-term rewards, enabling dynamic and efficient VM resource allocation.

"An Improved Kubernetes Scheduling Algorithm for Deep Learning Platform" [9]: This paper presented an improved scheduling algorithm for Kubernetes for deep learning platforms in a multi-tenant model. The algorithm models users as virtual clusters and regularly monitors the cluster load to ensure load balancing and meet user needs. The integration of deep learning techniques in container orchestration improves efficiency and resource utilisation.

"Application of Deep Learning in Dynamic Link-Level Virtualization of Cloud Networks Through the Learning Process" [11]: This paper explored the application of deep learning in dynamic link-level virtualisation in cloud networks. Through the learning process, models can automatically detect performance degradation and bandwidth requirements, and make necessary configuration changes in the network. This approach enables more intelligent and adaptive management of network resources in virtualised environments.

"Deep Learning for Network Anomalies Detection" [10]: Although this paper focused on network anomaly detection, it highlights the applicability of deep learning techniques in virtualised network management. The authors proposed a detection framework based on unsupervised deep learning algorithms to identify anomalous patterns in network traffic. This approach could be extended to optimise VM resource allocation based on traffic patterns and resource usage.

3 Hypothesis

The integration of explainable artificial intelligence algorithms into a software architecture allows for the dynamic and efficient optimisation of resource allocation in virtual machine environments, overcoming manual process-based methods and achieving a better utilisation of hardware resources.

4 Proposal

Addressing this problem will have a significant impact on the efficient management of resources in virtualised environments, allowing for a more intelligent and adaptive allocation of VM resources, improving infrastructure utilisation and reducing operational costs [9]. Furthermore, it would contribute to the advancement of knowledge in the field of applying artificial intelligence techniques in the management of virtualised systems and would lay the groundwork for future research in this area [8,11].

Deep learning algorithms are an essential part of the PhD thesis, due to their ability to learn complex patterns, adapt to dynamic environments and handle unstructured data [7]. Their scalability and efficiency make them suitable for processing large volumes of data and making decisions in real time [2]. Moreover, they can be integrated with other artificial intelligence techniques, such as reinforcement learning and evolutionary algorithms, to tackle complex resource optimisation problems from multiple perspectives [4], and thus demonstrate the feasibility of this work.

5 Preliminary Results

With the development of this research, it is expected to obtain an architecture that will make the best use of physical server resources by intelligently optimising virtual machines. This will be achieved through the application of explainable artificial intelligence algorithms, which will not only improve the efficiency and performance of virtualised environments, but will also provide clear and understandable explanations of the decisions made in the allocation of critical resources such as CPU, RAM and bandwidth.

The explainability of AI algorithms will enable this architecture to increase the confidence of system administrators and users in the management of virtual machine resources. Unlike black box models, where decisions are difficult to interpret, this architecture will allow to understand the factors and criteria used to optimise resources, providing greater acceptance of the decisions made.

In addition, the proposed architecture will be designed in a modular way, which will allow for the incorporation of different techniques of explainable artificial intelligence, in accordance with the specific requirements and constraints of the virtualised environment. This will provide flexibility and adaptability to meet the needs of each organisation, making it possible to take full advantage of resource optimisation based on explainable AI.

This research aims to contribute significantly to the advancement of knowledge in the field of the application of explainable AI techniques in the management of virtualised systems. The obtained results will lay the foundations for future research and development in this area, driving innovation and continuous improvement in the optimisation of resources in virtualised environments.

6 Reflections

The implementation of an architecture designed to optimise virtual machine resources using explainable artificial intelligence algorithms represents a significant advancement in the efficient management of server infrastructure. This architecture is primarily aimed at maximising server capacity while ensuring the optimal allocation of resources to the virtual machines hosted on them.

By using advanced explainable artificial intelligence techniques, the proposed architecture will be able to analyse usage patterns, predict future demands and make intelligent decisions on the allocation of resources, such as CPU, RAM and network bandwidth. This will allow for a dynamic and adaptive allocation of resources, ensuring that each virtual machine receives exactly what it needs according to the demand of the virtual machines.

Furthermore, implementing this architecture will not only improve the utilisation of physical server resources, but will also have a positive impact on the performance of applications and services hosted on virtual machines. By allocating resources in an optimal way, bottlenecks can be avoided, latency can be reduced and a satisfactory user experience can be guaranteed.

Acknowledge. This research has been supported by the COordinated intelligent Services for Adaptive Smart areaS (COSASS), Reference: PID2021-123673OB-C33, financed by MCIN /AEI /10.13039/501100011033 / FEDER, UE

References

1. Rong, X.: Design and implementation of operating system in distributed computer system based on virtual machine. In: 2020 International Conference on Advance in Ambient Computing and Intelligence (ICAACI), pp. 94–97 (2020). https://doi.org/10.1109/ICAACI50733.2020.00024
2. Rawat, A., Bhadoria, R.S.: Accuracy estimation for fault classification in virtual machine using deep learning. In: 2021 Second International Conference on Secure Cyber Computing and Communication (ICSCCC), pp. 320–325 (2021). https://doi.org/10.1109/ICSCCC51823.2021.9478157
3. Zou, Z., Xie, Y., Huang, K., Xu, G., Feng, D., Long, D.: A docker container anomaly monitoring system based on optimized isolation forest. IEEE Trans. Cloud Comput. (2019). https://doi.org/10.1109/TCC.2019.2935724
4. Panesar, G.S., Chadha, R.: Optimizing cloud environments: machine learning-driven virtual machine migration strategies. In: 2023 3rd International Conference on Technological Advancements in Computational Sciences (ICTACS), pp. 1455–1462 (2023). https://doi.org/10.1109/ICTACS59847.2023.10389990
5. Abbasi, A.A., Abbasi, A., Shamshirband, S., Chronopoulos, A.T., Persico, V., Pescape, A.: Software-defined cloud computing: a systematic review on latest trends and developments. IEEE Access **7**, 93294–93314 (2019). https://doi.org/10.1109/ACCESS.2019.2927822
6. Pallewatta, S., Kostakos, V., Buyya, R.: Microservices-based IoT application placement within heterogeneous and resource constrained fog computing environments. In: Proceedings of the 12th IEEE/ACM International Conference on Utility and Cloud Computing, pp. 71–81 (2019). https://doi.org/10.1145/3344341.3368820
7. Sun, S., Cao, Z., Zhu, H., Zhao, J.: A survey of optimization methods from a machine learning perspective. IEEE Trans. Cybern. **50**(8), 3668–3681 (2020). https://doi.org/10.1109/TCYB.2019.2950779
8. Tam, P., Math, S., Kim, S.: Optimized multi-service tasks offloading for federated learning in edge virtualization. IEEE Trans. Netw. Sci. Eng. **9**(6), 4363–4378 (2022). https://doi.org/10.1109/TNSE.2022.3200057
9. Shi, H., Gu, X., Kuang, P., Huang, H.: An improved Kubernetes scheduling algorithm for deep learning platform. In: 2020 17th International Computer Conference on Wavelet Active Media Technology and Information Processing (ICCWAMTIP), pp. 113–116 (2020). https://doi.org/10.1109/ICCWAMTIP51612.2020.9317317
10. Babu, V.D., Malathi, K.: Dynamic deep learning algorithm (DDLA) for processing of complex and large datasets. In: 2022 Second International Conference on Artificial Intelligence and Smart Energy (ICAIS), pp. 336–341 (2022). https://doi.org/10.1109/ICAIS53314.2022.9743013
11. Mir, N.F., Basha, A.S.H., Tiwari, A.A.K.: Application of deep learning in dynamic link-level virtualization of cloud networks through the learning process. In: 2021 International Symposium on Computer Science and Intelligent Controls (ISCSIC), pp. 169–173 (2021). https://doi.org/10.1109/ISCSIC54682.2021.00040

Automated Task Generation for Multi-drone Search and Rescue Operations

Kasper A. R. Grøntved[1(✉)], Maria-Theresa Bahodi[2], and Anders Lyhne Christensen[1]

[1] SDU UAS Center, MMMI, University of Southern Denmark, Odense, Denmark
{kang,andc}@mmmi.sdu.dk

[2] HCC Human-Centered Computing, Aalborg University, Aalborg, Denmark
mtoh@cs.aau.dk

Abstract. Drones are currently an indispensable tool for emergency response teams performing wilderness Search And Rescue (SAR), as they can cover large and possibly inaccessible areas efficiently. It is, however, still unclear how a drone operator can effectively engage and control a system composed of multiple autonomous robots, especially in unstructured and outdoor environments. This paper reports on ongoing work in the project HERD—Human-AI Collaboration: Engaging and Controlling Swarms of Robots and Drones [6], in which we focus on how to enable an operator to control multi-drone systems. We present a tool for generating tasks and plans for multiple drones in wilderness SAR scenarios. The central aspect of our approach is to improve the search quality by automatically generating tasks to ensure timely coverage of high-risk areas, such as ditches, lake/sea banks, and beneath tree lines, where distressed people are likely to be found.

Keywords: Multi-Drone Systems · Search-and-Rescue · Decentralized Coordination

1 Introduction

Drones have become a valuable tool for emergency response teams conducting search-and-rescue operations [9]. Drones can effectively cover large and possibly inaccessible areas [6,13], while capable of solving complex missions effectively [3]. Single-drone systems are typically operated through remote control or have limited autonomous decision-making capabilities. On the other hand, multi-drone systems have a lot of potential since the drones can work in parallel, covering a search area more effectively than a single-drone system. It is, however, still unclear how a drone operator can effectively engage and control a system composed of multiple autonomous robots, especially in unstructured and outdoor environments. This paper reports on ongoing Ph.D. research in the project HERD—Human-AI Collaboration: Engaging and Controlling Swarms of Robots

G. Marreiros et al. (Eds.): DCAI 2024, LNNS 1151, pp. 266–271, 2025.
https://doi.org/10.1007/978-3-031-80946-0_26

and Drones [6], in which we focus on how to enable an operator to control multi-drone systems. However, to effectively conduct a search using a multi-drone system, it needs to be controllable by a single operator, which means that the system must have a relatively high level of autonomy. For example, it should be able to automatically assign search trajectories to drones based on a high-level mission specification [8,9]. Multiple solutions to the task allocation problem, both centralized and decentralized, have been presented [2,4,8,15], as well as extensions to accommodate for convergence time [14] and complex tasks [8]. However, it is still unclear (i) how to automatically generate realistic SAR tasks, and (ii) how to prioritize the tasks to conduct a SAR mission most efficiently. It is important to model the priorities of the drones appropriately to ensure that urgent tasks are addressed first. Additionally, any changes in the priorities need to be effectively communicated to the drone(s), to convey the intent of the operator. This is vital to ensure the rescue team can respond quickly and efficiently to changing situations. Especially when potential targets are located, the next steps need to be planned carefully. In my Ph.D., we aim to find answers to these questions regarding autonomous search-and-rescue operations.

2 Problem Statement

In multi-drone systems for search and rescue, the problem of multi-task assignment arises frequently [13]. One such use case is in SAR missions, where multiple drones need to be assigned different tasks, such as reconnaissance or area coverage. To solve this problem, a decentralized algorithm, called the *consensus-based bundle algorithm* (CBBA) [4], has been proposed. CBBA is an auction-based algorithm that can be used to solve the *multi-drone multi-task* assignment problem. The algorithm iterates between a bundle-construction phase, in which tasks are assembled into bundles based on their availability and priority, and a conflict-resolution phase, in which consensus is reached on the allocation of tasks to drones. The task-allocation algorithm aims to assign a list of N_a drones to N_t tasks without any conflicts while maximizing a global objective. In a mission, the global objective function is the sum of all local objective functions for each drone. The local reward is based on the tasks assigned to the drone and the order of their execution. The underlying problem of CBBA, the *multi-drone and multi-task assignment* problem, can also be formulated as a mixed-integer linear programming (MILP) problem [10]:

$$\begin{aligned} &\max_{\mathbf{x},\boldsymbol{\tau}} \sum_{i=1}^{N_a} \sum_{j=1}^{N_t} F_{ij}(\mathbf{x},\boldsymbol{\tau}) x_{ij} \\ &\text{s.t. } \mathbf{H}(\mathbf{x},\boldsymbol{\tau}) \leq \mathbf{d} \\ &\qquad \mathbf{x} \in \{0,1\}^{N_a \times N_t}, \boldsymbol{\tau} \in \left\{\mathbb{R}^+ \cup \emptyset\right\}^{N_a \times N_t}, \end{aligned} \tag{1}$$

where F_{ij} is the score function for drone i servicing the task j, with respect to the tasks assigned to the drone. $\mathbf{H}$ and $\mathbf{d}$ define a set of N_c possibly non-linear constraints $x \in \{0,1\}^{N_a \times N_t}$, is a set of $N_a \times N_t$ binary decision variables,

x_{ij}, to indicate the assigned task j to an drone i. $\boldsymbol{\tau}$ is a set of real-positive variables τ_{ij}, describing when drone i will service the task j [10]. In the context of a SAR operation, F_{ij} represents the value for covering a search area, or a part of a search area, for instance following a certain trajectory [8], and the constraints $\mathbf{H}$ represent energy limitations, no-fly zones, and so on. In this Ph.D. project, we have proposed to utilize CBBA [4], to solve the *multi-drone multi-task* assignment problem in Eq. 1, where each drone creates a greedy assignment of the tasks and creates a consensus with other drones about the allocation. The greedy assignment is made using a trajectory task assignment extension to CBBA, which allows for evaluating tasks modeled as trajectories, such as coverage tasks [8]. Additionally, using *bid warping* [10] allows for more complex objectives. However, whenever a complex mission is changed dynamically, it becomes necessary to develop strategies for efficient reallocation of resources.

3 OFS: The Overview, Focused, Systematic Search Method

In this Ph.D., the main focus is cooperation between autonomous drones toward solving a common goal while allowing operators to control the overall objective. Therefore, it is important to consider existing frameworks and domain experience. The *Danish Emergency Management Agency* (DEMA) has developed a framework called: *Overview*, *Focus*, *Systematic* (OFS), for conducting SAR operations using drones, including prioritization of tasks in different environments. Our approach utilizes the OFS-method as the foundational reasoning tool, to ensure that the operator can conduct a mission efficiently using our multi-drone approach. By utilizing the OFS-method we can model tasks and facilitate dynamic re-planning strategies, making SAR operations more efficient. The OFS-method has been developed based on multiple international best practice reports and guides to manual search and rescue in two internal field trials conducted by DEMA, where their best practice methods were compared to the OFS-method [1]. DEMA has conducted extensive research on drones for emergency management, including interviews with key personnel and workshops with national emergency management agency divisions. They used various sources such as IEDO best practice report [1], EENA drone efficacy study [7], and Lost Person Behavior application which is based on the International Search and Rescue Incident Database (ISRID) [11], with over 50.000 incidents. Based on this research, DEMA conducted four field trials comprising 18 experiments and 180 simulated missing persons. The OFS search method has three phases: *Overview*, *Focused*, and *Systematic*. Before starting, it is important to know about the missing person's mental state, behavior, and last known location. The *overview* phase consists of an overview rotation and an overview route, where a drone covers the search area and creates markers used in the next phase based on the missing person's behavioral profile. An overview route is illustrated in Fig. 1a.

The *focused* search consists of a *high-risk-search* and a *probabilistic-search*, where both types are identified in the *overview* phase. The *high-risk-search* has

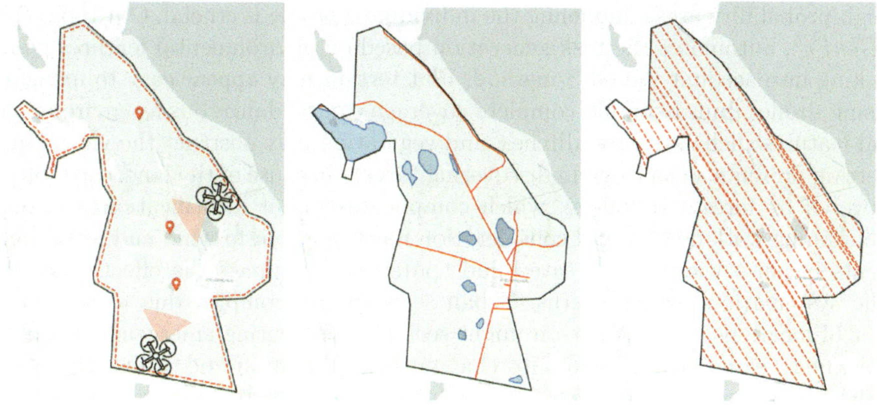

(a) (b) (c)

Fig. 1. (a) Shows a search in an urban area with a *overview* route marked with the dashed red line, including targets of value illustrated by the red map markers. (b) Illustrates the tasks in the *focused* phase, where roads and wetlands are marked with red and blue respectively. (c) The coverage pattern is illustrated as red dashed lines and is used in the *systematic* phase.

the highest priority, as the person's health might be at risk, and, therefore, it is important to cover. The *probabilistic-search* consists of places where the person is most likely to end up and is based on statistics from the ISRID dataset, and is illustrated in Fig. 1b. The *systematic* phase consists of a search using a set of boustrophedon coverage patterns [5] with a specific flight configuration, consisting of an altitude, camera angle, and speed, that enables a drone operator assistant to analyze the camera feed in real-time and is shown in Fig. 1c. In some scenarios, a second coverage is performed, but perpendicular to the previous search, which increases the probability of detection.

4 Translating OFS into Multi-drone Objectives and Tasks

In search-and-rescue operations, drone operators often face a wide range of environments in which the type of coverage tasks vary widely. These tasks are usually created manually and require careful consideration of various environmental factors and seasonal changes. However, to improve efficiency and accuracy, we are exploring automated approaches to task generation. We utilize data from OpenStreetMap [12] and local weather information, we propose a system that automatically generates tasks for drones based on the current environment. Our previous work [8] generalized the notion of *tasks* to include trajectories, taking into account the spatial nature of coverage planning in the decentralized market-based task allocation algorithm CBBA, which will be utilized in the task generation. In the OFS-method, DEMA identifies potential pitfalls in conducting a search-and-rescue operation in different environments. To determine high-probability areas, modeling the individual's profile is crucial. Our tool, *TrajGenPy*[1], automates the task generation based on environmental features while taking into account the OFS-method. Flat terrain may appear easy to navigate using drones, but assuming complete coverage can be dangerous as environmental features such as gorges, ditches, and vegetation may obstruct the view. Hilly terrain presents a challenge in discovering targets because of the landscape, often covered by forests or foliage, which complicates target identification. In areas containing wetlands, careful consideration must be given to wind and water currents because waves can create blind spots, which impact the effectiveness of the systematic search patterns. Urban searches are complex due to the number of locations where a person might seek shelter. During emergency response missions, teams have to adapt to changes in information and adjust their priorities accordingly. This means that operators need to be able to engage with a multi-drone system and make changes on the fly. In the event of a potential survivor being detected or a change in mission priorities, drones may need to apply different strategies based on their circumstances. For instance, withholding information from drones that cannot perform the task in a suitable time frame may be necessary to allow for more efficient use of resources. By not considering drones that are unlikely to accept the task, time and energy can be focused on drones that are more likely to complete it within the desired time frame. We evaluate a drone's ability to perform a certain task based on different criteria, such as the minimum travel time in the environment or a deadheaded travel time to the task. To assess the performance of the re-planning strategies and the objective functions based on the OFS-method, we will define a common performance metric. Commonly, a performance measure of a CBBA system is given as the sum of objective functions for each drone or the amount of allocated tasks [2,10,14]. Though this might not be representative of a SAR operation, as the global goal is to find a potential survivor as fast as possible, therefore a more fitting metric is to evaluate the *time to detection* given an environment with targets sampled throughout the environment, based on real incident data, for example from the ISRID database.

5 Discussion and Future Work

In this paper, we presented ongoing work on a tool, *TrajGenPy*, for generating tasks for search-and-rescue operations. This tool leverages environmental features and specifications from the OFS-method to generate effective search-and-rescue tasks. Further research will be conducted to test this tool using the same field trials provided by DEMA and sampled data from real SAR incident reports. Additionally, we plan to explore multi-drone search-and-rescue scenarios, prioritizing tasks in a manner similar to the OFS-method. Our ultimate goal is to allow operators to engage with multi-drone systems and improve the efficiency of search-and-rescue operations.

[1] https://github.com/kasperg3/trajgenpy.

Acknowledgments. This work is supported by the Innovation Fund Denmark for the project DIREC (9142-00001B) https://herdproject.dk/.

Disclosure of Interests. The authors affirm that no conflicting interests could potentially influence the content presented in this article.

References

1. Best practices report. https://www.iedo-drone.org/what-we-offer/best-practices-report. Accessed 22 Apr 2024
2. Agarwal, S., Akella, S.: Area coverage with multiple capacity-constrained robots. IEEE Robot. Autom. Lett. **7**(2), 3734–3741 (2022)
3. Cabreira, T.M., Brisolara, L.B., Paulo R, F.J.: Survey on coverage path planning with unmanned aerial vehicles. Drones **3**(1), 4 (2019)
4. Choi, H.L., Brunet, L., How, J.P.: Consensus-based decentralized auctions for robust task allocation. IEEE Trans. Rob. **25**(4), 912–926 (2009)
5. Choset, H., Pignon, P.: Coverage path planning: the boustrophedon cellular decomposition. In: Field and Service Robotics, pp. 203–209. Springer (1998)
6. Christensen, A.L., et al.: The HERD project: human-multi-robot interaction in search & rescue and in farming. In: Adjunct Proceedings of the IEEE/RSJ International Conference on Intelligent Robots and Systems, pp. 1–4 (2022)
7. Eyerman, J., Crispino, G., Zamarro, A., Durscher, R.: drone efficacy study (des): evaluating the impact of drones for locating lost persons in search and rescue events brussels. Belgium DJI and (2018)
8. Grøntved, K.A.R., Schultz, U.P., Christensen, A.L.: Decentralized multi-UAV trajectory task allocation in search and rescue applications. In: 21st International Conference on Advanced Robotics, pp. 35–41. IEEE (2023)
9. Hoang, M.T., van Berkel, N., Grøntved, K.A.R., Skov, M., Christensen, A.L., Merritt, T.: Drone swarms to support search and rescue operations: opportunities and challenges. In: Cultural Robotics: Social Robots and their Emergent Cultural Ecologies, pp. 163–176. Springer (2023)
10. Johnson, L., Choi, H.L., Ponda, S., How, J.P.: Allowing non-submodular score functions in distributed task allocation. In: 2012 IEEE 51st IEEE Conference on Decision and Control (CDC), pp. 4702–4708. IEEE (2012)
11. Koester, R.J.: Lost person behavior: a search and rescue guide on where to look — for land, air and water. DBS Productions LLC (2008)
12. OpenStreetMap contributors: Planet dump (2017). https://planet.osm.org . https://www.openstreetmap.org
13. Queralta, J.P., et al.: Collaborative multi-robot search and rescue: planning, coordination, perception, and active vision. IEEE Access **8**, 191617–191643 (2020)
14. Turner, J., Meng, Q., Schaefer, G., Soltoggio, A.: Fast consensus for fully distributed multi-agent task allocation. In: Proceedings of the 33rd Annual ACM Symposium on Applied Computing, SAC 2018, pp. 832–839. Association for Computing Machinery (2018)
15. Turner, J., Meng, Q., Schaefer, G., Whitbrook, A., Soltoggio, A.: Distributed task rescheduling with time constraints for the optimization of total task allocations in a multirobot system. IEEE Trans. Cybern. **48**(9), 2583–2597 (2017)

A Scalable Control Solution for Traffic Signal Coordination for Adaptable and Responsiveness Streets

Cristian Iza[1(✉)], Pedro Uribe-Chavert[1], Juan-Luis Posadas-Yagüe[2], and Jose-Luis Poza-Lujan[2]

[1] Doctoral School, Universitat Politècnica de València, 46022 Valencia, Spain
ciza@alumni.upv.es, pedurcha@doctor.upv.es

[2] Research Institute of Industrial Computing and Automatics, Universitat Politècnica de València, Camino de vera, sn, 46022 Valencia, Spain
jopolu@upv.es

Abstract. This paper proposes enhancing urban traffic management by integrating traffic signals into the SCATS system and developing a distributed control algorithm. The expected benefits include reduced congestion, improved road safety, and environmental sustainability. Scientifically, the research introduces innovative algorithms and integrates new features into existing systems, offering a scalable solution for urban traffic management. The study aims to optimise traffic flow and reduce travel times, ultimately improving quality of life in urban areas. The proposed approach promises efficient and sustainable urban mobility by dynamically adjusting to real-time traffic conditions. The system responds quickly to traffic conditions through collaborative coordination of traffic signals and autonomous decision-making, allowing effective urban traffic management. This research contributes to both practical applications and theoretical understanding, promising significant advances in urban traffic management practices.

Keywords: Traffic Management · Traffic Signals · Distributed Control

1 Problem Statement

This article will discuss how cities become congested and how traffic lights and signal systems can serve a dual purpose. The first problem is to prevent congestion, and the second is to resolve it once it has occurred. The most common way to address this is by regulating the duration of the traffic lights that control intersections. The duration of green light is typically increased or decreased on the basis of a control algorithm. However, city traffic can be regulated by other means as well. For example, it is possible to change speed limits to facilitate decongestion of an avenue or to allow circulation or turns in places that are not usually allowed to redirect traffic in the event of congestion or accidents. The main goal of the work is to develop a distributed control algorithm based on the

G. Marreiros et al. (Eds.): DCAI 2024, LNNS 1151, pp. 272–278, 2025.
https://doi.org/10.1007/978-3-031-80946-0_27

cooperation of agents located at control nodes, in which the input of the control is heterogeneous traffic data, depending on the available information. The output is the regulation of traffic lights and traffic signals.

2 Related Work

Various traffic control systems have been initially reviewed to determine the field of study and optimisation for the doctoral thesis. All traffic control systems are designed to manage and optimise traffic flow in the city through various methods. Firstly, systems such as CRONOS [2], LISA+ [1], OPAC [13], PRODYN [10], RHODES [12], SCOOT [6], and UTCS [11] focus on timing and cycle adjustment. These systems adjust traffic light timings based on various factors such as real-time traffic flow and demand. In particular, SYNCHRO[5] adjusts the duration of green light based on directional arrows. Second, traffic light synchronisation is a speciality of systems like PRODYN, SCATS [9], SCOOT, UTCS [11], and TRANSYT-7F & Traffic Lights [3]. These systems employ algorithms to coordinate the timings of traffic lights, thus improving the overall efficiency of traffic flow. Lastly, systems including CRONOS, LISA+, OPAC, SCATS, TRANSIT [8], UTCS, GRACE [7], and GREEN WAVE [4] consider special traffic situations. These systems ensure safe turns at intersections, safe crossings, smooth flow of traffic, priority for public transportation, and efficient speed recommendations. They also extend warning times and near construction sites and manage pedestrian traffic lights effectively. These systems optimise urban mobility by adjusting traffic light timings, synchronising traffic lights, or considering special traffic situations. The Table 1, discusses the acting mechanisms implemented by each of the systems studied.

The traffic signals mentioned for each algorithm in Table 1 are concrete examples of how urban traffic management algorithms are applied. Each of these signals uses algorithms to control and regulate vehicle flow, adapting to the specific needs of each situation. These algorithms are designed to optimise the coordination of traffic lights and improve traffic efficiency in urban environments.

Various algorithms are employed in different aspects of urban traffic management. For instance, CRONOS uses traffic, scheduling, and topology data to calculate green and red light intervals at traffic lights. GRACE provides speed recommendations based on traffic and topology. GREEN WAVE synchronises traffic lights on main roads to maintain traffic flow. LISA+ and SCOOT focus on light cycles at intersections, while OPAC adjusts light timings according to traffic and topology conditions. PRODYN dynamically adjusts light settings based on topology and traffic. RHODES and SCATS optimise light timings in real-time, and SYNCHRO coordinates light intervals at intersections. TRANSIT prioritises public transport vehicles, and TRANSYT-7F coordinates light timings based on traffic topology. UTCS also focuses on light coordination according to traffic topology.

Traffic management algorithms, such as, LISA +, PRODYN, RHODES, SCATS, SCOOT, use traffic data and algorithms to optimise mobility in urban

Table 1. Algorithm Implemented Signals Management Functionalities

Algorithm	Implemented Signals
CRONOS	Traffic Lights, Stop, Directional Arrows
LISA+	Traffic Lights, Pedestrians, Speed
OPAC	Traffic Lights, Yield, Turn
PRODYN	Traffic Lights, No Parking, One Way
RHODES	Traffic Lights, Priority, School Zone
SCATS	Traffic Lights, Warning, Exclusive Lane
SCOOT	Traffic Lights, Loading Zone, Crosswalk
TRANSIT	Traffic Lights, Bus, Stop
UTCS	Traffic Lights, Construction, Detour
TRANSYT-7F	Traffic Lights
SYNCHRO	Directional Arrows
GRACE	Suggested Speed Indicators
GREEN WAVE	Pedestrian Signals

environments. CRONOS and OPAC focus on traffic light green and red timings; GRACE provides speed recommendations; GREEN WAVE and SYNCHRO synchronize traffic lights to enhance traffic flow along corridors; TRANSIT and TRANSYT-7F prioritise public transportation vehicles; and UTCS coordinates traffic lights in urban areas to facilitate traffic. Table 2 presents a variety of control algorithms used in urban traffic management, highlighting their diversity in terms of control types and topological applications. A detailed analysis of these algorithms provides a comprehensive view of the strategies employed to address traffic challenges in different urban scenarios.

CRONOS system adjusts traffic signal timings for city avenues and intersections, GRACE harmonizes traffic speeds on highways. GREEN WAVE ensures effective synchronization of traffic lights along major thoroughfares to maintain continuous flow. Additionally, Lisa+, OPAC, and SCOOT offer adaptable solutions for different types of urban intersections and residential streets. On the other hand, PRODYN focuses on commercial areas, RHODES addresses congested avenues, and SCATS adapts to urban intersections. SYNCHRO and TRANSYT-7F coordinate intersection and avenue timings, respectively. TRANSIT prioritizes public transportation routes, while UTCS handles intersections in urban areas. Collectively, these algorithms work to enhance traffic efficiency and safety across diverse urban scenarios. Most of these algorithms are scalable and adaptable.

3 Hypothesis and Proposal

In our hypothesis and proposal, we address the following aspects. In the proposed work, the aim is to improve the reaction time of the control algorithm

Table 2. Algorithm Control Type Topologies Work Scopes

Systems	Control Type	Topologies
CRONOS	Adaptive Control	Avenues, urban intersections
GRACE	Adaptive Control	Highways, expressways
GREEN WAVE	Synchronized Adaptive Control	Main avenues, corridors
LISA+	Predictive Adaptive Control	Urban intersections, roundabouts
OPAC	Adaptive Control	Avenues, residential streets
PRDYN	Dynamic Adaptive Control	Intersections, commercial areas
RHODES	Adaptive Control	Avenues, congested areas
SCATS	Real-Time Adaptive Control	Intersections, urban areas
SCOOT	Predictive Adaptive Control	Urban intersections, main streets
SYNCHRO	Synchronized Adaptive Control	Intersections, main avenues
TRANSIT	Adaptive Control	Public transport routes
TRANSYT-7F	Adaptive Control	Intersections, main roads
UTCS	Adaptive Control	Intersections, urban areas

to reduce bottlenecks and resolve any bottlenecks that may have occurred more quickly. For the control algorithm we intend to use heterogeneous information, because each street may have different types of devices. As we have seen in the previous analysis, predictive control is the most widely used, so predictive control algorithms will be used which, in addition, allow congestion situations to be solved. From the previous analysis it can be deduced that the change of traffic control signals can go beyond traffic lights. In this work it is proposed to vary as many traffic control signals as possible. For example, not allowing parking if a large influx of vehicles is expected, or varying the permitted speed to accelerate the clearing of traffic congestion.

3.1 Hypothesis

The possibility of significantly enhancing urban traffic management is proposed by integrating traffic signals into traffic lights alongside a distributed control system. This integration allows for dynamic adjustment of traffic rules like speed and movement patterns in response to current traffic conditions. Consequently, it promises reduced congestion, improved traffic flow, and enhanced adaptability to specific intersection needs, fostering more efficient and sustainable urban mobility.

3.2 Proposal

SCATS is the ideal algorithm for implementing traffic signals in urban management due to its dynamic adjustment to real-time traffic conditions and adaptation to changes in network topology. This ensures smooth and safe transitions between signal phases, minimizing confusion and danger for drivers and pedestrians. Integrating IoT (Internet of Things) would further enhance the system by collecting real-time data from sensors and cameras, optimizing traffic signal synchronization more efficiently and effectively. Simulations will be conducted in SUMO of the proposed system to evaluate its performance under different traffic scenarios, validating and adjusting the necessary control parameters.

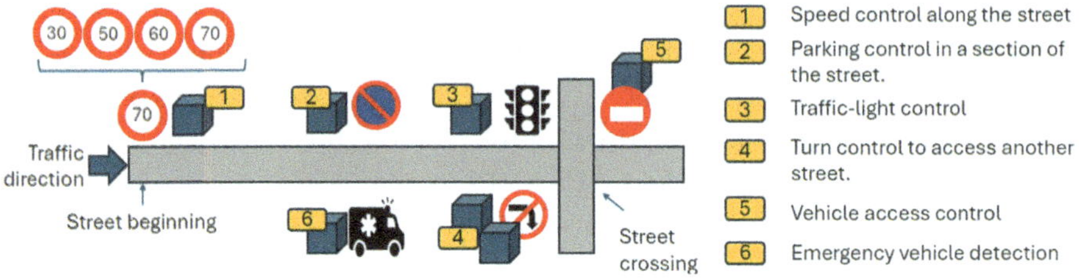

Fig. 1. Description of the environment and system proposal.

4 Preliminary Results and/or Evaluation Plan

Scientifically, SCATS stands as a notable advancement by introducing a groundbreaking method that adjusts dynamically to present traffic conditions and optimizes the layout of road networks. This structural design offers an effective and expandable resolution for managing urban traffic. Therefore, the proposal is to extend SCATS with the addition of comprehensive traffic signal control, in a collaborative way between streets. This leads to the concept of the "dynamic street" as proposed in this work in the Fig. 1. The notion of a "dynamic street" involves the incorporation of adaptable traffic signals into traffic lights to manage urban traffic more efficiently.

By combining advanced technologies and intelligent algorithms, the goal is to create a road environment that dynamically adjusts to real-time traffic conditions. This not only improves traffic flow and reduces congestion but also enhances road safety and user experience by enabling proactive and agile management of vehicular flow in urban areas. The extension of SCATS implies that the devices must coordinate with each other to make autonomous decisions regarding traffic control measures. This means that each device evaluates local conditions and adjusts traffic signals in its area.

Additionally, traffic signals are planned to be implemented to alter the algorithm's topology, allowing for a more dynamic adaptation to traffic conditions. Communication between devices enables a quick and efficient response to traffic conditions, thereby enhancing urban traffic management. Additionally, the integration of predictive analytics and machine learning into SCATS could optimize traffic control by better anticipating congestion and more effectively adjusting

5 Reflections

The outcome of the thesis is to optimize urban traffic management by incorporating traffic signals into the SCATS system and creating a distributed control algorithm. Social, economic, and environmental benefits are anticipated, including reduced traffic congestion, enhanced road safety, improved travel efficiency, cost and emissions reductions, effects on travel times, fuel savings, and broader effects on urban planning and land use. From a scientific perspective, the development of an innovative algorithm and the integration of new features into the existing system are the main contributions, providing a scalable and adaptable solution to improve urban traffic management. This research is expected to have a significant impact on both the practical and theoretical application of urban traffic management. There are plans to collaborate with urban planners and the public to ensure holistic improvements in urban living conditions.

Acknowledgements. Work supported by the Spanish Ministry of Science and Innovation MICINN Project: CICYT PRECON-I4: "Predictable and reliable information systems for Industry 4.0" TIN2017-86520-C3-1-R.

References

1. Barthauer, M., Friedrich, B.: Connecting microscopic traffic simulation and LISA+ external signal control. Transp. Res. Procedia **27**, 420–427 (2017)
2. Boillot, F., Vinant, P., Pierrelée, J.C.: On-field experiment of the traffic-responsive co-ordinated control strategy CRONOS-2 for under-and over-saturated traffic. Transp. Res. Part A Policy Pract. **124**, 189–202 (2019)
3. Cabrera Lara, G.: Metodología para la calibración del programa Transyt-7f para optimización de semáforos en áreas urbanas. Ph.D. thesis, Universidad Nacional de Colombia (2022)
4. Chou, L.D., Shen, T.Y., Tseng, Chia-Wei adn Chang, Y.J., Kuo, Y.W.: Esquema de gestión de semáforo virtual basado en ondas verdes con vanet. Revista Internacional de Computación Ad Hoc y Ubicua, pp. 22–32 (2017)
5. Hussein, N.A.: Synchro software-based alternatives for improving traffic operations at signalized intersections. Aro Sci. J. Koya Univ. **10**(1), 123–131 (2022)
6. Liu, K.: Design and application of real-time traffic simulation platform based on UTC/scoot and VISSIM. J. Simul. 1–18 (2023)
7. Ma, C., He, R.: Green wave traffic control system optimization based on adaptive genetic-artificial fish swarm algorithm. Neural Comput. Appl. **31**, 2073–2083 (2019)
8. Qadri, S.S.S.M., Gökçe, M.A., Öner, E.: State-of-art review of traffic signal control methods: challenges and opportunities. Eur. Transp. Res. Rev. **12**(1), 1–23 (2020). https://doi.org/10.1186/s12544-020-00439-1
9. Sun, J., Zhang, L.: Vehicle actuation based short-term traffic flow prediction model for signalized intersections. J. Central South Univ. **19**(1), 287–298 (2012)
10. Torabi, B., Zalila-Wenkstern, R., Saylor, R., Ryan, P.: Deployment of a plug-in multi-agent system for traffic signal timing. In: Proceedings of the 19th International Conference on Autonomous Agents and MultiAgent Systems, pp. 1386–1394 (2020)

11. Wang, Z.: Coordination and optimization of urban traffic control and path guidance based on multi-agent. In: Journal of Physics: Conference Series, vol. 1345, p. 042058. IOP Publishing (2019)
12. Yankevich, N.: Algorithm for control of traffic flow in an intelligent transport system. Athens J. Technol. Eng. **11**(1), 39–54 (2024)
13. Yilmaz, M., Smith, S.: Design perspectives of hardware-based cellular computational networks for traffic signal systems. In: 2024 IEEE Green Technologies Conference (GreenTech), pp. 242–247. IEEE (2024)

Evaluating AtomVM in LoRa-Based Industrial Use Cases

Dániel Ferenczi(✉) and Melinda Tóth

Faculty of Informatics, Eötvös Lóránd University, Pázmány Péter sétány 1/C, Budapest 1117, Hungary
danielf@inf.elte.hu, tothmelinda@elte.hu

Abstract. The BEAM ecosystem provides features for fault-tolerance, concurrency and live code upgrades for functional languages Erlang and Elixir. Apart from its standard distributions, it is available for IoT and Edge use cases through projects like Nerves, GRiSP and AtomVM. These use cases often require low-power use in general and long-range transmission, which makes communication protocols like LoRa appealing. We propose to explore the use of LoRa in BEAM-based IoT setups. In this work, we identify the criteria and steps for evaluating the details of using BEAM with Nerves and AtomVM. In particular, we are interested in how popular features of BEAM languages like scalability and robustness perform on IoT end devices.

Keywords: IoT · Erlang · AtomVM · BEAM · LoRa

1 Problem Statement

Internet of Things aims to connect a wide range of devices, including industrial devices, vehicles, sensors, and larger systems. The use of such setups ranges from home automation, water management, agriculture, manufacturing industry to healthcare to name a few examples. Depending on the particular application, end devices in IoT systems may require fault-tolerance, wireless connectivity ranging kilometres or autonomous operation through many years, in some cases on battery power [12]. As IoT use cases are often composed of a large number of nodes, on-site maintenance, in general, should happen seldom and ideally, the software should be fault-tolerant. In battery-powered setups choices concerning software implementations or communication protocols also affect maintenance through power consumption. In our work, we propose the use of the Erlang [8] programming language running on AtomVM [2] nodes and the LoRa [7] communication protocol for IoT use cases and identify the steps for evaluating these technologies. We take as a base the inherent features of these technologies, like the ease of development of concurrent applications, fault-tolerance, low power use, and long-range communication and propose an evaluation plan to explore their limits and identify particular scenarios where they can be used.

G. Marreiros et al. (Eds.): DCAI 2024, LNNS 1151, pp. 279–285, 2025.
https://doi.org/10.1007/978-3-031-80946-0_28

In the following paragraphs, we present Erlang, its distributions for IoT devices and the LoRa protocol. Section 2 outlines work on industrial applications of Erlang and LoRa. In Sect. 3 we propose using Erlang and LoRa together in industrial IoT. Evaluation of this hypothesis is presented in Sect. 4. We detail future steps and research areas in Sect. 5.

1.1 Erlang

Erlang is a general-purpose, dynamically typed functional programming language created with fault-tolerance in mind [6]. Developed originally for Ericsson's telecommunication switches in the '80 s, by now it is used for fintech software, message brokers, chat applications and much more [9]. Erlang programs are executed on a runtime, called BEAM, that provides the features to build reliable, fault-tolerant, concurrent, soft real-time distributed systems. It also supports upgrading running code without any noticeable impact on the end-user. These features are part of the language and do not need additions to the application's software stack. This has a clear benefit in resource-constrained environments where developers want to minimize the number of software components.

While typically Erlang is deployed on server-grade computers, there were three projects of note in the past decade to bring BEAM to lower-powered devices: GRiSP [4], Nerves [5] and AtomVM [2]. These projects bring BEAM-based high-level languages to devices typically running as IoT end nodes. Nerves focuses on providing a software platform and the tooling for writing Elixir [3] applications for systems like the Raspberry Pi or BeagleBone computers. Elixir is a functional language also running on BEAM and also suitable for building scalable, fault-tolerant software. GRiSP, compatible only with specific GRiSP boards, provides hard real-time event handling by virtue of its tight integration with it. Due to running BEAM directly on its hardware it also provides better performance than counterparts using an operating system as well. Both GRiSP and Nerves bring a feature-complete BEAM to relatively powerful devices, that typically require continuous power. AtomVM on the other hand targets microcontrollers and can be run on battery-powered devices. The downside is, that the port of BEAM it includes is not feature-complete and lacks some features, notably bignums, the tooling and protocol required for distributed Erlang, a REPL and hot code loading amongst others [1]. However, it provides the constructs used to develop reliable applications: language idioms for building concurrent applications and supervision trees used for managing Erlang processes.

1.2 LoRa

LoRa [20], short for Long Rage is a radio communication system developed by the company Cycleo, later acquired by Semtech. The technique is popular in scenarios where end devices are located up to 10 kilometres away from their gateways and require a low data rate. LoRa modems have low power consumption, making the technology popular with battery-powered devices [21]. LoRa operates on the unlicensed radio spectrum. The technique uses "chirps", linear changes to the frequency over time, to encode the data [19]. LoRa devices allow developers to customize their applications "Spreading Factor": a low value increases the data rate and has a lower time-on-air, albeit has worse sensitivity for the receiver.

LoRaWAN is a medium access control (MAC) and network layer protocol based on LoRa, maintained by the LoRa Alliance [19]. Its specification describes end device classes depending on data rate and power consumption-related requirements and roles for gateways and network servers. These allow for building systems consisting of multiple authenticated devices sending their data to gateways, that are coupled together under network servers that ensure deduplication and connectivity to applications. Many industrial applications require scheduling of real-time data-flow, which LoRaWAN by default does not support [15].

2 Related Work

Using BEAM-based languages in IoT setups has been the subject of past research. The CAEFL [22] project investigated Elixir's use in federated learning. They defined agents capable of enhancing a global model with local information specific to their environment. The setup made use of hot code loads to redefine existing environments, create new ones, and even add or remove sensors on an agent level without downtime.

With regards to the industrial use of Nerves, efficient, autonomous communication of end nodes has been researched by Takase [11]. Kikuchi [23] researched its use for control in smaller water plants. Achlys' [13] authors investigated Erlang's use in IoT. In their work, they relied on GRiSP devices. The research looked aimed at finding alternatives to the centralization prevalent in IoT systems. The authors managed to build nodes that were capable of building network connections and clusters automatically with each other. They also made use of Erlang's feature for fault tolerance to create a setup where the individual node's availability is uncertain.

Applications of LoRa and LoRaWAN have been surveyed in past research [12]. The largest range covered by the included use cases is 5 kilometres. In this instance, the technology was used to monitor the water level in troughs. With regards to LoRaWAN its indoor industrial use for monitoring has been explored [17]. Their test simulation assumed several nodes ranging between 10 and 1000 with a coverage radius fixed to 200 m. They concluded that the technology performs well concerning reliability, timeliness and energy consumption. Further research [18] evaluated the technology in multi-gateway setups. They concluded that the technology is suitable for monitoring, although it requires a precise understanding of the business need and monitored metrics to avoid overprovisioning.

Research has been made to support LoRa's adoptions in scenarios with stricter timeliness-related requirements. One such project, Industrial LoRa [15] designed a MAC layer protocol in place of LoRaWAN, that can provide through their protocol's Contention Access Period soft-real time guarantees for a wider range of industrial applications. Their proposal also includes a Contention-Free Period to support scenarios that require high scalability without the need for real-time communication. However, this period does not provide high reliability. The authors made further contributions to improve non-real-time reliability, improve real-time bandwidth efficiency [14] and add multi-hop capabilities [16].

More recent research [24] modified the role of traditional LoRaWAN gateways to ensure real-time properties. In standard setups, it is the role of network servers to acknowledge messages sent from end devices. This poses a delay in communication, as these acknowledgements have to be relayed through the gateways. In the modified setup the authors added acknowledgement and node management to the gateways to include confirmed, real-time messaging applicable for emergency and control scenarios. Additionally, shifting this role from the network server reduced its load and power consumption.

3 Hypothesis and Proposal

Industrial IoT has many applications for low-powered, reliable systems. BEAM has built-in capabilities for building scalable, robust systems and has been ported to lower-powered devices suitable for use in industrial scenarios, like AtomVM. Apart from the operational benefits, using BEAM on IoT also has the advantage of being able to program with high-level, functional languages and simplify development by offering pattern-matching for binary data. Erlang, from its root as a programming language for telecommunication switches, is designed for building soft real-time systems, and can thus be used in industrial applications where timely responses are of importance.

Wireless connectivity allows use in setups where a wired connection is unavailable or expensive to build. LoRa in particular is a good candidate for cases where data has to be sent over large distances. We hypothesise that industrial applications can benefit from the features of BEAM-based languages and LoRa communication. As a result of our evaluation, we wish to identify those particular industrial use cases where these technologies can be efficiently applied.

We primarily focus on power-constrained setups based on AtomVM and Erlang to see, how such systems handle concurrency and communication concerning soft real-time responsiveness and power consumption. For this task, we aim to build a prototype system consisting of multiple AtomVM nodes.

4 Evaluation Plan

To evaluate our hypothesis we wish to test the following points:

1. evaluate concurrency performance of AtomVM nodes
2. test fault-tolerance of AtomVM nodes
3. test performance in distributed computing scenarios
4. measure power consumption across different uses of these features

5. evaluate tests on different combinations of hardware and AtomVM and standard Erlang releases

In our first two objectives, we wish to develop experiments that would stress test the nodes' responsiveness under different load levels and measure the velocity at which supervision trees manage processes, e.g. restart them in the event of failures. In our third step, we want to add our nodes to a LoRa-based industrial setup and see how the distance between sources and targets affects the timeliness of distributed applications. Investigating overall reliability, and implications on development in scenarios with varying distances would also be a topic of interest. We are currently discussing the details of such experiments and non-functional requirements in general with an industrial partner. Although AtomVM already includes support for the LoRa layer, adding support for LoRaWAN still requires development effort. For this task, we plan to reuse code to read and write LoRaWAN frames that have been already implemented in Erlang [10].

Power consumption is an important subject in IoT systems, given the high number of deployed devices and their limited access to power sources. To see how AtomVM and LoRa fit such setups, we also wish to measure how different loads, code structures and networking setups affect power consumption.

We also plan to evaluate our measurements on different releases of AtomVM and Erlang/OTP running on a variety of hardware. Our goal is to help assess performance differences between releases, developer and deployment target hardware, and distributions.

5 Reflections and Future Steps

With our evaluation plan, we wish to explore the constraints of AtomVM and LoRa-based setups in IoT. Afterwards, with the aim of an industrial partner, we aim to investigate the limits of these setups in particular scenarios.

With regards to tooling, Erlang applications typically require changes when targetting AtomVM. Investigating how static analysis and refactoring tools can help this endeavour would be of interest.

Researching Nerves's application would also be of interest. Although devices running Nerves require more power, the distribution offers the complete capabilities of BEAM, including hot code loading and distributed Erlang/Elixir. Applying hot code loading of applications on devices where the typical upgrade paths involve powering off the device and manually copying firmware, would improve the developer workflow, and perhaps even the operation of the entire system.

Disclosure of Interests. The authors have no competing interests to declare that are relevant to the content of this article.

References

1. Atomvm limitations (2024). https://www.atomvm.net/doc/main/programmers-guide.html#limitations. Accessed 19 Apr 2024
2. Atomvm project (2024). https://www.atomvm.net/. Accessed 19 Apr 2024
3. Elixir language (2024). https://elixir-lang.org. Accessed 25 Apr 2024
4. Grisp project (2024). https://www.grisp.org/. Accessed 19 Apr 2024
5. Nerves project (2024). https://www.nerves-project.org/. Accessed 19 Apr 2024
6. Armstrong, J.: Making reliable distributed systems in the presence of software errors. Ph.D. thesis (2003)
7. Augustin, A., Yi, J., Clausen, T., Townsley, W.M.: A study of lora: long range & low power networks for the internet of things. Sensors **16**(9), 1466 (2016)
8. Cesarini, F., Thompson, S.: Erlang Programming: A Concurrent Approach to Software Development. O'Reilly Media (2009)
9. Cesarini, F.: Which companies are using erlang, and why? (2019). https://www.erlang-solutions.com/blog/which-companies-are-using-erlang-and-why-mytopdogstatus/. Accessed 24 Apr 2024
10. Helium: Erlang lorawan (2023). https://github.com/helium/erlang-lorawan. Accessed 25 Apr 2024
11. Hideki, T.: On the way to achieve autonomous node communication elixir (2022). https://www.youtube.com/watch?v=Y4IASAU4Bjo. Accessed 24 Apr 2024
12. Khutsoane, O., Isong, B., Abu-Mahfouz, A.M.: IoT devices and applications based on lora/lorawan. In: IECON 2017-43rd Annual Conference of the IEEE Industrial Electronics Society, pp. 6107–6112. IEEE (2017)
13. Kopestenski, I., Van Roy, P.: Erlang as an enabling technology for resilient general-purpose applications on edge IoT networks. In: Proceedings of the 18th ACM SIGPLAN International Workshop on Erlang, pp. 1–12 (2019)
14. Leonardi, L., Battaglia, F., Bello, L.L.: RT-lora: a medium access strategy to support real-time flows over lora-based networks for industrial IoT applications. IEEE Internet Things J. **6**(6), 10812–10823 (2019)
15. Leonardi, L., Battaglia, F., Patti, G., Bello, L.L.: Industrial lora: a novel medium access strategy for lora in industry 4.0 applications. In: IECON 2018-44th Annual Conference of the IEEE Industrial Electronics Society, pp. 4141–4146. IEEE (2018)
16. Leonardi, L., Bello, L.L., Patti, G.: MRT-lora: a multi-hop real-time communication protocol for industrial IoT applications over lora networks. Comput. Commun. **199**, 72–86 (2023)
17. Luvisotto, M., Tramarin, F., Vangelista, L., Vitturi, S., et al.: On the use of lorawan for indoor industrial IoT applications. Wirel. Commun. Mob. Comput. **2018** (2018)
18. Magrin, D., Capuzzo, M., Zanella, A., Vangelista, L., Zorzi, M.: Performance analysis of lorawan in industrial scenarios. IEEE Trans. Industr. Inf. **17**(9), 6241–6250 (2020)
19. Reddy, G.P., Kumar, Y.P.: Demystifying lora wireless technology for IoT applications: concept to experiment. In: 2021 4th International Symposium on Advanced Electrical and Communication Technologies (ISAECT), pp. 1–6. IEEE (2021)
20. Semtech: Lora and lorawan (2024). https://lora-developers.semtech.com/documentation/tech-papers-and-guides/lora-and-lorawan/. Accessed 25 Apr 2024
21. Senol, N.S., Rasheed, A.: A testbed for lora wireless communication between IoT devices. In: 2023 11th International Symposium on Digital Forensics and Security (ISDFS), pp. 1–6. IEEE (2023)

22. Xu, R., Michala, A.L., Trinder, P.: Caefl: composable and environment aware federated learning models. In: Proceedings of the 21st ACM SIGPLAN International Workshop on Erlang, pp. 9–20 (2022)
23. Yutaka, K.: An elixir app to a small hydropower plant with nerves (2022). https://www.youtube.com/watch?v=2vbQLT005zc&t. Accessed 24 Apr 2024
24. Zhong, C., Nie, X.: A novel single-channel edge computing lora gateway for real-time confirmed messaging. Sci. Rep. **14**(1), 8369 (2024)

Anomaly Detection for Trust Management in Internet of Things Systems

Sahibzada Saadoon Hammad[1,2](✉) and Sergio Trilles[1,3]

[1] Institute of New Imaging Technologies, Universitat Jaume I, Avda. Sos Baynat, s/n, 12071 Castelló de la Plana, Spain
{hammad,strilles}@uji.es

[2] ValgrAI - Valencian Graduate School and Research Network of Artificial Intelligence, Camí de Vera s/n, 46022 Valencia, Spain

[3] Facultat de Ciencias y Tecnología, Universidad Isabel I, Calle Fernán González, 76, 09003 Burgos, Spain

Abstract. The fast growth of Internet of Things systems has transformed the way policy decisions are made based on the data produced by these devices. These data are prone to errors and anomalies and can present privacy and security issues which can affect policy decisions. Therefore, it is necessary that, at each layer of Internet of Things architecture, the data undergo a process to ensure its quality. This paper presents a framework for building a trust management system for Internet of Things devices using anomaly detection in different layers of IoT architecture and formulation of a reputation index.

Keywords: IoT · Anomaly Detection · TinyML · Reputation Index · Trust Management

1 Introduction

The Internet of Things (IoT) has witnessed a significant surge in recent years, with projections estimating a staggering 29.42 billion IoT devices by 2030 [18]. These devices, including sensors, generate time series data that serve as a critical resource for analysis and form the basis for numerous policy-oriented decisions. The data captured by these sensors come from a diverse range of applications, including, but not limited to, smart cities [20], agricultural practices [22], health monitoring systems, and industrial operations [8]. The time series data generated by IoT devices are susceptible to anomalies, necessitating robust mechanisms for anomaly detection [11]. However, traditional anomaly detection techniques often fail to efficiently encapsulate the intricate relationships within the data,

Supported by ValgrAI - Valencian Graduate School and Research Network for Artificial Intelligence (Generalitat Valenciana).

G. Marreiros et al. (Eds.): DCAI 2024, LNNS 1151, pp. 286–291, 2025.
https://doi.org/10.1007/978-3-031-80946-0_29

leading to suboptimal performance. With the advent of technological advancements and enhanced computational capabilities, Deep Learning (DL) methods such as Convolutional Neural Networks (CNN) [16], Long Short-Term Memory (LSTM) [9], and Autoencoders [2] have gained widespread acceptance in anomaly detection applications. Concurrently, Graph Neural Networks (GNNs) [17] have emerged as potent techniques for modeling intricate relationships within data, capturing correlations between different time series, and modelling spatiotemporal dependencies. The correlation between different sensors can help develop trust management which is critical to addressing issues like privacy and reliability of interconnected devices in a sensor network [25].

In this paper, we propose a technological framework for the development of a trust management system for IoT devices in a sensor network. We explored the literature to identify current trends in anomaly detection on the edge and applied anomaly detection algorithms in urban noise sensor networks to detect anomalies.

2 Concepts and Related Work

2.1 Anomaly Detection

Anomaly detection techniques include statistical methods, Machine Learning (ML), and DL algorithms. The authors of [10] employ Recurrent Neural Networks (RNN) to identify irregularities in data gathered from environmental sensors using Gated Recurrent Units (GRU) and Long Short-Term Memory (LSTM). For the identification of point, contextual, and collective anomalies in smart home sensors LSTM, One-Class Support Vector Machine (OCSVM), Isolation Forest and Convolutional Neural Networks (CNN) have been used [15]. The authors in [24] propose an anomaly detection system for air quality monitoring data using LSTM based on the reconstruction error between the original time series and the predicted series. Smart agriculture systems have employed LSTM and Multivariate Linear Regression (MLR) to identify anomalies. Experiments have been carried out using time series data from environmental sensors including air temperature, humidity, pressure, soil temperature, soil moisture, and wind speed/direction. The study findings indicate that the LSTM model outperformed the MLR approach [4].

2.2 TinyML

The IoT architecture consists of three layers: the edge, fog, and cloud. The edge is the layer where the data is produced and contains sensors and smart devices such as microcontrollers. The fog layer has better computing capabilities than the edge layer, and the cloud layer is where high-end computations and analytics are performed on the data to make decisions. In Edge computing, these smart devices with limited computational power are used to process data at the same place where are generated. Edge computing helps cut down data transmission costs by processing data at the edge, resulting in low latency and better data security and privacy [6]. In recent years, a concept called 'TinyML' has been introduced, which refers to running ML algorithms on low-power, resource-constrained devices [19]. Applications of TinyML have been found in many recent works. Using an on-board wristband, CNN as TinyML was able to achieve 96.4% accuracy in hand gesture detection. An Arduino Nano with 256 KB of RAM and 1 MB of flash memory was used to implement the tiny model [3]. The authors in [13] use an Arduino Nano 33 BLE for detecting anomalies in balanced and unbalanced laundry loads using two neural networks.

2.3 Reputation and Trust Management

In the realm of IoT, trust management systems are defined as intelligent information processing systems that can handle anomalies and threats by taking advantage of trust-based data. These systems use trust models to determine trust values and categorize items in an Internet of Things network as trustworthy or untrustworthy [23]. Different approaches are used to develop a trust management system. The authors in [1] deploy an LSTM for the identification of malicious nodes. A behavior model is proposed using trust properties of network Quality of Service (QoS) with a DL algorithm. The authors quantified the 'trust' in the system by comparing the similarity between real and predicted network behaviors [14]. Perceptron Detection (PD) computes the trust values of IoT nodes and identifies malicious nodes based on those values by combining the perceptron and K-means methods [12]. A lightweight trust management for wireless sensor networks based on Bayesian and Entropy methods calculates the trust values of a node. It uses adaptive decay factors and effective history records to derive direct trust values, which lowers energy usage and increases algorithm accuracy. The entropy method is used to unbiasedly assign weights to various trust values, improving flexibility and evaluating results [5].

3 Proposed Methodology

The proposed research aims to develop a trust management system for IoT devices in the sensor network by the application of anomaly detection on the edge and fog layer. The methodological steps defined to develop the proposed system are as follows: the first step in the proposed methodology is to identify state-of-the-art anomaly detection techniques in IoT systems. For this purpose an explorative literature review has been carried out to investigate the following: a) the techniques/algorithms in anomaly detection on the edge b) edge devices used for running the algorithms c) application areas of edge computing and TinyML. This step lays the foundation for the technological framework of study. It comprises the potential IoT datasets, DL techniques for anomaly detection, and the design of a trust management system. In this framework, the first step is to develop a novel anomaly detection technique and deploy a lightweight anomaly detection model on edge devices (microcontrollers) to evaluate the efficiency of resource-constrained devices. The next step in the technological framework is to consider the spatiotemporal relationship between the sensors in a geographic location and, based on the correlation between their spatial and temporal characteristics, detect anomalies in the fog layer. GNN will be used to construct a spatiotemporal graph structure to capture inter-sensor relationships and find malicious nodes in the network. There are many variants of GNNs such as Graph Convolutional Networks (GCNs), GraphSAGE, and Graph Attention Networks (GATs), which will be applied and compared. This part of the framework will use information from the edge layer such as a) data quality (anomalies, duplicates, and missing values) and b) intrinsic properties of the sensor (sensor data resolution and accuracy) which will help in building the reputation index. In the next phase, the findings from the fog layer will be transferred to the cloud layer to develop a reputation index and trust management system based on the reputation of each node from the fog layer (Fig. 1).

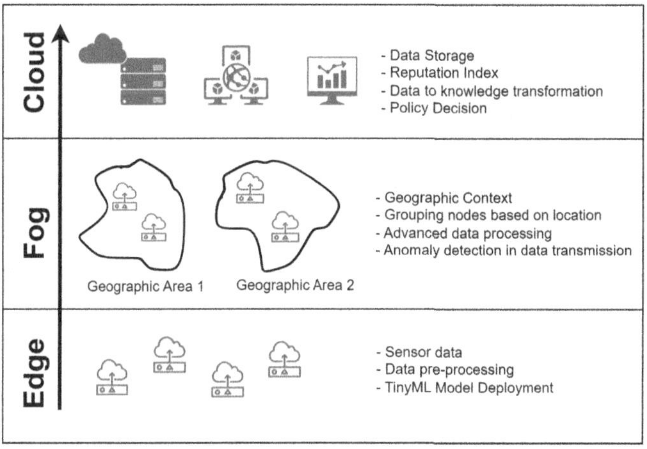

Fig. 1. Proposed Technological Framework.

4 Preliminary Results

Some of the preliminary results after the literature review indicate that detecting anomalies using TinyML is an emerging field with significant potential to exploit the gaps in this domain [21]. Out of 18 articles in the review study, CNN is the most used algorithm, followed by LSTM and Autoencoder. Raspberry Pi is the most used microcontroller in 8 out of 18 studies, followed by Arduino and ESP32, each used in 3 studies. There is an increasing trend in the number of studies related to TinyML in anomaly detection over the years. While the preliminary findings suggest great potential they also highlight several challenges, such as the lack of relevant labeled datasets for anomaly detection. We applied anomaly detection in the context of urban noise data using two LSTM autoencoder models trained with normal data and tested to detect point and collective anomalies. The models are quantized and converted to TensorFlow Lite models for deployment on an ESP-32 microcontroller. The results in terms of accuracy, heap memory consumption, and inference time are evaluated [7]. The power consumption of the microcontroller during inference is $0.2693\,\text{W} \pm 0.039$ and $0.3268\,\text{W} \pm 0.015$, while the inference time is 4 ms.

5 Conclusion

This paper outlines the doctoral research plan and methodology for tackling the challenges in the rapidly changing field of anomaly detection in IoT systems to enhance trust management. Our proposed research seeks to contribute to the field of anomaly detection and trust management in IoT ecosystems. While in the methodology there is mention of data quality and sensor properties for the development of the reputation index more such indicators need to be identified to develop a robust trust management system.

Acknowledgements. This publication is part of the project PID2022-141813OB-I00 funded by MCIN/AEI/10.13039/501100011033 and by ERDF/EU. Sahibzada Saadoon Hammad has been funded by the ValgrAI - Valencian Graduate School and Research Network for Artificial Intelligence (Generalitat Valenciana).

References

1. Alghofaili, Y., Rassam, M.A.: A trust management model for IoT devices and services based on the multi-criteria decision-making approach and deep long short-term memory technique. Sensors **22**(2), 634 (2022)
2. Baldi, P.: Autoencoders, unsupervised learning, and deep architectures. In: Proceedings of ICML Workshop on Unsupervised and Transfer Learning, pp. 37–49 (2012)
3. Bian, S., Lukowicz, P.: Capacitive sensing based on-board hand gesture recognition with tinyml. In: Adjunct Proceedings of the 2021 ACM International Joint Conference on Pervasive and Ubiquitous Computing and Proceedings of the 2021 ACM International Symposium on Wearable Computers, pp. 4–5 (2021)
4. Catalano, C., Paiano, L., Calabrese, F., Cataldo, M., Mancarella, L., Tommasi, F.: Anomaly detection in smart agriculture systems. Comput. Ind. **143**, 103750 (2022)
5. Che, S., Feng, R., Liang, X., Wang, X.: A lightweight trust management based on Bayesian and entropy for wireless sensor networks. Secur. Commun. Netw. **8**(2), 168–175 (2015)
6. Dutta, L., Bharali, S.: Tinyml meets IoT: a comprehensive survey. Internet Things **16**, 100461 (2021)
7. Hammad, S.S., Iskandaryan, D., Trilles, S.: An unsupervised tinyml approach applied to the detection of urban noise anomalies under the smart cities environment. Internet Things **23**, 100848 (2023)
8. Hassija, V., Chamola, V., Saxena, V., Jain, D., Goyal, P., Sikdar, B.: A survey on IoT security: application areas, security threats, and solution architectures. IEEE Access **7**, 82721–82743 (2019)

9. Hochreiter, S., Schmidhuber, J.: Long short-term memory. Neural Comput. **9**(8), 1735–1780 (1997)
10. JaeMyoung, K., Cho, Y.W., Kim, D.H.: Anomaly detection of environmental sensor data using recurrent neural network at the edge device. In: 2020 International Conference on Information and Communication Technology Convergence (ICTC), pp. 1624–1628. IEEE (2020)
11. Li, D., Deng, L., Liu, W., Su, Q.: Improving communication precision of IoT through behavior-based learning in smart city environment. Futur. Gener. Comput. Syst. **108**, 512–520 (2020)
12. Liu, L., Ma, Z., Meng, W.: Detection of multiple-mix-attack malicious nodes using perceptron-based trust in IoT networks. Futur. Gener. Comput. Syst. **101**, 865–879 (2019)
13. Lord, M., Kaplan, A.: Mechanical anomaly detection on an embedded microcontroller. In: 2021 International Conference on Computational Science and Computational Intelligence (CSCI), pp. 562–568. IEEE (2021)
14. Ma, W., Wang, X., Hu, M., Zhou, Q.: Machine learning empowered trust evaluation method for IoT devices. IEEE Access **9**, 65066–65077 (2021)
15. Majib, Y., Barhamgi, M., Heravi, B.M., Kariyawasam, S., Perera, C.: Detecting anomalies within smart buildings using do-it-yourself internet of things. J. Ambient. Intell. Humaniz. Comput. **14**(5), 4727–4743 (2023)
16. O'shea, K., Nash, R.: An introduction to convolutional neural networks. arXiv preprint arXiv:1511.08458 (2015)
17. Scarselli, F., Gori, M., Tsoi, A.C., Hagenbuchner, M., Monfardini, G.: The graph neural network model. IEEE Trans. Neural Networks **20**(1), 61–80 (2009)
18. Transforma insights, & exploding topics (1 July 2023). Number of internet of things (IoT) connected devices worldwide from 2019 to 2023, with forecasts from 2022 to 2030 (in billions) (2023). https://www.statista.com/statistics/1183457/iot-connected-devices-worldwide/. Accessed 13 May 2024
19. Sudharsan, B., et al.: Tinyml benchmark: executing fully connected neural networks on commodity microcontrollers. In: 2021 IEEE 7th World Forum on Internet of Things (WF-IoT), pp. 883–884 (2021)
20. Trilles, S., Calia, A., Belmonte, Ó., Torres-Sospedra, J., Montoliu, R., Huerta, J.: Deployment of an open sensorized platform in a smart city context. Future Gener. Comput. Syst. **76**, 221–233 (2017)
21. Trilles, S., Hammad, S.S., Iskandaryan, D.: Anomaly detection based on artificial intelligence of things: a systematic literature mapping. Internet Things 101063 (2024)
22. Trilles, S., et al.: Development of an open sensorized platform in a smart agriculture context: a vineyard support system for monitoring mildew disease. Sustain. Comput. Inform. Syst. **28**, 100309 (2020)
23. Tyagi, H., Kumar, R., Pandey, S.K.: A detailed study on trust management techniques for security and privacy in IoT: challenges, trends, and research directions. High-Confidence Comput. 100127 (2023)
24. Wei, Y., Jang-Jaccard, J., Xu, W., Sabrina, F., Camtepe, S., Boulic, M.: LSTM-autoencoder-based anomaly detection for indoor air quality time-series data. IEEE Sens. J. **23**(4), 3787–3800 (2023)
25. Yan, Z., Zhang, P., Vasilakos, A.V.: A survey on trust management for internet of things. J. Netw. Comput. Appl. **42**, 120–134 (2014)

Enhancing Recruitment with LLMs and Chatbots

Liliana Novais[1,2,3(✉)], Vitor Rocio[1,4], and Jorge Morais[1,4]

[1] Universidade Aberta, Lisbon, Portugal
2302227@estudante.uab.pt
[2] Instituto Superior de Engenharia do Porto, Porto, Portugal
[3] SEG Automotive Portugal, Vila do Conde, Portugal
[4] INESC TEC, Porto, Portugal

Abstract. Traditional approaches in the competitive recruitment landscape frequently encounter difficulties in effectively identifying exceptional applicants, resulting in delays, increased expenses, and biases. This study proposes the utilisation of contemporary technologies such as Large Language Models (LLMs) and chatbots to automate the process of resume screening, thereby diminishing prejudices and enhancing communication between recruiters and candidates. Algorithms based on LLM can greatly transform the process of screening by improving both its speed and accuracy. By integrating chatbots, it becomes possible to have personalised interactions with candidates and streamline the process of scheduling interviews. This strategy accelerates the hiring process while maintaining principles of justice and ethics. Its objective is to improve algorithms and procedures to meet changing requirements and enhance the competitive advantage of talent acquisition within organisations.

Keywords: Recruitment · Large Language Models · Chatbots · Automation · Efficiency · Bias Mitigation

1 Introduction

In order to completely comprehend the Recruitment Process, it is important to have a comprehensive awareness of its diverse character, which includes both theoretical principles and actual application. Over the years, this process has been carefully analysed and enhanced, as demonstrated by the extensive scholarly research and literature dedicated to its different aspects [5,7,15]. Acquiring a thorough comprehension of the different methods of recruitment is crucial for tailoring strategies that result in effective talent acquisition. The main classifications of recruitment, such as internal and external recruitment, encompass various sub-categories including direct, indirect, passive, campus, and diversity recruiting. These different avenues are designed to attract talent and serve different organisational needs and goals. The ultimate objective is to achieve optimal alignment between job requirements and candidate qualifications [6,19,21].

G. Marreiros et al. (Eds.): DCAI 2024, LNNS 1151, pp. 292–297, 2025.
https://doi.org/10.1007/978-3-031-80946-0_30

Nevertheless, the Selection Process encounters obstacles, particularly bias and referral pressure, that have the potential to distort decision-making. Organisations are using organised and transparent selection processes to limit the dangers of referral pressure on merit-based selection and diversity, due to the presence of conscious and unconscious biases. This is done to assure fairness and impartiality. Ensuring ethical practices in the selection process is of utmost importance, as it guarantees fairness, a wide range of perspectives, and the inclusion of diverse individuals. By acknowledging and addressing unconscious bias, and by using specific measures to lessen its influence, organisations can prevent discriminatory behaviours and foster a more complete and equitable approach to evaluating candidates [6,8]. In order to create a positive environment for hiring talented individuals and achieving organisational success, it is essential to have a thorough awareness of the Recruitment Process, along with ethical concerns and proactive strategies to address any challenges that may arise.

The project will be developed in SEG Automotive Portugal, Unipessoal Lda, according to the HR department an average of 170 CVs are received per vacancy [14].

2 Problem

Currently, the recruitment procedure encompasses several challenges. Conventional manual approaches frequently lead to inefficiencies, as human participation in activities such as resume screening and candidate sourcing can result in errors and subjectivity, especially when dealing with a large number of applications. Furthermore, the labor-intensive nature of manual processes might overwhelm HR workers, causing them to be diverted from more strategic pursuits. In addition, the presence of unconscious prejudices in the hiring process poses a substantial threat, as it perpetuates inequities and leads to the exclusion of highly skilled individuals who do not align with preconceived conceptions or criteria.

3 Related Work

Recent advancements in artificial intelligence offer promising solutions to challenges in automated curriculum analysis, as evidenced by studies such as those by Harsha et al. and Sudha et al. [9,22]. Harsha et al. introduce the "Resumé Screening" web application, utilizing NLP and ML algorithms to efficiently analyze resumes and enhance screening efficiency [9]. Additionally, studies by Mridha et al. [13] and Kinge et al. [11] explore the effectiveness of CNNs and machine learning algorithms like Random Forest and Naive Bayes in resume screening, aiming to optimize the recruitment process by identifying qualified candidates [11,13]. Moreover, Raghavan et al. [17] and Alina et al. [12] delve into the ethical and legal considerations of automated recruitment processes, emphasizing fairness, transparency, and monitoring metrics to mitigate biases and ensure responsible implementation. Automated recruiting systems hold promise in efficiently identifying competent individuals, leading to time and resource savings for organizations. However, addressing ethical and legal concerns is crucial to ensure fairness and transparency in the recruitment process. Continued research and development efforts are necessary to refine algorithms, minimize biases, and adapt to evolving ethical and legal standards [12,17]. Organizations must remain vigilant and implement responsible strategies to fully leverage the benefits of automated recruitment tools while ensuring their long-term effectiveness. Researchers highlight crucial factors in the utilization of large language models (LLMs), emphasizing contextual understanding, transparency, and acknowledgment of limitations [3,4,10]. Despite their potential in supporting diverse language skills and facilitating research, challenges and risks associated with LLM applications underscore the importance of addressing biases and educating about responsible use [3,10]. Novais' thesis [14] presents a proof of concept demonstrating the effectiveness of automated curriculum classification using NLP, validating its performance against manual methods.

4 Hypothesis

The objective of the project is to update recruitment methods by utilising LLMs and chatbots, thereby transforming the process of resume screening and candidate interaction. LLMs effectively classify resumes, while chatbots provide initial evaluations, transmit job specifics, and customise interactions [20]. Automating duties enables recruiters to dedicate their attention to nurturing relationships with top applicants. By including pre-interview evaluations, we may identify highly qualified individuals at an early stage. Continuous improvement fosters adaptability, mitigates prejudices, and amplifies fairness, so optimising hiring procedures and bolstering organisational efficacy.

5 Proposal

The PhD thesis aims to revolutionize traditional recruitment practices by leveraging LLMs and chatbots, offering a multifaceted approach to optimization. LLMs play a pivotal role in resume categorization, autonomously extracting crucial information and expediting candidate screening. Chatbots, equipped with natural language processing skills, engage with candidates early on, providing job details and conducting initial assessments with tailored questions, enhancing efficiency and accuracy. Moreover, chatbots facilitate personalized communication between recruiters and candidates, automating repetitive tasks and allowing recruiters to prioritize cultivating relationships with top candidates for an exceptional candidate experience. Integrating pre-interview evaluations into the chatbot interface enables recruiters to systematically assess candidates' skills and cultural fit, identifying promising individuals early in the process. Continuous learning and improvement of LLMs and chatbots are essential to adapt to changing recruitment needs, minimize biases, and ensure fairness in candidate selection, ultimately streamlining hiring processes, enhancing candidate engagement, and improving organizational effectiveness.

6 Evaluation Plan

Evaluating chatbots, especially those utilizing Language Models (LLMs), is crucial in contemporary recruitment. Examining candidate interactions with these sophisticated chatbots compared to conventional approaches provides valuable insights on the future of recruitment. Research can investigate various aspects of user experience, including trust, bias, efficacy, and candidate participation. Gaining a comprehensive understanding of these factors might lead to the restructuring of recruitment techniques in order to improve effectiveness and contentment.

Integrating chatbots powered by LLM technology into the recruitment process can enhance efficiency and satisfaction. Essential components comprise of conducting usability testing to guarantee an interface that is easy to use, interaction that is seamless, and responses that are precise. Effectiveness will be assessed using performance measures such as response speed, screening accuracy, and interview scheduling success. User feedback will evaluate the level of happiness, equity, and general effectiveness of the recruitment procedures. Comparative assessments will emphasize the advantages of chatbots, including their time efficiency, decreased need for manual labor, and enhanced compatibility with applicants. Conducting bias detection and scalability tests will guarantee impartiality and flexibility. Iterative enhancements will maintain industry standards, promote efficiency, optimize applicant satisfaction, and cultivate creativity.

7 Reflections

Examining the incorporation of advanced technologies like LLMs and chatbots into recruiting procedures provides valuable understanding of the advantages and difficulties linked to automated talent acquisition. The research conducted by Harsh et al. and Sudha et al. highlights the effectiveness of NLP and ML algorithms in improving the efficiency and effectiveness of resume screening [9,22]. Implementing automation for processes such as resume processing and early candidate screening can result in time and resource savings for organisations, while also guaranteeing a more uniform and unbiased evaluation process [9,22]. Nevertheless, the ethical and legal issues raised by Raghavan and Alina regarding algorithmic biases and discrimination in recruitment procedures emphasise the significance of responsible implementation and ongoing monitoring to maintain fairness and transparency in candidate selection [12,17]. In addition, the application of NLP and ML techniques, as evidenced in the research referenced as [11,13], shows potential in improving the accuracy and precision of resume analysis [11,13]. CNNs and ML algorithms like Random Forest and Naive Bayes have demonstrated effectiveness in detecting pertinent talents and experiences from resumes, assisting in the process of rating and selecting candidates [11,13]. However, it is crucial to continue conducting research and development in order to tackle problems like algorithmic biases and to constantly improve algorithms to meet changing recruitment needs and candidate preferences [11,13]. Moreover, the successful applications described by Pimpalkar et al., Sanyalet al., Alsharefet al., and Alsaif et al. highlight the capacity of NLP and ML approaches to enhance recruiting processes in different industries [1,2,16,18]. These research provide evidence of the efficacy of several algorithms and models, such as text-similarity measures, named entity identification, and bi-directional recommendation systems, in improving the efficiency and accuracy of candidate selection [1,2,16,18]. Nevertheless, the adoption of automated recruitment methods requires thorough examination of ethical, legal, and technical aspects, as highlighted by Alina and Raghavan [12,17]. Continual research and development are crucial for enhancing algorithms, reducing biases, and guaranteeing adherence to changing ethical and legal norms in recruitment processes [12,17].

References

1. Alsaif, S.A., et al.: NLP-based bi-directional recommendation system: towards recommending jobs to job seekers and resumes to recruiters. Big Data Cogn. Comput. **6**(4) (2022). https://doi.org/10.3390/bdcc6040147. ISSN 2504-2289
2. Alsharef, A., et al.: Exploring the efficiency of text-similarity measures in automated resume screening for recruitment. In: 2023 10th International Conference on Computing for Sustainable Global Development (INDIACom), pp. 36–42 (2023)
3. Amirova, A., et al.: Framework-based qualitative analysis of free responses of Large Language Models: Algorithmic fidelity. PLoS ONE **19**(3), e0300024 (2024). https://doi.org/10.1371/journal.pone.0300024
4. Balcerak, A., Woźniak, J.: ICT-based recruitment and selection tools: the recruiters' perspective. Sci. Papers Silesian Univ. Technol. **178**, 1–25 (2023). https://doi.org/10.29119/1641-3466.2023.178.2
5. Cambridge English Dictionary. https://dictionary.cambridge.org/dictionary/english/. Accessed 29 Jan 2023
6. Cardoso, A.A.: Recrutamento & Selecção de Pessoal. Ed. by Editora RH (2005)
7. Chiavenato, I.: Gestão de Pessoas. Productivity Press (2010)
8. Consul, N., et al.: Mitigating unconscious bias in recruitment and hiring. J. Am. Coll. Radiol. **18**(6), 769–773 (2021). https://doi.org/10.1016/j.jacr.2021.04.006. https://www.sciencedirect.com/science/article/pii/S1546144021003227. Focus on Private Practice. ISSN 1546-1440
9. Harsha, T.M., et al.: Automated resume screener using natural language processing (NLP). In: 2022 6th International Conference on Trends in Electronics and Informatics (ICOEI), pp. 1772–1777 (2022). https://doi.org/10.1109/ICOEI53556.2022.9777194
10. Hosseini, M., et al.: An exploratory survey about using ChatGPT in education, healthcare, and research. PLOS ONE (2023). https://doi.org/10.1371/journal.pone.0292216
11. Kinge, B., et al.: Mitigating unconscious bias in recruitment and hiring. Int. J. Sci. Res. Comput. Sci. Eng. Inf. Technol. **18**, 769–773 (2022)
12. Köchling, A., Wehner, M.C.: Discriminated by an algorithm: a systematic review of discrimination and fairness by algorithmic decision making in the context of HR recruitment and HR development. Bus. Res. **13**(3), 795–848 (2020). https://doi.org/10.1007/s40685-020-00134-w

13. Mridha, M.F., et al.: A machine learning approach for screening individual's job profile using convolutional neural network. In: 2021 International Conference on Science & Contemporary Technologies (ICSCT), pp. 1–6 (2021). https://doi.org/10.1109/ICSCT53883.2021.9642652
14. Novais, L.: Automated Resume Screening System Using Natural Language Processing (2023)
15. Oxford dictionary. https://www.oxfordlearnersdictionaries.com/. Accessed 29 Jan 2023
16. Pimpalkar, A., et al.: Job applications selection and identification: study of resumes with natural language processing and machine learning. In: 2023 IEEE International Students' Conference on Electrical, Electronics and Computer Science (SCEECS), pp. 1–5 (2023). https://doi.org/10.1109/SCEECS57921.2023.10063010
17. Raghavan, M., et al.: mitigating bias in algorithmic hiring: evaluating claims and practices. In: Proceedings of the 2020 Conference on Fairness, Accountability, and Transparency, FAT* 2020, Barcelona, Spain, pp. 469–481. Association for Computing Machinery (2020). https://doi.org/10.1145/3351095.3372828. ISBN 9781450369367
18. Sanyal, S., et al.: Resume parser with natural language processing. Ph.D. thesis, March 2017. https://doi.org/10.13140/RG.2.2.11709.05607
19. Several. Recruitment and Selection Revised Edition. Routledge, Amsterdam (2009)
20. Skondras, P., Zervas, P., Tzimas, G.: Generating synthetic resume data with large language models for enhanced job description classification. Future Internet **15**, 363 (2023). https://doi.org/10.3390/fi15110363. https://www.mdpi.com/journal/futureinternet
21. Society for Human Resource Management. https://www.shrm.org/. Accessed 29 Jan 2023
22. Sudha, G., et al.: Personality prediction through cv analysis using machine learning algorithms for automated E-recruitment process. In: 2021 4th International Conference on Computing and Communications Technologies (ICCCT), pp. 617–622 (2021). https://doi.org/10.1109/ICCCT53315.2021.9711787

Towards Non-invasive Detection of Gastric Intestinal Metaplasia: A Deep Learning Approach Using Narrow Band Imaging Endoscopy

Sílvio Capela[1(✉)], Jorge Lage[2], and Vítor Filipe[1,3]

[1] Universidade de Trás-os-Montes e Alto Douro (UTAD), 5000-081 Vila Real, Portugal
silvioc.25@hotmail.com, vfilipe@utad.pt

[2] Serviço de Gastrenterologia, Unidade Local de Saúde de Trás-os-Montes e Alto Douro (ULSTMAD), 5000-508 Vila Real, Portugal

[3] INESC TEC – INESC Technology and Science, 4200-465 Porto, Portugal

Abstract. Gastric cancer, ranking as the sixth most prevalent cancer globally and a leading cause of cancer-related mortality, follows a sequential progression known as Correa's cascade, spanning from chronic gastritis to eventual malignancy. Although endoscopy exams using Narrow Band Imaging are recommended by internationally accepted guidelines for diagnostic Gastric Intestinal Metaplasia, the lack of endoscopists with the skill to assess the NBI image patterns and the disagreement between endoscopists when assessing the same image, have made the use of biopsies the gold standard still used today. This proposal doctoral thesis seeks to address the challenge of developing a Computer-Aided Diagnosis solution for GIM detection in NBI endoscopy exams, aligning with the established guidelines, the Management of Epithelial Precancerous Conditions and Lesions in the Stomach. Our approach will involve a dataset creation that follows the standardized approach for histopathological classification of gastrointestinal biopsies, the Sydney System recommended by MAPS II guidelines, and annotation by gastroenterology experts. Deep learning models, including Convolutional Neural Networks, will be trained and evaluated, aiming to establish an internationally accepted AI-driven alternative to biopsies for GIM detection, promising expedited diagnosis, and cost reduction.

Keywords: Gastric Intestinal Metaplasia · Narrow Band Imaging · NBI · Endoscopy · Gastric Cancer · Deep Learning · Computer-Aided Diagnosis · Biopsies

1 Introduction

In recent years, the concept of Artificial Intelligence (AI) has been a heavily discussed topic, growing interest not only in the research and scientific fields but also in the media, being this interest especially propelled by the resurgence of deep learning [1]. This, together with the amount of data produced every day, impacts the way we live our daily

G. Marreiros et al. (Eds.): DCAI 2024, LNNS 1151, pp. 298–303, 2025.
https://doi.org/10.1007/978-3-031-80946-0_31

lives, with more and more artificial intelligence systems influencing our behavior based on our preferences. The same can happen in the health sector, where a huge amount of data is produced every day and can also be used to develop AI systems to support decision making or diagnose diseases [2].

Gastric cancer ranks as the sixth most prevalent cancer globally and stands as the fourth leading cause of cancer-related mortality worldwide [3]. The onset of stomach cancer is usually a sequential process that was first described by Pelayo Correa [4] and is now known as the Correa's cascade, represented in Fig. 1.

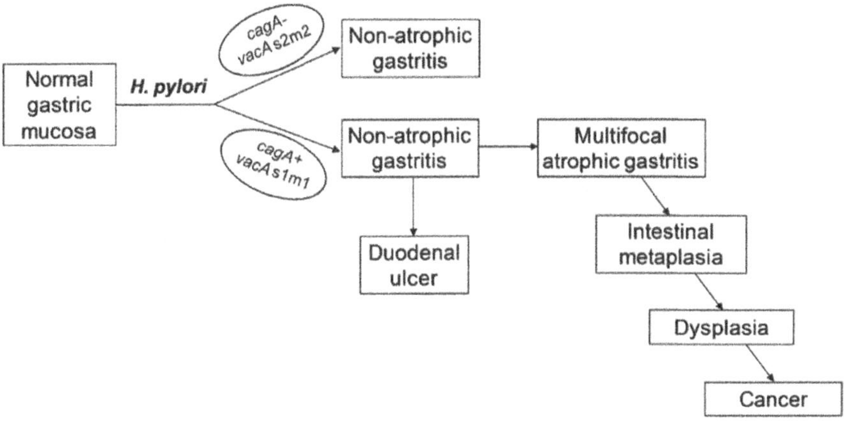

Fig. 1. Correa's cascade [4]

The process described by Pelayo Correa consists of five phases:

1. Chronic Gastritis: This refers to stomach inflammation primarily caused by Helicobacter pylori (HP) infection, although other factors like smoking or alcohol consumption can contribute.
2. Gastric Atrophy: In this phase, due to HP infection, there's a reduction of the normal stomach glandular cells. This atrophy weakens the stomach lining, a reduces acid secretion and a promotes a favorable environment for the Helicobacter pylori bacteria.
3. Intestinal Metaplasia: Cells in the stomach lining start acquiring characteristics resembling those of intestinal cells.
4. Dysplasia: Intestinal metaplasia may progress to dysplasia, where abnormal cells in the stomach lining have the potential to become cancerous.
5. Initial Gastric Cancer: Dysplastic cells can evolve into cancer cells, leading to the formation of a malignant tumor in the stomach.

A set of guidelines called Management of Epithelial Precancerous Conditions and Lesions in the stomach (MAPS) was proposed in 2012 by the European Society of Gastrointestinal Endoscopy (ESGE), the European Helicobacter and Microbiota Study Group (EHMSG), the European Society of Pathology (ESP) and the Portuguese Society of Digestive Endoscopy (SPED). This document was the first international guideline on the management of precancerous conditions and lesions in the stomach. Later in 2019, an update to the MAPS guideline was proposed, the MAPS II, developed according to evidence in this area published after 2010 [5].

This document defines guidelines for diagnosing gastric cancer's pre-malignant conditions and also a set of surveillance guidelines for patients with any of the pre-malignant conditions. This guideline concluded that the use of Narrow Band Imaging (NBI) for the detection and diagnosis of Gastrointestinal Intestinal Metaplasia (GIM) provides better overall results when compared with the classic White-Light Endoscopy (WLE). NBI technology uses narrowband filters to restrict the emitted light to just two narrow wavelength bands, blue and green, making it possible to see the blood vessels in the gastric mucosa in greater detail, which allows pre-malignant conditions such as metaplasia to be identified more easily without the need for biopsies [6], but the lack of endoscopists familiarized with advanced image patterns makes the need for biopsies still recommended in the present and it should be taken according to the original Sydney System, two biopsies from the antrum (lesser and greater curvature) and two from the corpus (lesser and greater curvature) [5].

The updated Sydney System is the most accepted protocol for gastritis classification and grading, it recommends five biopsies, two from the antrum (lesser and greater curvature), two from the corpus (lesser and greater curvature), and a final one from incisura. However, MAPPS II guidelines suggest the use of the original Sydney System, as a large study [12] proved that the original Sydney System had the highest yield for the diagnosis of GIM, stating that the inclusion of an incisura biopsy does not add much information to the diagnosis with a bigger cost [5, 12]. Figure 2 presents a comparison between WLE and NBI images on a patient diagnosed with GIM. According to [6], in WLE it is only possible to detect features suggesting the presence of GIM on the angle but when using NBI also lesser curvature presents features that indicate the presence of GIM, meaning that NBI provided more information than WLE.

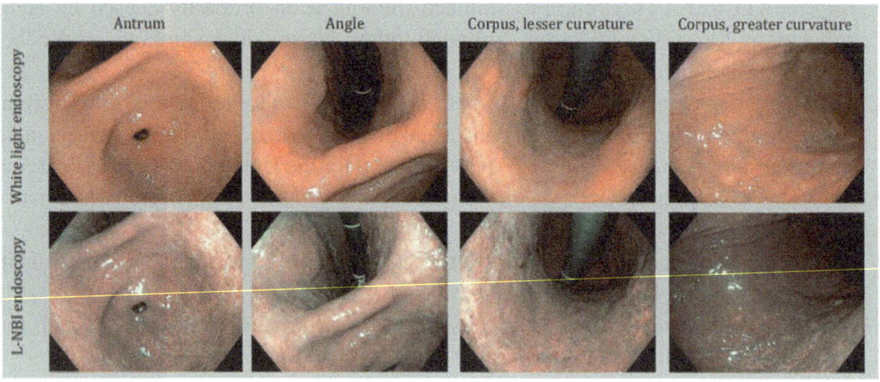

Fig. 2. Endoscopic images examples used by [6]. White Light Endoscopy Images (first line), NBI images from same stomach location (second line)

Computer Aided Diagnosis (CAD) for gastrointestinal diseases has been a widely researched area with multiple solutions being proposed in this area. The deep learning-based methods, mainly the Convolutional Neural Networks (CNNs), have achieved revolutionary results in many medical image problems [7, 8], meaning that potentially they easily detect image patterns and thus make the use of biopsies unnecessary. This could represent an important advance in the diagnosis of GIM, meaning easier and faster diagnosis contributing to an improved overall cost-effectiveness.

2 Related Work

Despite the results obtained by deep learning methods in the medical field in recent years, especially CNNs, the literature review shows that there is a lack of studies that combine the use of this is technology with NBI. From the ones that use these two technologies, it is highlighted the good results obtained by the deep learning models when compared with endoscopists [8]. Tao Yan et al. [9] proposed an Intelligent Diagnostic System (IDS) for GIM detection, composed of three deep learning models for classification tasks, one of them to distinguish NBI frames from Magnifying Endoscopy combined with Narrow Band Imaging (ME-NBI) frames and two separated models based on CNNs to classify these frames based on the endoscopic technique that they were collected. The EfficientNet-b4 model was trained and evaluated with the ME-NBI dataset, obtaining an average 89% accuracy on a five-fold cross-validation. The final IDS diagnosis results were 82.7%, 88.6% and 86.2% accuracy on the NBI images dataset, ME-NBI dataset and both datasets combined, respectively. Similarly, a research compared five CNN models (ResNet-50, VGG-16, Inception-v3, DenseNet-121, EfficentNet-b4) for classification task, on a small dataset combining both NBI and ME-NBI images into a total of 125 images [10]. Image flip data augmentation technique was used to improve the quantity of images available for training the models and the results obtained were between 72.1% and 82.4% accuracy, from EfficientNet-b4 and VGG-16, respectively. Contrary to the results obtained in the study conducted by [9], the EfficientNet-b4 had the worst performance in this case.

From another perspective of the subject of this research, deep learning models have been used to evaluate GIM grading and Operative Link on Gastritis Assessment based on Intestinal Metaplasia (OLGIM) staging using ME-NBI images collected according to the Sydney System, from antrum and body stomach locations [11]. The authors proposed the use of the Faster-RCNN detection and classification model for metaplasia detection and grade classification. The performance of the proposed was not statistically different when compared with three junior and three senior endoscopists but higher than three untrained endoscopists, meaning that the model could not be used alone for diagnosis but could assist endoscopists on this task.

After a literature review, it is possible to identify that there are some interesting results when combining deep learning-based methods with NBI endoscopic images. Therefore, it can be observed that almost no study took into account the MAPS II recommendation, so there is only one dataset from our review so far that collected endoscopic images following the Sydney System guideline but used it for grading GIM, meaning no healthy images present. As MAPS II, are the internationally accepted guidelines for managing precancerous lesions in the stomach, there is an unexplored research opportunity for further studies to provide a solution that can be presented for acceptance and integration in these recommendations.

3 Research Question and Proposal Work

According to the research and analysis carried out to date, the main research question (RQ) on which we intend to carry out studies is:

RQ: It is possible to detect precancerous conditions of stomach cancer, especially metaplasia, using artificial intelligence and obtain significant results that justify the removal of the need for taking biopsies?

At this point, the study raised some concerns:

- There are no public datasets that use NBI endoscopy exams, meaning that a data collection process will be needed to conduct this study.
- The possible lack of data can limit the study results.
- Obtaining data for this research may encounter ethical challenges regarding privacy concerns.

However, we propose the creation of a CAD solution for the detection of GIM in NBI endoscopy exams. For the development of a proposal solution, we intend to conduct research in a distinctive manner:

1. Create a dataset of NBI endoscopy images according to the Sydney System, with four images per patient, from both greater and lesser curvature of the stomach antrum and body locations. Each image will represent a biopsy site and the ground truth will be the histological confirmation obtained from that biopsy.
2. Annotate regions of interest on NBI images, according to doctors specialized in gastroenterology and endoscopic exams.
3. Apply deep learning methods to the dataset created, compare multiple models, and explore classical data augmentation techniques and new augmentation approaches, such as Generative Adversarial Networks (GANs) and diffusion models, to improve the amount of data and the models' results.
4. Compare the models' predictions with the histological findings to assess the efficiency of the models. This evaluation will involve analyzing the Sensitivity, Specificity, Accuracy and overall diagnostic performance of the models in comparison to histopathological results.

By creating a dataset that takes into account the MAPS II guidelines and the use of the Sydney System in the selection of NBI endoscopy exam frames, we aim for a CAD solution that can remove the existing need for biopsies, meaning faster diagnosis and a cost reduction, and that can be internationally accepted as it matches the current MAPS II recommendations for this precancerous condition.

4 Final Reflections

The advances in endoscopy exam technologies like NBI have contributed to improving early diagnosis of gastric cancer [8] but the lack of endoscopists with the skills to analyze and comprehend the image patterns shown by NBI images, limit the use of it on endoscopic exams. On the other hand, Artificial Intelligence, mainly deep learning methods like CNNs have achieved revolutionary results in multiple subjects in the medical image area and their use in this type of medical image can solve this problem. The objective of this scientific research is to explore the use of AI models on NBI endoscopic images, in order to do so we propose the creation of a dataset that can represent real endoscopic exams where each image will have a histological validation from a biopsy, and they will be collected following the MAPS II guidelines. The research challenge of this thesis is to present an AI model that can be accepted internationally as an alternative to biopsies on detecting GIM in endoscopy exams, meaning less costs and faster diagnosis.

References

1. Chollet, F.: DEEP LEARNING with Python. Manning Publications Co., Shelter Island (2018)
2. Bohr, A., Memarzadeh, K.: Chapter 2 - The rise of artificial intelligence in healthcare applications. In: Artificial Intelligence in Healthcare, pp. 25–60 (2020). https://doi.org/10.1016/B978-0-12-818438-7.00002-2
3. World Health Organization. https://www.who.int/en/news-room/fact-sheets/detail/cancer. Accessed 11 Apr 2023
4. Correa, P., Piazuelo, M.B.: The gastric precancerous cascade. J. Digest. Diseases **13**(1), 2–9 (2012). https://doi.org/10.1111/j.1751-2980.2011.00550.x
5. European Society of Gastrointestinal Endoscopy. https://www.esge.com/management-of-epithelial-precancerous-conditions-and-lesions-in-the-stomach-maps-ii-esge-esp-sped-guideline-update-2019. Accessed 11 Apr 2023
6. Lage, J., et al.: Light-NBI to identify high-risk phenotypes for gastric adenocarcinoma: do we still need biopsies? Scand. J. Gastroenterol. **51**, 501–506 (2015)
7. Litjens, G., et al.: A survey on deep learning in medical image analysis. Med. Image Anal. (2017). https://doi.org/10.1016/j.media.2017.07.005
8. Horiuchi, Y., Hirasawa, T., Fujisaki, J.: Application of artificial intelligence for diagnosis of early gastric cancer based on magnifying endoscopy with narrow-band imaging. Clin. Endosc. **57**(1), 11–17 (2024). https://doi.org/10.5946/ce.2023.173
9. Yan, T.: Intelligent diagnosis of gastric intestinal metaplasia based on convolutional neural network and limited number of endoscopic images. Comput. Biol. Med., **126** (2020). https://doi.org/10.1016/j.compbiomed.2020.104026
10. Martins, M., et al.: Diagnostic performance of deep learning models for gastric intestinal metaplasia detection in narrow-band images. In: Annual International Conference of the IEEE Engineering in Medicine & Biology Society (2023). https://doi.org/10.1109/EMBC40787.2023.10340055
11. Niu, W., et al.: A deep learning model based on magnifying endoscopy with narrow-band imaging to evaluate intestinal metaplasia grading and OLGIM staging: a multicenter study. Dig. Liver Dis. (2024). https://doi.org/10.1016/j.dld.2024.02.001
12. Lash, J.G., Genta, R.M.: Adherence to the Sydney System guidelines increases the detection of Helicobacter gastritis and intestinal metaplasia in 400 738 sets of gastric biopsies. Aliment. Pharmacol. Ther. **38**, 424–431 (2013). https://doi.org/10.1111/apt.12383

Reflecting on how to Analyze Poisoning Attacks in Gossip Learning

Alexandre Pham(✉)

Sorbonne Université, CNRS, LIP6, 75005 Paris, France
Alexandre.Pham@lip6.fr

Abstract. Gossip Learning enables the automatic learning of a model without using a central server. This distributed control prompts the study of the possibility of malicious attacks by the participants themselves. While poisoning attacks on other forms of machine learning have been extensively studied, their effects on collective learning via gossip did not receive a similar level of attention yet. Our work is the first to propose a methodology for evaluating poisoning attacks in this context in scenarios with or without churn. Additionally, to assess our methodology on a representative case study, we extended an existing simulator with a poisoning attack injection module.

Keywords: Decentralized Federated Learning · Gossip Learning · Attacks

1 Introduction

In traditional machine learning (ML) systems, a central entity manages both the model and the data collected for model training. Such data centralization poses a problem when the data used to train the model is sensitive and needs to be kept private, or simply to comply with regulations. Federated Learning, introduced by Google in 2016 [4], maintains a central entity that coordinates training across multiple machines scattered throughout the network, without the training data being communicated to the server.

Despite its reduced role, the existence of a central server constitutes a point of failure and an obvious target for attack. Gossip learning (GL) aims to perform Federated Learning without relying on a central server, using peer-to-peer communications.

In this work, we evaluate the impact of attackers (also called Byzantine or malicious nodes) attempting to corrupt the models of honest participants in a GL system. Our reference GL algorithm was proposed by Hegedűs et al. [3]. Close to our work, Giaretta et al. [2] studied the applicability of GL, highlighting problems and proposing solutions when the data distribution is correlated either with the degree distribution or the communication speed. However, they do not consider attacks in their work.

G. Marreiros et al. (Eds.): DCAI 2024, LNNS 1151, pp. 304–309, 2025.
https://doi.org/10.1007/978-3-031-80946-0_32

Our Contributions. This work is a summary of a technical report [5] and conference paper, where our contribution is threefold. Firstly, we propose a methodology to evaluate the effects of poisoning attacks in a system running a GL algorithm. Secondly, to evaluate our methodology, we implemented a public extension[1] of the popular simulator *gossipy*[2] to simulate poisoning attacks. Finally, we apply our methodology to Hegedűs et al.'s GL algorithm [3] to evaluate its resilience to poisoning attacks based on several factors such as topology, Byzantine node distribution, and churn. Our analysis paves the way for the design of countermeasures against poisoning attacks.

2 Methodology

In this section, we present the various elements of our methodology: the choice of the GL algorithm to be tested, the choice of topologies, the dataset, and the machine learning algorithm, the attack, churn parameters, and the metrics used to evaluate the impact of the attacks.

GL Algorithm. Recently, Hegedűs et al. [3] proposed a GL algorithm that exhibits the best overall performance when considering communication resources. Their algorithm employs two interesting mechanisms. The first is a compression technique, by not transmitting all parameters in each exchange, but only a part called *Partition*, and we denote by S the total number of partitions the system uses, with a higher S implying smaller messages. The second mechanism balances the sending of periodic messages and the sending of messages in response to an event, here following a local update after receiving a partition.

Topologies for the GL Infrastructure. We study the following topologies[3] with n nodes: 20 fan-out graph, as initially used by Hegedűs et al. [3]; random 20-regular graphs; Watts-Strogatz graphs for their small-world property; Erdős-Rényi graphs for their balanced property; Zipf-based graphs, where each node has at least one neighbor. For the last two topologies, we restrict our simulations to instances where the graph is connected.

ML Dataset. We use the MNIST dataset as our dataset. It consists of 2 sets $\mathcal{D}_{\text{training}}$ and $\mathcal{D}_{\text{test}}$. An instance of this dataset is a pair $(x, y) \in \mathbb{R}^{784} \times [\![0, 9]\!]$, where x represents a 28×28 pixel image with the digit y written on it. Each node k will have as its local training set $\mathcal{D}^k_{\text{training}}$ of size 250, and such that $\bigcap_{k=1}^{n} \mathcal{D}^k_{\text{training}} = \emptyset$.

ML Model. Each node k trains a logistic regression (multinomial) model to correctly classify the images from the aforementioned dataset. We set $\eta = 1$ and $\lambda = 0$, these values offering the best performance in the Erdős-Rényi case

[1] https://gitlab.lip6.fr/apham/data-poisoning-attacks-in-gossip-learning.
[2] https://github.com/makgyver/gossipy/.
[3] Here, we present only a subset of the results from our technical report [5].

without churn, but the choice of these 2 values is not trivial and must be studied for all topologies as done by Hegedűs et al. [3].

Byzantine Attacks. In this study, Byzantine nodes insert a trigger pattern of 9 pixels into 20% of the images in their dataset, which they then label as 0. Their goal is to make honest nodes classify the marked images as 0, without decreasing classification performance on clean data (i.e., without the trigger pattern). As Byzantine nodes attack the GL system by manipulating their dataset to disrupt the network, this is called a *data poisoning attack*, and as their goal is to introduce a hidden objective, this type of attack is also referred to as a *backdoor attack*. To evaluate the performance and impact of the attack on honest nodes, we draw 2000 images from the test set that we divide into two equal-sized sets. The first half will be used to evaluate performance on the usual classification task. For the second half, we remove all (x, y) where $y = 0$, and apply the same modifications to x as those done by the Byzantine nodes on the remaining samples. These two sets are called *test set* and *backdoor set*.

To place Byzantine nodes, we will use two strategies: *classic* and *random*. In the *classic* strategy, we first select nodes with the highest degree. In the *random* strategy, we select nodes by randomly sampling without replacement.

Metrics. We are interested in the average *accuracy* of honest nodes on the test and backdoor sets. *Accuracy* is defined as the number of correct predictions over the total number of predictions. For the first, the higher it is, the better: (honest) nodes should perform well on data that hasn't been marked with the pattern, and for the second, it's the opposite: nodes should perform poorly on tampered images.

3 Simulation Results

Let us define $n \in \{100, 150\}$ as the number of nodes and f as the number of Byzantine nodes in a simulation. Each curve presented here is an average over at least 10 random executions. In the following, we present two scenarios, with and without churn, starting with the churn-free scenario, i.e., when nodes remain online throughout the execution.

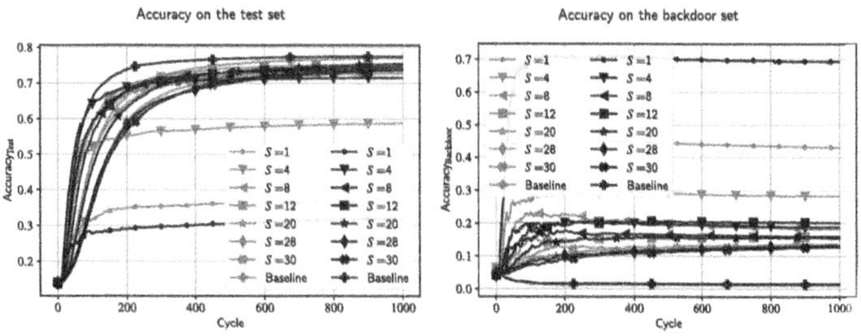

Fig. 1. Accuracy on the test set (left) and the backdoor set for the 20 fan-out topology for $n = 100$ (grey) and $n = 150$ (black) with the random placement strategy with $f = 0.3n$. Baselines curves are the best results on malicious-free simulations among the S studied.

Churn-Free Scenario. In Fig. 1, we represent the performance on the 2 sets defined in Sect. 2 when $n = 100$ and $n = 150$ with $f = 0.3n$ Byzantine nodes placed according to the *random* strategy.

We can observe that when S is low, the attack is (too) powerful for honest nodes, as they misbehave on the test set and classify altered inputs very well compared to the considered reference (best S studied here in simulations without Byzantine on the test set). This can be explained by the fact that a partition likely contains parameters that interact with the trigger pattern. As S increases, we notice that the attack is less powerful, as honest nodes behave closely to the reference on the test set. Here, parameters that interact with the trigger pattern are too diluted among all partitions, and thus less likely to be updated. But we notice that Byzantine nodes still manage to introduce undesirable behavior (although not as much as they would like).

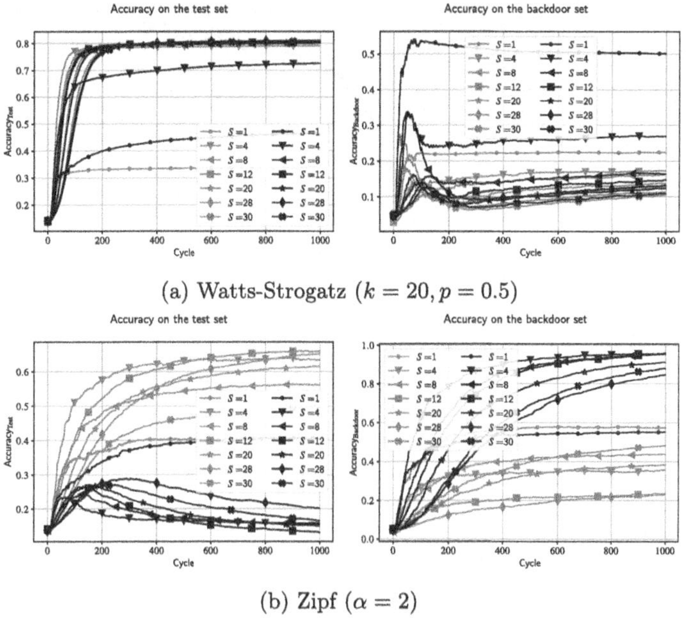

Fig. 2. Accuracy on the test set and backdoor set for different topologies with $n = 150$ and $f = 0.3n$ with the random placement strategy (grey) and classical placement strategy (black).

In Fig. 2, we compare the *classical* and *random* strategies for Byzantine nodes, in graphs where the degree distribution is not balanced. We notice that when $n = 150$, nodes placed in a Watts-Strogatz network and Byzantine nodes are placed randomly, honest nodes behave similarly to what was shown in Fig. 1 when S is high (and perform better when $S = 1$). Interestingly, with the backdoor, when Byzantine nodes are placed randomly, honest nodes placed in a Watts-Strogatz topology handle the attack better than when placed in a Zipf-based network. We assume this result is due to local parts of the graph where honest nodes react strongly to the backdoor, whereas this is probably not the case for the Watts-Strogatz topology due to its small-world property. When Byzantine nodes are placed in a classical manner, honest nodes perform much better on the backdoor than on the test, this is because it is easier for them to optimize with tampered data, and when sending partitions, their solutions have a strong impact.

Churn Scenario. We now study the case where nodes can disconnect and reconnect and have a 20% probability of being online[4].

In Fig. 3, we study the effects of the placement strategy and the number of malicious nodes in a Zipf-based topology. Relative to the number f, unsurprisingly, the classical method is more powerful than the random strategy, however, given enough nodes placed randomly, it can outperform the classical strategy when there are less malicious nodes.

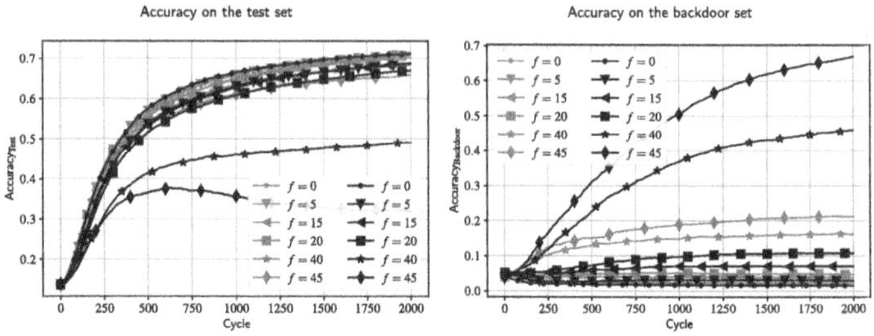

Fig. 3. Accuracy on the test and backdoor set for $n = 150$, $S = 8$ for $f \in \{0, 5, 15, 20, 25, 40, 45\}$ with the random (gray) and classical (black) placement strategy when nodes degree follow a Zipf law.

In Fig. 4, we study $n = 100$ and 150, with $f = 0.3n$ Byzantine nodes placed randomly. We observe that even though $S = 1$ is special, and although honest nodes are not resilient against the attack initially, they manage to reach a much better solution in the end, close to the other values of S studied. This can be explained by the fact that nodes are offline most of the time, including Byzantine nodes, so malicious partitions are less likely to be processed. We notice that for the other values of S studied, the result is very close to what was shown in Fig. 1 despite the difference in topology, and can be explained by the uniform distribution of data: honest nodes having the same (type of) learning material, learn to accomplish the same task (but more slowly than when they are always present).

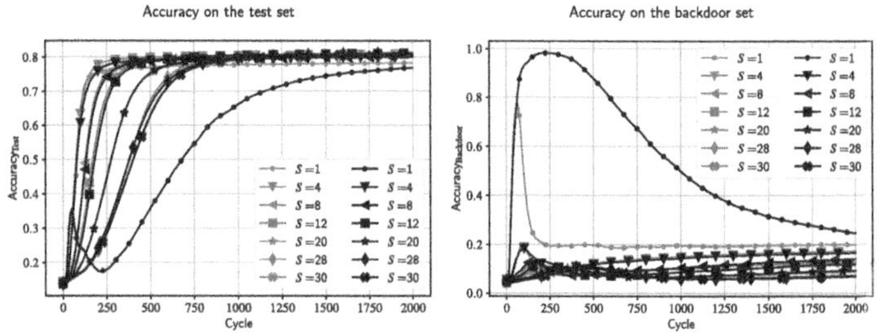

Fig. 4. Accuracy on the test set and backdoor set with $n = 100$ (grey) and 150 (black) in a 20-random regular graph with $f = 0.3n$.

The fact that the accuracy in the backdoor set can be a growing function with respect to the number of rounds or the difference in the churn-free and churn scenario when S is low are interesting differences to study.

[4] To approximate the churn scenario used in [3].

4 Reflections and Discussions

It would be interesting to use the presented method to study other GL or more broadly Decentralized FL (DFL) algorithms, and we refer to [1] for an overview of DFL.

Besides, as GL involves Distributed Computing and Machine Learning, we can analyze and reflect about the summarized work under both perspectives.

Distributed Computing. On this side of this work, it would be interesting to further this study by studying other topologies, for example those found in IoT context, and to study more malicious behavior than the limited behavior studied here, as nodes here followed the same behavior as honest nodes besides tampering with their dataset.

Machine Learning. On this side of this work, we can closely follow what is being studied in the ML (models...) and FL (aggregation...) literature and adapt it to DFL or GL instances.

5 Conclusion

We have proposed a methodology to study the resilience of Gossip Learning in the presence of poisoning attacks. As a case study, we target the state-of-the-art GL algorithm, and study its resilience across a wide range of topologies in scenarios with and without churn. Our results show that communication optimizations and the choice of underlying topology are not always favorable to honest nodes. Furthermore, the distribution of Byzantine nodes also has a significant impact on the accuracy of the results.

Acknowledgments. The work presented in this document has received funding from the EU Horizon Europe research and innovation Programme under Grant Agreement No. 101070118.

References

1. Beltrán, E.T.M., et al.: Decentralized federated learning: fundamentals, state of the art, frameworks, trends, and challenges. IEEE Commun. Surv. Tutor. **25**, 2983–3013 (2023)
2. Giaretta, L., Girdzijauskas, Š.: Gossip learning: off the beaten path. In: 2019 IEEE International Conference on Big Data (Big Data), pp. 1117–1124 (2019)
3. Hegedűs, I., Danner, G., Jelasity, M.: Decentralized learning works: an empirical comparison of gossip learning and federated learning. J. Parallel Distrib. Comput. **148**, 109–124 (2021)
4. McMahan, H.B., Moore, E., Ramage, D., Hampson, S., y Arcas, B.A.: Communication-efficient learning of deep networks from decentralized data (2016)
5. Pham, A., Potop-Butucaru, M., Tixeuil, S., Fdida, S.: Data poisoning attacks in gossip learning. Technical report, Sorbonne Université, January 2024

Advancing Wildlife Monitoring in Gregarious Species with Drone Swarms

Edouard G. A. Rolland[1](✉), Lucie Laporte-Devylder[2,3], and Anders Lyhne Christensen[1]

[1] UAS Drone Center, MMMI, University of Southern Denmark (SDU), Campusvej 55, 5230 Odense M, Denmark
{edr,andc}@mmmi.sdu.dk
[2] WIPSEA, Rennes, France
lucie.laporte-devylder@wipsea.com
[3] Department of Biology, University of Southern Denmark (SDU), Campusvej 55, 5230 Odense M, Denmark

Abstract. The widespread adoption of autonomous multi-drone missions in wildlife research remains limited by the absence of readily available automated solutions. In this paper, we present an application of drone swarms to perform monitoring of gregarious species. As part of a PhD project within the interdisciplinary WildDrone consortium, we describe the potential of utilizing drones to gather multi-view data about animals for posture analysis, individual identification, and 3D reconstruction. We introduce our strategy to tackle the deployment of drone swarms for such a task, as well as our methodology to conduct our first real-world tests in January 2025 in Kenya.

Keywords: Drone Swarms · Wildlife Monitoring · WildDrone Project

1 Introduction

This paper presents a PhD project, which is part of the EU-funded MSCA Doctoral Network, WildDrone [1]. WildDrone is a research consortium consisting of 13 different thesis topics, focused on exploring the applications of drone technologies and machine vision in the field of nature conservation. The relevance of this initiative arises from the alarming decline in biodiversity over the past decades. It has become evident that there is an urgent need to develop new conservation methods that are cost-effective, time-efficient, and less labour-intensive [4]. For scientists, data collection is of paramount importance as it enables monitoring the health, behaviour, and populations of animal and plant species. The continuous advancements in the field of aerial robotics have opened up new opportunities for biologists, offering them a versatile experimental platform for conducting aerial surveys. In the scientific literature, drones have already been identified as effective tools in the realm of animal conservation, see [5,19] for examples.

G. Marreiros et al. (Eds.): DCAI 2024, LNNS 1151, pp. 310–316, 2025.
https://doi.org/10.1007/978-3-031-80946-0_33

As part of the WildDrone consortium, the PhD project presented in this paper aims to develop new methods for controlling drone swarms to extend the possibilities of data collection over large areas and among gregarious species, which tend to live in groups or communities. Coordinating multiple drones to collect data simultaneously will allow us to increase the efficiency of data collection and its robustness, as swarm systems are known to be highly error-resilient [20]. To the best of our knowledge, limited research has focused on real-world testing of drone swarms for wildlife conservation. In this project, we aim to move drone swarms out of simulations and laboratories, building on research outcomes from other WildDrone PhDs in biology and computer vision.

2 Problem Statement

Drone swarms can be applied in various contexts within species monitoring. To refine the focus of this PhD project, we have aligned its objectives with those of other members of the consortium who specialize in biology.

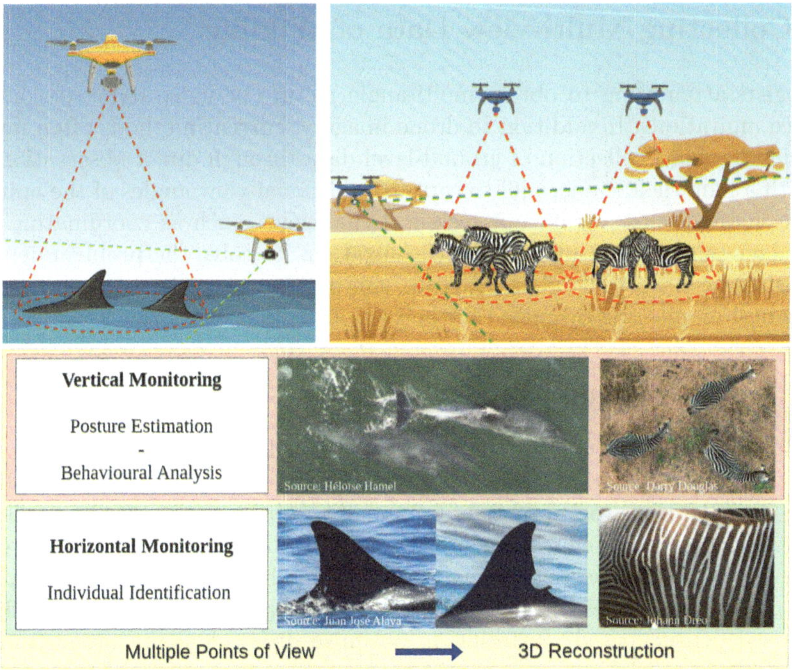

Fig. 1. Illustration of the problem and expected outcomes of multi-view monitoring of gregarious species

The ability to simultaneously collect data from different angles opens up new avenues that can significantly enhance the accuracy and efficiency of wildlife studies. While top-view monitoring can enable the acquisition of information used in wildlife census [3], behavioural studies [10], and posture analysis [11], the addition of lateral views would greatly improve and extend the possibilities. For instance, it would allow for individual identification by recognizing unique and distinctive physical features, such as fin notches or markings for marine mammals [14], and fur coat patterns such as zebra stripes [13], as depicted in Fig. 1. Using multiple viewpoints would also facilitate the creation of detailed 3D reconstructions of individuals and their environments, providing invaluable insights into habitat structure and animal behaviour.

In line with these objectives, this PhD project aims to develop swarm intelligence algorithms for dispatching drones over a group of animals to perform horizontal and vertical monitoring while respecting certain constraints, such as the closest distance to an individual and the spatial resolution of the pictures acquired. Moreover, the drones' positions must be chosen to maximize the number of animals accurately monitored in each configuration. The monitoring quality will be assessed using dedicated metrics defined by the consortium's biologists.

3 Collecting Multi-view Data of Wildlife

Biologists attempting to obtain multi-angle picture or video datasets continue to face limitations. In addition to drone imagery, current methods often involve the simultaneous collection of ground-level data through direct observation [10] or with a hand-held camera [6] to capture additional view angles of the animals. These methods pose significant technical challenges, such as coordinating multiple devices and ensuring data synchronization. To solve the problem depicted in Fig. 1, we have divided it into two sub-problems: vertical monitoring and horizontal monitoring discussed in the sections below.

3.1 Vertical Monitoring

The first sub-problem deals with monitoring animals from a vertical perspective. To date, tracking animal movements and behaviour is primarily accomplished using a single drone, often operated manually. Subsequently, the data collected by the drone is processed using computer vision techniques [12,17]. The potential of drone swarms for vertical animal observations has been previously explored with gregarious species in livestock monitoring [2]. In their study, Li et al. [15] employed density-based clustering to optimize drone deployment, maximising animal coverage while minimizing the average distance between drones and animals. More broadly, the optimization problem of covering a maximum number of targets with a minimum number of drones has also been addressed using bio-inspired algorithms. For instance, the elephant herding optimization algorithm [21] and artificial bee colony algorithms [22] have been utilized for this purpose.

3.2 Horizontal Monitoring

The use of drones for identification of individual animals in biological studies has been reported in the literature [6,7]. However, existing protocols for conducting photo identification flight surveys are currently defined qualitatively, with drones being manually flown. These protocols do not establish a standard procedure for positioning the drone relative to the species being studied [6,7]. Rather, the distance between the animals and the drone is selected to ensure sufficient resolution to identify their individual physical characteristics while ensuring the animals are not disturbed [14]. A step towards automation would be to define, through a model, the likelihood of correct animal identification based on the relative position of the drone and the animal. Such a model could be based on the empirical experience of biologists, but also on specific criteria that have been defined to assess the quality of a photo with respect to identification, such as those described in [8,14]. This model could thus be used to evaluate the performance of drones conducting the horizontal monitoring, and subsequently to build an objective function. The optimization could be carried out by employing methods used around the *camera placement optimization problem (CPO problem)* defined in [9]. This problem involves optimally placing a certain number of cameras in designated regions to monitor as much of the ground area as possible. A similar approach could be adopted to solve the animal identification problem, this time considering our custom-made objective function. To solve the CPO problem, researchers have explored various strategies, including bio-inspired algorithms like binary Particle Swarm Optimization [16] and a clustering-based multi-resolution optimization method [18] aimed at enhancing the speed of problem resolution.

4 Evaluation Strategy, and Further Work

The WildDrone consortium aims to conduct real-world assessments of drone technologies for wildlife conservation in different natural reserve parks, such as the Wadden Sea in Denmark and the Ol-Pejeta Conservancy in Kenya. The first field trip is planned to be in January 2025 in Kenya. To ensure that we meet the deadline for conducting real-world experiments, we have decided to solve the problem outlined in this paper under stricter assumptions. We define a fixed camera tilt and altitude for each type of drone (vertical or horizontal). Additionally, we assume that the animals are stationary and that their spatial distribution is known. We are currently attempting to solve the optimization problems using a centralized approach and a simulated environment. This environment, depicted in Fig. 2, allows us to utilize 3D models of animals and emulate drone camera views. This initial simplified approach will enable us to quickly transition to the implementation of the hardware solution, ensuring our operational readiness by January 2025.

The experience gained during the first trip and the evaluation of our initial methods will allow us to highlight the limitations of our approach and its potential. The next step will be to introduce a more complex model for the animals, including a spatial distribution based on real data and their potential movement. This will involve making our algorithms responsive to a dynamic environment, thus focusing on a decentralized algorithmic solution to unlock the potential of a fully decentralized system.

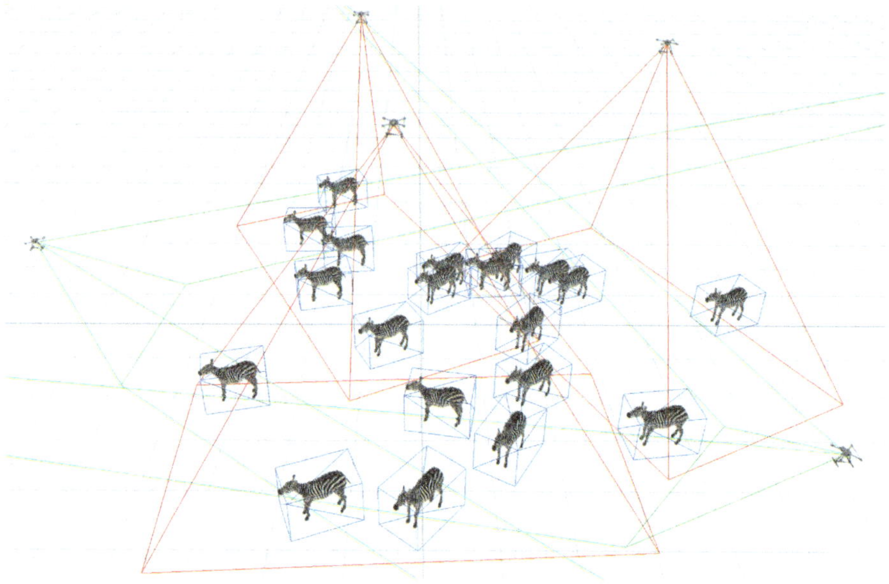

Fig. 2. Drones performing the vertical (red) and horizontal (green) monitoring tasks in the simulator

Acknowledgments. This work is supported by the WildDrone MSCA Doctoral Network funded by EU Horizon Europe under grant agreement no. 101071224.

References

1. Autonomous Drones for Nature Conservation Missions | WildDrone | Project | Fact sheet | HORIZON | CORDIS | European Commission. Online (2023). https://cordis.europa.eu/project/id/101071224. Accessed 10 Feb 2023
2. Alanezi, M.A., Shahriar, M.S., Hasan, M.B., Ahmed, S., Yusuf, A., Bouchekara, H.R.: Livestock management with unmanned aerial vehicles: a review. IEEE Access **10**, 45001–45028 (2022)
3. Attard, M.R., et al.: Review of satellite remote sensing and unoccupied aircraft systems for counting wildlife on land. Remote Sens. **16**(4), 627 (2024)

4. Ceballos, G., Ehrlich, P.R., Dirzo, R.: Biological annihilation via the ongoing sixth mass extinction signaled by vertebrate population losses and declines. Proc. Natl. Acad. Sci. **114**(30), E6089–E6096 (2017)
5. Corcoran, E., Winsen, M., Sudholz, A., Hamilton, G.: Automated detection of wildlife using drones: synthesis, opportunities and constraints. Methods Ecol. Evol. **12**(6), 1103–1114 (2021)
6. Degollada, E., Amigó, N., O'Callaghan, S.A., Varola, M., Ruggero, K., Tort, B.: A novel technique for photo-identification of the fin whale, balaenoptera physalus, as determined by drone aerial images. Drones **7**(3), 220 (2023)
7. Durban, J.W., Fearnbach, H., Barrett-Lennard, L., Perryman, W., Leroi, D.: Photogrammetry of killer whales using a small hexacopter launched at sea. J. Unmanned Veh. Syst. **3**(3), 131–135 (2015)
8. Elliser, C.R., van der Linde, K., MacIver, K.: Adapting photo-identification methods to study poorly marked cetaceans: a case study for common dolphins and harbor porpoises. Mammalian Biol. **102**(3), 811–827 (2022)
9. Hoffmann, F.: On the rectilinear art gallery problem. In: Paterson, M.S. (ed.) ICALP 1990. LNCS, vol. 443, pp. 717–728. Springer, Heidelberg (1990). https://doi.org/10.1007/BFb0032069
10. Inoue, S., Yamamoto, S., Ringhofer, M., Mendonça, R.S., Pereira, C., Hirata, S.: Spatial positioning of individuals in a group of feral horses: a case study using drone technology. Mammal Res. **64**, 249–259 (2019)
11. Kholiavchenko, M., et al.: KABR: in-situ dataset for Kenyan animal behavior recognition from drone videos. In: Proceedings of the IEEE/CVF Winter Conference on Applications of Computer Vision (WACV) Workshops, pp. 31–40. IEEE (2024)
12. Koger, B., Deshpande, A., Kerby, J.T., Graving, J.M., Costelloe, B.R., Couzin, I.D.: Quantifying the movement, behaviour and environmental context of group-living animals using drones and computer vision. J. Anim. Ecol. **92**(7), 1357–1371 (2023)
13. Lahiri, M., Tantipathananandh, C., Warungu, R., Rubenstein, D.I., Berger-Wolf, T.Y.: Biometric animal databases from field photographs: identification of individual zebra in the wild. In: Proceedings of the 1st ACM International Conference on Multimedia Retrieval, pp. 1–8 (2011)
14. Landeo-Yauri, S.S., Ramos, E.A., Castelblanco-Martínez, D.N., Niño-Torres, C.A., Searle, L.: Using small drones to photo-identify antillean manatees: a novel method for monitoring an endangered marine mammal in the Caribbean sea. Endanger. Species Res. **41**, 79–90 (2020)
15. Li, X., Xing, L.: Reactive deployment of autonomous drones for livestock monitoring based on density-based clustering. In: 2019 IEEE International Conference on Robotics and Biomimetics (ROBIO), pp. 2421–2426. IEEE (2019)
16. Morsly, Y., Aouf, N., Djouadi, M.S., Richardson, M.: Particle swarm optimization inspired probability algorithm for optimal camera network placement. IEEE Sens. J. **12**(5), 1402–1412 (2011)
17. Perryman, R.J., Brown, C., Pasian, N., Ward, A.J., Kent, M.: Investigating manta ray collective movements via drone surveys. bioRxiv pp. 2022–07 (2022)
18. Puligandla, V.A., Lončarić, S.: A multiresolution approach for large real-world camera placement optimization problems. IEEE Access **10**, 61601–61616 (2022)
19. Schad, L., Fischer, J.: Opportunities and risks in the use of drones for studying animal behaviour. Meth. Ecol. Evol. **14**, 1864–1872 (2022)
20. Schranz, M., et al.: Swarm intelligence and cyber-physical systems: concepts, challenges and future trends. Swarm Evol. Comput. **60**, 100762 (2021)

21. Strumberger, I., Bacanin, N., Tomic, S., Beko, M., Tuba, M.: Static drone placement by elephant herding optimization algorithm. In: 2017 25th Telecommunication Forum (TELFOR), pp. 1–4. IEEE (2017)
22. Zhang, C., Zhang, L., Zhu, L., Zhang, T., Xiao, Z., Xia, X.G.: 3d deployment of multiple UAV-mounted base stations for UAV communications. IEEE Trans. Commun. **69**(4), 2473–2488 (2021)

Security Considerations for 5G Network Slicing

Carlos D. Aguilar-Mora[1(✉)] and Rodolfo P. Merino-Vivanco[2]

[1] Universidad Nacional de Loja, Av. Pío Jaramillo Alvarado, Loja, Ecuador
cdaguilar@usal.es

[2] Centro de Investigación, Innovación, Desarrollo de Prototipos Tecnológicos y Pruebas Electroenergéticas I2TEC, Universidad Nacional de Loja, Av. Pío Jaramillo Alvarado, Loja, Ecuador
rodolfo.merino@unl.edu.ec

Abstract. 5G network slicing is an emerging technology that enables the sharing of physical resources for various purposes, used in multiple sectors, such as robotics, medicine, agriculture, mining, media, as well as applications such as IoT, Industry 4.0, among others. This article presents an overview of the potential security threats in the field of Network Slicing, the components that make up its physical infrastructure (RAN, CORE, and transport network slicing), as well as its lifecycle, which consists of the preparation, commissioning, operation, and decommissioning phases. Due to the risk of denial-of-service (DoS) attacks, misconfiguration attacks and man-in-the-middle (MITM) attacks, it is of paramount interest to propose and apply artificial intelligence-based frameworks as the main proposal to mitigate the impact of the constant threats that the 5G network could have.

Keywords: 5G · Network Slicing · 5G Security

1 Introduction

5G network technology is the fifth generation of mobile technology, designed to provide a faster, more reliable, and adaptive network, compared to previous generations of 4G and LTE. Some key features of this technology include much faster download speed, reduced latency, an improved architecture designed to handle more simultaneously connected devices, as well as a wider range of applications.

To deliver a faster, more reliable, and adaptive network that provides enhanced benefits, 5G proposes, in addition to its physical architecture, a different service-based architecture that consists of separating the control plane from the data plane through the use of software. Therefore, 5G is based on softwarization through network virtualization, comprising three main technologies: software-defined networking (SDN), network functions virtualization (NFV) and cloud computing. Network Slicing (NS), on the other hand, configures network functions so that logical end-to-end networks identified as "network slices" can be assigned, ensuring that mobile networks can accommodate the diverse network requirements of 5G applications [1].

G. Marreiros et al. (Eds.): DCAI 2024, LNNS 1151, pp. 317–322, 2025.
https://doi.org/10.1007/978-3-031-80946-0_34

Although network slicing offers significant advantages in terms of performance, adaptability, and capabilities, it is not devoid of vulnerabilities, therefore, along with segmentation, latency and throughput requirements, security is a fundamental requirement in the design and deployment of network slicing. Consequently, it is necessary to analyse these requirements to best address the security considerations for network slicing.

The main objective of this study is to understand NS, as well as its main vulnerabilities, basic security requirements and mitigation strategies.

2 5G Network Slicing

Network slicing or 5G network fragmentation, first introduced by the next generation mobile network (NGMN), is defined by the NGMN as a logical network slice running on a common underlying infrastructure (physical or virtual), mutually isolated, with independent control and management that can be created on demand. A network slice may consist of separate domain components in the same or different administrations, or components applicable to the access network, the transport network, the core network and edge networks.

2.1 Physical Infrastructure

A network slice can comprise several network domains used by a network slice provider (NSP), as shown in Table 1. It is composed of resources that can be dedicated or shared in terms of functionality, processing, storage and bandwidth [2].

Table 1. Network slicing domains.

Domain	Key features	References
RAN slicing	Selects an appropriate network slice and applies the appropriate configuration to handle specific traffic Types: Distributed-RAN (D-RAN), Centralised-RAN (C-RAN), Open-RAN (O-RAN), Fog-RAN (F-RAN) Operates as a gateway for all user equipment (UE) and is constantly growing	[3–5]
Core network slicing	It includes both the control plane and the user plane with shared and unshared functions between slices	[6]
Transport network slicing	Programmable end-to-end data path connectivity, across all network segments (fronthaul, midhaul, backhaul)	[7, 8]

2.2 Lifecycle Management

Multiple network slices enable the provision of a variety of communication services. Network slice management involves, at a minimum, the following concepts [9]:

- Network slices are generally composed of network slice subnetworks (e.g. network slice RAN, Core and transport network slicing).
- Network slice subnetworks are composed of network functions (NF).
- Network functions can be performed by a combination of virtualised network functions (VNF) and/or physical network functions (PNF).
- Communication services can be supported by one or more network slices.

Network slices must be managed throughout their lifecycle. A typical lifecycle includes the following phases: preparation, refers to the preparation of the network environment, from design to on-boarding; commissioning, refers to the creation of instances, through the allocation of network resources and implementation of configurations necessary to meet slice requirements; operation, concerns the activation, modification and deactivation of instances, with important activities such as monitoring and generation of operation reports; decommissioning, with the termination of network slice instances it is necessary to recover the allocated resources as well as the configuration of the dependent resources [1, 9]. The lifecycle of a network slice is shown in Fig. 1.

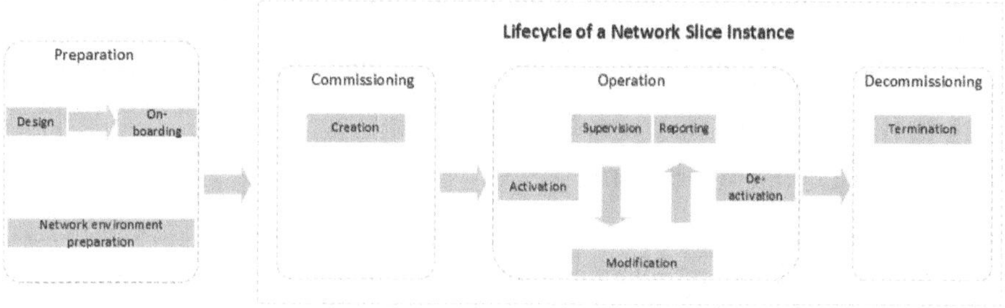

Fig. 1. Lifecycle of a network slice instance [9].

The lifecycle phases require special consideration when it comes to security, as it is a vulnerable area which could be broadly exploited.

3 Potential Security Threats in Network Slicing

Many threat vectors affect network slicing, among which, denial of service (DoS) attacks, misconfiguration attacks and man-in-the-middle (MITM) attacks pose a high risk [2].

These attacks directly affect the cornerstones of security - confidentiality, integrity, and availability.

Denial of Service (DoS) Attacks. These attacks directly compromise the availability and functionality of systems, including loss of access to 5G infrastructure. A DoS attack can result in several attacks such as hardware tampering, sensor errors, malware and distributed DoS (DDoS) [6, 10].

Misconfiguration Attacks. These attacks occur when malicious actors exploit the control settings of a system, and they may also include the inadvertent disabling of security functions or the inadvertent disabling of monitoring services. Such attacks affect the confidentiality, integrity and availability of a network slice [10].

Man-in-the-Middle (MITM) Attacks. This type of attack involves the attacker relaying and possibly altering communications between two points, compromising the availability, confidentiality, and integrity of a network slice. MITM attacks exploit ARP spoofing to intercept data. Additionally, confidentiality can be compromised when information or data is intercepted or exposed [5, 10].

Network slicing brings with it a wide range of services to cover the diverse needs of multiple sectors, which opens up many attack possibilities:

- **Lifecycle of a slice:** of the four phases of the lifecycle mentioned above, in the preparation phase the template configuration for slices is established, poor design, incorrect procedures, lack of latest security patches, can produce risks in the preparation phase, leading to data leakage and even exposure of sensitive information [11, 12]
- **Communication Type:** In general, to provide services, slices use two types of communication: Inter-Slice communication and Intra-Slice communication. The possible threats include denial-of-service, session hijacking, malicious mobile node flooding, man-in-the-middle, and redirection attacks [12, 13].
- **Technologies used**: NFV and SDN are two crucial technologies that come to the forefront when implementing segmentation capabilities in 5G network architectures. Some attacks on the technology used include: data plane layer flooding attacks, DoS attacks, poisoned applications in the application layer [13], among the most common.

4 Securing Network Slicing

Network Slicing brings with it several security challenges, which need to be addressed during the lifecycle. Several solutions have been proposed to address network slicing security issues, all of them related to prevention, mitigation and detection strategies to protect the network against potential threats, which can also be classified into three main categories: RAN, Core and General strategies [6].

Beyond prevention, mitigation, and detection strategies, it is necessary to think about mechanisms that lead to effective network slicing protection, which is why, among other mechanisms, artificial intelligence (AI) is proposed as the most suitable technology for tackling security problems. To this end, it is necessary to develop and apply frameworks in the field of network slicing that integrate artificial intelligence with 5G networks [14].

According to [15], in view of the diverse threats and challenges associated with the security of network slicing, it is necessary to prioritise the mitigation of these threats. This is especially crucial given that 5G is used in multiple sectors, such as robotics, medicine, agriculture, mining, media, as well as applications such as IoT, Industry 4.0 and others; the security of network slicing is of paramount concern. Therefore, the authors also identified the following research challenges: 5G architecture and AI-based network slicing, secure authentication, secure integration of services, secure and continuous connectivity, front-haul/backhaul/xHaul security, secure deployment, performance metrics and optimisation.

5 Conclusion

In view of the risk to which sectors and applications using 5G networks may be exposed, a broad scope of research into network slicing infrastructure has emerged, covering its components, lifecycle management, potential security threats and, above all, mitigation strategies. The use of artificial intelligence techniques is considered to be a key approach to responding to the security needs of the different environments in which 5G is used.

References

1. De Alwis, C., Porambage, P., Dev, K., Gadekallu, T.R., Liyanage, M.: A survey on network slicing security: attacks, challenges, solutions and research directions. IEEE Commun. Surv. Tutor. **26**, 534–570 (2024). https://doi.org/10.1109/COMST.2023.3312349
2. National Security Agency (NSA): Cybersecurity and Infrastructure Security Agency (CISA): 5G Network Slicing: Security Considerations for Design, Deployment, and Maintenance (2023). https://www.cisa.gov/news-events/alerts/2023/07/17/nsa-cisa-release-guidance-security-considerations-5g-network-slicing
3. ESF 5G NETWORK SLICING-SECURITY CONSIDERATIONS FOR DESIGN, DEPLOYMENT, AND MAINTENANCE_FINAL (1).pdf
4. Azimi, Y., Yousefi, S., Kalbkhani, H., Kunz, T.: Applications of machine learning in resource management for RAN-slicing in 5G and beyond networks: a survey. IEEE Access. **10**, 106581–106612 (2022). https://doi.org/10.1109/ACCESS.2022.3210254
5. Bjerre, S.A., Klæbel Blomsterberg, M.W., Andersen, B.: 5G attacks and countermeasures. In: 2022 25th International Symposium on Wireless Personal Multimedia Communications (WPMC), pp. 285–290 (2022). https://doi.org/10.1109/WPMC55625.2022.10014962
6. Salahdine, F., Liu, Q., Han, T.: Towards secure and intelligent network slicing for 5G networks. IEEE Open J. Comput. Soc. **3**, 23–38 (2022). https://doi.org/10.1109/OJCS.2022.3161933
7. 5G Americas: Envolving 5G security for the cloud (2022). https://www.5gamericas.org/wp-content/uploads/2022/09/Evolving-5G-Security-for-the-Cloud-2022-InDesign.pdf
8. Mohammed, C.M., Kheder Shaikhah, S.: A survey and analysis of architecture and models of network slicing in 5G. In: 2022 8th International Engineering Conference on Sustainable Technology and Development (IEC), pp. 192–198 (2022). https://doi.org/10.1109/IEC54822.2022.9807564
9. Soveri, M.C., et al.: 5G Network slice management (2023). https://www.3gpp.org/technologies/slice-management
10. National Security Agency (NSA): Cybersecurity and Infrastructure Security Agency (CISA): ESF Potential Threats to 5G Network Slicing (2022). https://media.defense.gov/2021/May/10/2002637751/-1/-1/0/POTENTIAL%20THREAT%20VECTORS%20TO%205G%20INFRASTRUCTURE.PDF
11. Wu, T.-Y., Jie, T.F.: 5G network slicing security. In: Nicopolitidis, P., Misra, S., Yang, L.T., Zeigler, B., Ning, Z. (eds.) Advances in Computing, Informatics, Networking and Cybersecurity: A Book Honoring Professor Mohammad S. Obaidat's Significant Scientific Contributions, pp. 755–780. Springer, Cham (2022). https://doi.org/10.1007/978-3-030-87049-2_28
12. Singh, V.P., Singh, M.P., Hegde, S., Gupta, M.: Security in 5G network slices: concerns and opportunities. IEEE Access. **12**, 52727–52743 (2024). https://doi.org/10.1109/ACCESS.2024.3386632
13. Sajjad, M.M., Bernardos, C.J., Jayalath, D., Tian, Y.-C.: Inter-slice mobility management in 5G: motivations, standard principles, challenges, and research directions. IEEE Commun. Stand. Mag. **6**, 93–100 (2022). https://doi.org/10.1109/MCOMSTD.0001.2000025

14. Alanazi, M.H.: Machine learning-based secure 5G network slicing: a systematic literature review. Int. J. Adv. Comput. Sci. Appl. **14** (2023). https://doi.org/10.14569/IJACSA.2023.0141239
15. Dangi, R., Jadhav, A., Choudhary, G., Dragoni, N., Mishra, M.K., Lalwani, P.: ML-based 5G network slicing security: a comprehensive survey. Future Internet **14**, 116 (2022). https://doi.org/10.3390/fi14040116

Graffiti Identification Using Color Analysis: An Approach Based on K-Means Clustering

Miguel García García[1(✉)], Guillermo Hernández[1,2], Angélica González-Arrieta[1,2], and Sara Rodríguez González[1,2]

[1] Department of Computer Science and Automation, Faculty of Science, University of Salamanca, Plaza de los Caídos s/n, 37008 Salamanca, Spain
{restauramiguel,guillehg,angelica,srg}@usal.es
[2] BISITE Research Group, University of Salamanca, Edificio I+D+I, Calle Espejo 2, 37007 Salamanca, Spain

Abstract. This article presents a method for graffiti identification based on the analysis of its dominant colors using artificial intelligence. The method employs a K-Means clustering algorithm to group the pixels of a graffiti image according to their color values in the Red, Green, Blue or The Lab color space. From this clustering, the dominant colors of each group are extracted and ranked by their prevalence. This approach enables the characterization of the predominant colors in graffiti, facilitating their subsequent localization and analysis in a database.

Keywords: Graffiti · K-Means · Artificial Intelligence · color · unsupervised classification

1 Problem Statement

The proliferation of graffiti in cities represents a challenge for local authorities and organizations in charge of managing public space. These elements can have a negative impact on urban aesthetics, public safety, and the quality of life of citizens. The identification and analysis of graffiti is essential for its subsequent removal, prevention and control. The trend is to demonstrate that the same colors used by the same author can be related and these colors are automatically extracted with artificial intelligence (AI).

There are several traditional methods for identifying graffiti, however, these methods have several limitations:

- Visual inspection: It is a time-consuming and expensive method that requires trained personnel and can be susceptible to human error.
- Satellite imagery: While they can provide an overview, their resolution may be insufficient to identify graffiti accurately.
- Pattern recognition: It requires an extensive and accurate training dataset, and its performance can be affected by variability in the appearance of graffiti.

G. Marreiros et al. (Eds.): DCAI 2024, LNNS 1151, pp. 323–328, 2025.
https://doi.org/10.1007/978-3-031-80946-0_35

2 Related Work

2.1 Graffiti Identification System Using Low-Cost Sensors

Previous studies present the possibility of studying the authorship of graffiti using a colorimeter developed with Arduino hardware technology according to the Do It Yourself (DIY) philosophy [1]. Through the Red, Green, Blue (RGB) data obtained, the aim is to study and compare the information extracted from each of the graffiti present on different walls. The same color can be found in different parts of the same graffiti, but also in other graffiti that a priori could be of different authorship. However, graffiti can be related and it is possible to group graffiti artists and "gangs" working together. The methodology followed for the construction of the colorimeter and its actual application in a case study is described in four case studies.

2.2 Object Detection Using Color Dissimilarity Based Segmentation Method

Associated with object detection, the image segmentation method based on color difference is able to give good results. This segmentation method splits the image based on the dissimilarity of the R, G, and B values in the adjacent image pixels. The object detection results displayed by this method are very good [2].

2.3 Batik Motif Classification Using Color-Texture-Based Feature Extraction and Back Propagation Neural Network

This research develops software to automatically identify batik image motifs using feature extraction based on color textures and a back propagation neural network [3]. The color and texture characteristics of the batik image are extracted by a combination of the color co-occurrence matrix, the difference between pixels of the scan pattern, and the color histogram for the K-Means methods.

3 Hypothesis

Color analysis using K-Means grouping techniques[1] it makes it possible to accurately and efficiently identify the dominant colors in a graffiti, facilitating their subsequent location and analysis in a database.

The hypothesis is based on the assumption that graffiti exhibits distinctive color characteristics that can be captured and analyzed using image processing techniques. The K-Means clustering algorithm is a suitable tool to identify these characteristics, as it allows pixels of the image to be grouped according to their color values, revealing the dominant colors present in the graffiti.

The hypothesis will be verified by conducting a series of experiments that will evaluate the effectiveness of the proposed method for the identification of graffiti. The

[1] K-means is an unsupervised classification algorithm (clustering) that groups objects into k clusters based on their characteristics. The clustering is performed by minimizing the sum of distances between each object and the centroid of its cluster [6].

experiments will be carried out using a database of graffiti images and various aspects will be analyzed, such as the accuracy in identifying dominant colors, the efficiency of the algorithm and the scalability of the method to images of different sizes and lighting conditions.

If the hypothesis is confirmed, the proposed method will have a significant impact on graffiti identification and analysis. This can lead to better management of public space, a reduction in the costs associated with graffiti removal, and an improvement in the quality of life of citizens.

4 Proposal

The proposal consists of using a K-Means clustering algorithm to group the pixels of a graffiti image based on their color values in the RGB[2] or The Lab Color Space (CIELAB[3]). From this grouping, the dominant colors of each group are extracted and ordered according to their prevalence. The approach makes it possible to characterize the predominant colors in graffitis made by the same author, facilitating their subsequent location and analysis in a database.

The work consists of the following stages:

- Image pre-processing: The graffiti image is pre-processed to remove noise and improve image quality.
- Color representation: Each pixel in the image is represented as a point in the RGB or CIELAB color space.
- K-Means clustering: The K-Means clustering algorithm is used to group pixels into a predefined number of clusters.
- Dominant color extraction: The dominant color of each group is extracted by taking the midpoint of the group.
- Dominant color sorting: Dominant colors are sorted according to the number of pixels they contain in each group.

Advantages of the proposed method:

- Accuracy: It is able to accurately identify the dominant colors in graffiti.
- Efficiency: It is computationally efficient and can be applied to images of different formats.
- Flexibility: It is flexible and can be adapted to different color spaces.
- Profitability: It's a cost-effective method that doesn't require specialized hardware or software.

[2] RGB: the term RGB refers to a color space that reproduces colors visible to humans through the additive mixing of the three primary colors. The RGB color space is named after these three colors: red, green, and blue [6].

[3] CIELAB, is currently one of the most popular and uniform color spaces used for evaluating the color of an object. This color space is widely used because it correlates numerical color values consistently with human visual perception. Researchers and manufacturers use it to assess color attributes, identify inconsistencies, and precisely convey their results to others in numerical terms [6].

5 Evaluation Plan

In order to evaluate in depth the performance of the proposed method, an evaluation plan will be carried out that includes the following aspects:

- Same colors for the same author: The results obtained in different graffiti attributed to the same author will be compared to verify that the method can be effective in this aspect.
- Assessing accuracy in a larger database: A larger and more diverse database of graffiti will be used to assess the accuracy of the method in different scenarios.
- Comparison with other methods: The performance of the proposed method will be compared with other methods of graffiti identification, such as visual inspection or pattern recognition.
- Robustness assessment: The robustness of the method against different lighting conditions, noise and image distortions will be evaluated.
- Scalability analysis: The scalability of the method to process images of different sizes and resolutions will be evaluated.

To evaluate the performance of the proposed method for identifying graffiti by color analysis and K-Means clustering, experiments are performed using a database of x images of graffiti with known colors (see Fig. 1, Fig. 2, Fig. 3). The experiments are carried out using two color spaces: RGB and CIELAB.

Fig. 1. Color extraction according to graffiti reference 1 based on dominant colors (4 and 6).

6 Reflections

The proposed method for the identification of graffiti by color analysis and grouping by K-Means has proven to be a valuable tool for the management of public space. The method is accurate, efficient, flexible and cost-effective, and can be applied to a wide range of applications.

This method has the potential to significantly improve graffiti identification and analysis. This can lead to better management of public space, a reduction in the costs associated with graffiti removal, and an improvement in the quality of life of citizens.

There are several areas of future research related to the proposed method:

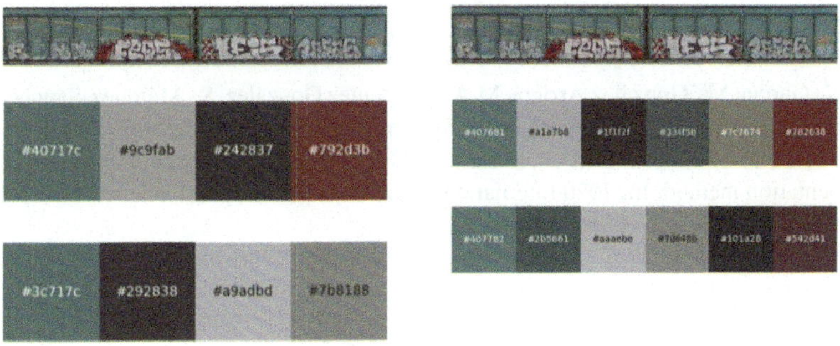

Fig. 2. Color extraction according to graffiti reference 2 based on dominant colors (4 and 6).

Fig. 3. Color extraction according to graffiti reference 3 based on dominant colors (4 and 6).

- Development of more advanced clustering algorithms: Other clustering algorithms, such as hierarchical clustering or density-based clustering, could be explored to improve the accuracy and efficiency of the method.
- Integration with graffiti management systems: The proposed method could be integrated into graffiti management systems to facilitate the identification, location and removal of graffiti.
- Application to other types of images: The method could be adapted to the identification of other objects in images, such as road signs or street furniture.

Color analysis is a powerful tool for identifying and analyzing objects in images. The proposed method for the identification of graffiti by color analysis and clustering by K-Means is an example of how this technique can be used to address a real problem and improve the quality of life in cities.

Acknowledgments. This research has been supported by the project "COordinated intelligent Services for Adaptive Smart areaS (COSASS)", Reference: PID2021-123673OB-C33, financed by MCIN /AEI /https://doi.org/10.13039/501100011033 / FEDER, UE.

References

1. García García, M., González Arrieta, M.A., Rodríguez González, S., Márquez-Sánchez, S., Da Silva Ramos, C.F.: Graffiti Identification System Using Low-Cost Sensors (2024)
2. Karma, I.G.M., Sulastra, I.M.D.J., Susanti, J.: Object detection using color dissimilarity based segmentation method. In: 2020 International Conference on Applied Science and Technology (iCAST), pp. 366–370. IEEE, October 2020
3. Suciati, N., Pratomo, W.A., Purwitasari, D.: Batik motif classification using color-texture-based feature extraction and backpropagation neural network. In: 2014 IIAI 3rd International Conference on Advanced Applied Informatics, pp. 517–521. IEEE, August 2014
4. Konica Minolta, June 2023. https://sensing.konicaminolta.us/mx/blog/entendiendo-el-espacio-de-color-cie-lab/
5. Ionos, January 2024. https://www.ionos.es/digitalguide/paginas-web/diseno-web/colores-rgb/#:~:text=RGB%3A%20el%20t%C3%A9rmino%20RGB%20se,red%2C%20green%2C%20blue)
6. Universidad de Oviedo: K-means clustering (n.d.). https://www.unioviedo.es/compnum/laboratorios_py/kmeans/kmeans.html. Accessed 18 June 2024

Author Index

G. Marreiros et al. (Eds.): DCAI 2024, LNNS 1151, pp. 329–330, 2025.
https://doi.org/10.1007/978-3-031-80946-0

The manufacturer's authorised representative in the EU is Springer Nature Customer Service Centre GmbH, Europaplatz 3, 69115 Heidelberg, Germany. If you have any concerns regarding our products, please contact ProductSafety@springernature.com

Printed and bound by CPI Group (UK) Ltd, Croydon, CR0 4YY
30/09/2026
02225561-0009